Springer Theses

Recognizing Outstanding Ph.D. Research

Aims and Scope

The series "Springer Theses" brings together a selection of the very best Ph.D. theses from around the world and across the physical sciences. Nominated and endorsed by two recognized specialists, each published volume has been selected for its scientific excellence and the high impact of its contents for the pertinent field of research. For greater accessibility to non-specialists, the published versions include an extended introduction, as well as a foreword by the student's supervisor explaining the special relevance of the work for the field. As a whole, the series will provide a valuable resource both for newcomers to the research fields described, and for other scientists seeking detailed background information on special questions. Finally, it provides an accredited documentation of the valuable contributions made by today's younger generation of scientists.

Theses are accepted into the series by invited nomination only and must fulfill all of the following criteria

- They must be written in good English.
- The topic should fall within the confines of Chemistry, Physics, Earth Sciences, Engineering and related interdisciplinary fields such as Materials, Nanoscience, Chemical Engineering, Complex Systems and Biophysics.
- The work reported in the thesis must represent a significant scientific advance.
- If the thesis includes previously published material, permission to reproduce this must be gained from the respective copyright holder.
- They must have been examined and passed during the 12 months prior to nomination.
- Each thesis should include a foreword by the supervisor outlining the significance of its content.
- The theses should have a clearly defined structure including an introduction accessible to scientists not expert in that particular field.

More information about this series at http://www.springer.com/series/8790

Ricardo Puebla

Equilibrium and Nonequilibrium Aspects of Phase Transitions in Quantum Physics

Doctoral Thesis accepted by
the Universität Ulm, Ulm, Germany

 Springer

Author
Dr. Ricardo Puebla
Centre for Theoretical Atomic,
    Molecular and Optical Physics
Queen's University Belfast
Belfast, UK

Supervisor
Prof. Dr. Martin B. Plenio
Institut für Theoretische Physik
    and IQST
Universität Ulm
Ulm, Germany

Tag der Promotion (date of the Ph.D. defense): 13.03.2018

ISSN 2190-5053   ISSN 2190-5061   (electronic)
Springer Theses
ISBN 978-3-030-13138-8   ISBN 978-3-030-00653-2   (eBook)
https://doi.org/10.1007/978-3-030-00653-2

This Springer imprint is published by the registered company Springer Nature Switzerland AG
The registered company address is: Gewerbestrasse 11, 6330 Cham, Switzerland

Supervisor's Foreword

Phase transitions are an experience of everyday life that has fascinated scientists for centuries. The quest to understand them has led to many Nobel Prize-worthy developments ranging from the renormalization group in statistical mechanics to fundamental aspects of quantum field theory. Freezing and boiling water constitute simple examples of first-order phase transitions for which the density, a first derivative of the free energy, changes discontinuously. But these are by no means the only phase transitions. Indeed, in many respects, more interesting are second-order phase transitions such as the ferromagnetic phase transition. Here, the magnetization increases continuously from zero as the temperature is lowered below the transition temperature, and it is the magnetic susceptibility, the second derivative of the free energy, that changes discontinuously. Second-order phase transitions are particularly interesting as they exhibit the remarkable phenomenon of spontaneous symmetry breaking. When there is no external magnetic field and the setting is therefore completely symmetric, the ferromagnetic material can become magnetized in one of two possible directions which are chosen at random; i.e., in each repetition of the experiment, there may be a new direction of magnetization. Although the physics was completely symmetric, the resulting state is not and we speak of a spontaneously broken symmetry. The mechanism of spontaneous symmetry breaking underlies, for example, the theory of the Higgs boson in the standard model.

In quantum mechanics, a second-order quantum phase transition in a quantum system describes an abrupt change in the ground-state properties upon a change of a system parameter and separates quantum phases of matter, whose physical properties cannot be analytically connected. According to the traditional understanding, the emergence of phase transitions can be explained only by invoking the concept of the thermodynamic limit of infinitely many system components. Indeed, up until recently, it has been widely assumed that a phase transition can only occur in this limit. The reason for this assumption can be understood from the consideration of the partition function. A partition function can be represented as a sum of exponential functions, and the number of terms to be summed over is equal to the dimension of the corresponding Hilbert space. As the sum of finite number of exponential functions, which are individually analytic functions, can never become non-analytic, the possibility of any non-analyticity in the partition function and thus the possibility of phase transition only arise when this sum is over infinitely many terms. This simple observation rules out the existence of a phase transition for quantum systems where each system component has a finite Hilbert space dimension, such as in the case of spin systems. In the folklore of the research community that has been concentrated, not correctly as we will see, in the belief that only infinitely extended systems can exhibit a phase transition.

In his dissertation, Ricardo Puebla shows that this belief is not true. Indeed, in a seminal contribution, he shows that this no-go type of principle does not apply to a system composed of unbounded quantum systems residing in an infinite dimensional Hilbert space such as a harmonic oscillator. In this case, the partition function is a sum over an infinite number of exponential functions and therefore the possibility of non-analytic behavior is not necessarily ruled out by the finite number of system components. Building on this insight, he then proceeds to construct explicit examples of quantum phase transitions and critical phenomena in a minimal quantum system, comprising a single two-level system and a bosonic mode. He shows that in a suitable limit of the system's parameters rather than in the conventional thermodynamic limit in the particle number, this system exhibits all the phenomenology of phase transitions. This novel notion, dubbed finite-component size quantum phase transition, challenges the well-established idea of phase transitions which emerge due to an infinite number of degrees of freedom. Remarkably, it is shown that by investigating and exploring systems featuring this novel phase transition, which do not require large number of constituents, one can learn about the critical traits of a truly quantum many-body system—typically more complicated to realize experimentally—in the same manner as one studies simplified Ising models to understand more complex systems which belong to the same universality class. Ricardo does not only show that the ground-state properties exhibit a phase transition but also extends these observations to time-dependent phenomena. Indeed, when traversing a symmetry breaking second-order phase transition at a finite rate, different parts of the system cannot interact fast enough and will choose their symmetry broken phase at random and independent of each other. Hence, different spatial regions may choose different symmetries, and the boundary between two such regions forms a domain wall or a defect. In the traditional picture of phase transitions, remarkably, the number of these domains is a function of the rate at which one traverses the phase transition with exponents that are, somewhat surprisingly, determined by the equilibrium properties of the phase transition. This is the celebrated Kibble–Zurek scaling, and it was by no means clear that this phenomenology would also be realized in the finite system phase transition, but Ricardo was able to show this too by assuming that quantum excitations play the role of defects. Beyond the actual conceptual novelty, a detailed analysis of a potential experimental realization unveils its feasibility based on dynamical decoupling techniques to cope with different and realistic noise sources. Experiments of this type are now underway with laser-cooled trapped ions.

This thesis not only offers a detailed account for the aforementioned scientific results, but also places them into context by reviewing previous findings and discussing prospects directions for the future work. In addition, the introduction firms down the necessary general concepts for the understanding of the subsequent chapters and discussion. This thesis is therefore an excellent read for those interested in the realm of critical phenomena and Kibble–Zurek physics as well as in the novel notion of finite-component system quantum phase transitions.

Ulm, Germany   Prof. Dr. Martin B. Plenio
July 2018

Abstract

Phase transitions are inherent to many-body systems and constitute a paradigmatic example of an emergent phenomenon, which stems from an infinite number of microscopic constituents. Different phases of matter, that are characterized by having distinct macroscopic properties, are separated by critical values of certain external magnitudes at which a change of phase takes place, namely when a phase transition occurs. As matter varies its collective properties in many and disparate manners, phase transitions or critical phenomena appear also in the quantum realm. In this regard, our understanding and the subsequent exploitation of distinct phases of matter crucially rely on the comprehension of phase transitions and of their underlying mechanisms. Continuous phase transitions deserve special attention as they allow to investigate same physics across systems that are a priori utterly distinct. This remarkable property is known as universality, and it provides a powerful tool for the inspection of critical phenomena by analyzing simple models that reproduce the main traits of more complex systems. However, while these notions are well established in a static or equilibrium case, much less it is known about how universal features emerge or are inherited in nonequilibrium scenarios. In this context, we find the Kibble–Zurek mechanism, a theory aiming to explain defect formation promoted by finite-rate ramps across a phase transition. This mechanism establishes a relation between universal equilibrium critical exponents and nonequilibrium features of the dynamics involving a phase transition. In this thesis, we analyze continuous phase transitions in different low-dimensional systems, addressing their equilibrium and nonequilibrium aspects, as well as paying special attention not only to the Kibble–Zurek mechanism but also to other universal traits of the dynamics. Moreover, we show that a model with only two constituent elements undergoes a phase transition in a suitable parameter limit rather than in the conventional thermodynamic limit and thus challenging the common notion about phase transitions.

The Kibble–Zurek mechanism evidenced that universal traits also apply to nonequilibrium scenarios. However, since Kibble–Zurek mechanism is based on properties only valid in the thermodynamic limit, their examination and ultimate verification in finite-size systems must be done prudently. Here, we discuss how universality of phase transitions can be attained and extended by exploiting and combining finite-size scaling theory with a Kibble–Zurek problem. This is illustrated in a Ginzburg–Landau model as well as in a realistic linear to zigzag phase transition in a Coulomb crystal, where in addition an approximate method is proposed to gain insight into the dynamics. Furthermore, a brief exploration of a feasible quantum many-body system suggests that the quantum Kibble–Zurek mechanism may be observed within current ion-trap technology.

A relevant part of this thesis has been devoted to investigate the quantum Rabi model, which describes the interaction between a qubit and a single bosonic mode. Despite its apparent simplicity, it is demonstrated that the quantum Rabi model undergoes a quantum phase transition when the ratio of qubit over bosonic frequency becomes infinite, while keeping finite the number of system constituents—a novel manner of attaining criticality that we have dubbed finite-component system phase transition. In this limit, the exact effective low-energy solution reveals a spontaneous symmetry breaking and the emergence of ground-state singularities at the critical point. Furthermore, for finite, yet large, ratio values, an equivalent formulation of finite-size scaling theory can be applied to this quantum phase transition. This allows us to disclose that the quantum Rabi model falls in the same universality class as that of the Dicke model. We extend our study of this critical system to a nonequilibrium scenario, where among another findings, we show that Kibble–Zurek physics can actually be observed in a zero-dimensional system, contrary to the common belief prior this work. Hence, the quantum Rabi model serves as an excellent test bed for the examination of different physics related to quantum phase transitions. We complete our analysis by proposing a potential experimental implementation of the quantum Rabi model using a single trapped ion to probe the dynamics of a superradiant quantum phase transition, while at the same time coping with realistic noise sources.

List of Publications

Parts of this thesis are based on or have been taken from material first published in the following peer-reviewed journals or as preprint (arXiv) versions:

[P1] Fokker-Planck formalism approach to Kibble-Zurek scaling laws and nonequilibrium dynamics R. Puebla, R. Nigmatullin, T. E. Mehlstäubler, and M. B. Plenio Phys. Rev. B **95**, 134104 (2017) arXiv:1702.02099

[P2] Probing the dynamics of a superradiant quantum phase transition with a single trapped ion R. Puebla, M.-J. Hwang, J. Casanova, and M. B. Plenio Phys. Rev. Lett. **118**, 073001 (2017) arXiv:1607.03781

[P3] A robust scheme for the implementation of the quantum Rabi model in trapped ions R. Puebla, J. Casanova, and M. B. Plenio New J. Phys. **18**, 113039 (2016) arXiv:1608.07434

[P4] Excited-state quantum phase transition in the Rabi model R. Puebla, M.-J. Hwang, and M. B. Plenio Phys. Rev. A **94**, 023835 (2016) arXiv: 1607.00283

[P5] Quantum phase transition and universal dynamics in the Rabi model M.-J. Hwang, R. Puebla, and M. B. Plenio Phys. Rev. Lett. **115**, 180404 (2015) ℘arXiv:1503.03090

The following publications are not covered in this thesis:

[P6] Equivalence among generalized nth order quantum Rabi models J. Casanova, R. Puebla, H. Moya-Cessa, and M. B. Plenio Submitted arXiv:1709.02714

[P7] Protected ultrastrong coupling regime of the two-photon quantum Rabi model with trapped ions R. Puebla, M.-J. Hwang, J. Casanova, and M. B. Plenio Phys. Rev. A **95**, 063844 (2017) arXiv:1703.10539

[P8] Magnetic field fluctuations analysis for the ion trap implementation of the quantum Rabi model in the the deep strong coupling regime R. Puebla, J. Casanova, and M. B. Plenio J. Mod. Opt. **65**, 745 (2018) arXiv:1704.07303

[P9] Irreversible processes without energy dissipation in an isolated Lipkin-Meshkov-Glick model R. Puebla, and A. Relaño Phys. Rev. E **92**, 012101 (2015) arXiv:1404.6146

[P10] Phase space determination of electron beam for intraoperative radiation therapy from measured dose data E. Herranz, J. L. Herraiz, P. Ibáñez, M. Pérez-Liva, R. Puebla, J. Cal-González, P. Guerra, R. Rodríguez, C. Illana, and J. M. Udías Phys. Med. Biol. **60**, 375 (2015)

[P11] Non-thermal excited-state quantum phase transitions R. Puebla, and A. Relaño Europhys. Lett. **104**, 50007 (2013) arXiv:1305.3077

[P12] Excited-state phase transition leading to symmetry-breaking steady states in the Dicke model R. Puebla, A. Relaño, and J. Retamosa Phys. Rev. A **87**, 023819 (2013) arXiv:1209.5320

Acknowledgements

This thesis has been possible thanks to the help, commitment, and support of several people, to whom I would like to express my gratitude. I start by thanking Martin B. Plenio, the Director of the Institute of Theoretical Physics at Ulm University, who offered me the opportunity to become a member of his outstanding research group. He has been always accessible and willing to help. Besides having encouraged and proposed exciting directions of research, he has always welcomed discussions about new ideas. His guidance during these last years has been fundamental, and I wish that his profound knowledge as well as his passionate and excellent way of doing science have left a mark on me. I am really grateful to him and feel fortunate to have had the occasion to have worked with, but specially to have learned from him.

I am also most grateful to Jorge Casanova, Myung-Joong Hwang, and Ramil Nigmatullin, who became my day-to-day supervisors in different projects and from whom I have learned enormously, not only about physics. Their dedication, knowledge, and patience have inspired and motivated me every day. It has been a pleasure and an honor working with them. I would like also to thank Tanja E. Mehlstäubler, for her willingness and encouragement, and for her hospitality during my visit to the PTB in Braunschweig, as well as to Ramil for his friendliness during my short stay in Oxford.

In addition, during the course of this thesis, I have benefited from discussions and exchange of ideas with different members of the group. In particular, I would like to thank Mehdi Abdi, Jorge Casanova, Felipe Caycedo-Soler, Dario Egloff, Pelayo Fernández-Acebal, Myung-Joong Hwang, Oliver Marty, Polichronis Muratidis, Ramil Nigmatullin, Santiago Oviedo-Casado, Julen S. Pedernales, Javier Prior, Jochen Rau, Andrea Smirne, Alejandro D. Somoza, Dario Tamascelli, Zhengyu Wang, Bobo Wei as well as to external collaborators, Lapo Bogani, Héctor Moya-Cessa, Kihwan Kim, Christopher Monroe, Jacob Smith, Jiehang Zhang, and Jing-Ning Zhang. In the different conferences and workshops I have attended, I had the opportunity to meet and learn from Stefan Schütz, Pedro Pérez-Fernández, Michael Klotz, Francisco Pérez-Bernal, José Enrique García

Ramos, Armando Relaño, Lea F. Santos, Ángel Rivas, Simone Felicetti, Iñigo Arrazola, among others. Special thanks to Barbara Bender-Palm and Gerlinde Walliser for their tireless efforts, indispensable help, kindness, and willingness. Also, special thanks to the diligent and thorough proofreaders Jorge Casanova, Felipe Caycedo-Soler, Myung-Joong Hwang, Ramil Nigmatullin, Julen S. Pedernales, Cecilia Puebla, and Andrea Smirne. Your revisions and feedback have certainly improved both readability and content of this thesis. Thank you all!

My period in Ulm comes now to an end. I feel fortunate and indebted to the wonderful people I have met during this time, whom I will resist to say good-bye: Alex, Andrea M. and Andrea S., Caco, Dario, Edgar, Felipe, Johnny, Julen, Kike, Moha, Lina, Oli, Pelayo and Raquel. Thank you to the GastroClub and the unforgettable moments we have lived together. Also to Kike and Marín, hope to meet you soon. Thank you all, because your friendship transformed Ulm into *our* home. Thank you to Samu and Ana, and at the same time my apologies for not having visited you as often as I should have, which also applies to the people listed below. I cannot forget my cherished Madrid—where good friends, family, and warm memories weave an almost unreal, yet cozy, feeling. Good memories as MH41 (Adri, Marco, Juli, Samu), the friends from always (Alex, Carlitos, Chuso, Fer, Jaime, Isma, Nacho, Rodri) and from the undergraduate (Ali, Belén, Gato, Ignacio, Laura, Miguel, Puy, Rafa), the GFN (Esther, Edu, Jacobo, Leti, Mailyn, Marie, Vadym, Vicky), and my mentors during the MSc. (José Manuel, Armando and Joaquín), now all spread over the world but whom I miss every day. I am grateful to Ayala's and Rubio's family, in special to Mari Carmen, Fernando, and Zaira for their attention and for finding always nice words that cheered me up. Thank you to my family, always supportive and understanding the long periods without showing up or replying, in particular to my beloved parents and sister, Dora, Cefe, and Ceci, because they have always encouraged me, no matter what, and from who I have learned priceless lessons. Mónica, although you know it better than anyone, I am indebted to you, for all the years we have been together; there are simply no words to express all my gratitude.

July 2018   Ricardo Puebla

Cooperations and Funding

This work has been carried out at the Institute of Theoretical Physics ("Institut für Theoretische Physik") at the Ulm University under the supervision of Prof. Dr. Martin B. Plenio. The research presented in Chap. 2 has been conducted in collaboration with Dr. Ramil Nigmatullin, currently in the University of Sidney, and Dr. Tanja E. Mehlstäubler from the Physikalisch-Technische Bundesanstalt (PTB). Chapters 3 and 4 contain original research results obtained in collaboration with Dr. Myung-Joong Hwang from Ulm University. Chapter 5 presents different results carried out in collaboration with Dr. Myung-Joong Hwang and Dr. Jorge Casanova from Ulm University, while the results shown in Chap. 6 have been performed in collaboration with Oliver Marty. Private communications from Dr. Kihwan Kim are acknowledged regarding the experimental conditions on the considered setup in Chap. 5, Dr. Christopher Monroe, Dr. Jiehang Zhang, and Dr. Jacob Smith, on the experimental setup for the results presented in Chap. 6.

Financial support is acknowledged to an Alexander von Humboldt Professorship, the EU STREPs PAPETS and QUCHIP, the EU Integrating Project SIQS and the ERC Synergy grant BioQ.

Contents

Chapter 1
Introduction

Our understanding of nature is built on the precise knowledge of the interaction between distinct constituents of matter. Certainly, the description of macroscopic properties arising from microscopic details is one of the fundamental problems in modern science, and of primarily importance in condensed matter physics and statistical mechanics [1]. In this regard, enlarging the degrees of freedom of a system typically provokes the emergence of qualitatively different behavior, leading to the appearance of macroscopic manifestation of microscopic interactions—cooperative or emergent phenomena that cannot be grasped by few interacting elements, as P. W. Anderson pointed out in his famous essay [2]. Examples are ubiquitous in nature, such as the existence of matter in diverse thermodynamic *phases* despite comprising the same atomic composition. These phases account for diverse macroscopic properties, while it is possible to transform matter into another phase by changing external conditions, an issue that has enthralled humankind throughout history and has been addressed by diverse means. An explanation of this phenomenon dates back to the ancient Greek philosophers Leucippus and Democritus, but it was not until the arrival of statistical mechanics in the 19th century when a precise and formal theory was introduced. It was then when the term *phase transition* was coined in relation to the transitions between different phases of matter. Not surprisingly, phase transitions appear at every scale of nature, from cosmology to the quantum world, thus representing an omnipresent phenomenon in physics, since matter varies its properties in a number of manners, breaking and restoring symmetries at expense of variations on external conditions. Among the vast number of known phase transitions, the emblematic case of water is perhaps the best illustration of this phenomenon. Despite the fact that water is made of the same constituents, it appears in distinct phases, namely, solid, liquid and gas. By simply varying the temperature one can easily obtain different macroscopic properties, like crystalline order in ice.

The advent of quantum mechanics irrevocably modified our view of nature and the underlying laws that govern it [3, 4]. This theory successfully addressed previously unsolved problems, while pointing out novel phenomena, that led to the discovery

© Springer Nature Switzerland AG 2018
R. Puebla, *Equilibrium and Nonequilibrium Aspects of Phase Transitions in Quantum Physics*, Springer Theses, https://doi.org/10.1007/978-3-030-00653-2_1

of new and exotic phases of matter. For example, the fact that Bose–Einstein condensation or superconductivity cannot be explained in classical terms reveals the
richer and striking properties that matter can acquire when quantum mechanical
effects become dominant. These emergent phases of quantum matter have attracted
the attention among the scientific community and their inspection assuredly has
become an active field of research during the last decades, not only due to our insa-
tiable desire to push the frontiers of knowledge, but also for practical reasons. In this
respect, for example, the quest for combining quantum mechanical effects and statistical mechanics has ultimately led to the discovery of phase transitions triggered by
quantum fluctuations, called quantum phase transitions (QPTs), which separate different exotic ordered phases of matter in the same manner as the more conventional
phase transitions [5, 6].

The great technological advances in quantum technologies have made possible
to study individual quantum entities with an unprecedented degree of precision,
isolation and control, bringing closer the compelling idea of Feynman [7]. Currently
it is feasible to explore the fundamental ingredients of matter and their interaction in
laboratories, an inconceivable task not long ago, as E. Schrödinger made clear: *"[...]
in thought-experiments we sometimes assume that we do (work with single atoms or
molecules); this invariably entails ridiculous consequences"* [8]. The manipulation
of mesoscopic quantum systems has made possible to inquire nature about the well-
grounded principles of thermodynamics and statistical mechanics and how they break
down when dealing with single interacting particles—a fascinating and fundamental
issue.

Either classical or quantum, phase transitions can be classified in two subgroups,
namely first-order and continuous [5, 9, 10]. This classification has its roots in
the work of P. Ehrenfest [11] and took firm ground by 1940s, less than a decade
after being proposed. Motivated by contemporary experiments on the superfluid
transition of helium (the λ transition) , Eherenfest argued that whenever the Gibbs
free energy becomes discontinuous, there is a phase transition. Depending on the
order n of the derivative in which the discontinuity appears, the transition is then
considered to be of first ($n = 1$), second order ($n = 2$), or higher ($n \geq 3$). Ironically,
this classification did not explain the λ transition, nor the logarithmic divergence
featured by a two-dimensional Ising model [12]. Thus, the Eherenfest classification
was patently incomplete and had to be either extended or amended. Later, it resulted
in the so-called modern classification, which only distinguishes between first-order
and continuous phase transitions [9, 10]. However, for historical reasons, the terms
second-order and continuous phase transitions are sometimes used indistinguishably.
Spontaneous symmetry breaking lies at the core of phase transitions, in which one of
the phases exhibits an order after a critical value (called critical point) of an external
parameter, such as temperature or magnetic field intensity.[1]

---

[1]Note that there are continuous phase transitions that do not break any symmetry, as it is the case
of the Kosterlitz–Thouless phase transition [13]. However, this special type of phase transition is
not covered in this thesis.

On top of equilibrium properties, the dynamics of systems featuring phase transitions or subject to symmetry breaking is of relevance, which is typically embraced under the generic name of nonequilibrium phenomena [9, 14–17]. Nonequilibrium dynamics addresses the broad question of how a particular system behaves when driven out of equilibrium, as, for example, how thermalization is attained after a sudden change in the external parameters or how a system reacts when driving across a phase transition. In addition, such effects can be of great relevance in other assignments, such as quantum annealing in the realm of quantum information processing [18, 19]. These issues, despite of their great relevance in nature, are much less understood than their equilibrium or static counterparts. In particular, in the context of dynamics of phase transitions we find the paradigmatic Kibble–Zurek (KZ) mechanism of defect formation [20]. This mechanism establishes an elegant and acclaimed relation between static and nonequilibrium properties of the phase transition in terms of the rate at which it is traversed.

This thesis, however, does not attempt to provide an analysis of the disparate critical phenomena emerging on diverse physical systems. Nonetheless, we will steer the reader through different topics, always within the framework of continuous phase transitions, starting in Chap. 2 with the paradigmatic Ginzburg–Landau model, to then move into the quantum realm. In particular, Chaps. 3, 4 and 5 examine the quantum Rabi model (QRM), a model in which we will unveil a novel type of critical behaviors, and study its static and nonequilibrium physics, as well as a potential experimental implementation. In addition, in the search for a robust implementation of the QRM, we analyze a suitable strategy to cope with certain noise sources. In Chap. 6 we consider a quantum many-body system, namely, an Ising model, which will serve us as a testbed for the inspection of the quantum Kibble–Zurek (QKZ) mechanism. The work presented in this thesis extends our knowledge about phase transitions and their associated dynamics, and may help to elucidate the rich physics uncovered by small quantum systems and their emergent phenomena. In addition, the main outcomes within this thesis may encourage experimental inquiries into these issues, which ultimately will allow us to understand and explore the intriguing features of quantum matter out of equilibrium. Finally, in Chap. 7 we will present the main conclusions and prospects for future work. Parts of this thesis have been first published in [21–25], see List of publications in the front matter. Although we have not introduced phrases or figures directly from these references, this thesis is based on these works, and therefore the content may be similar to that of these references, while citations are often omitted.

In this thesis we will focus on continuous phase transitions, and on their nonequilibrium dynamics, that is, on the features of systems when phase transitions participate or are involved in the dynamics. As aforementioned, in this context we find the KZ mechanism, which will be examined in detail in distinct scenarios, both classical and quantum. As continuous phase transitions constitute the central topic of this thesis, it is convenient to provide an overview of these critical phenomena, introducing the main concepts we will make use in the subsequent chapters. This introductory part is devoted to this task, where we exemplify these aspects reviewing the celebrated Ising model. In addition, we discuss the nonequilibrium scenario we will consider, together with an explanation of the KZ mechanism.

1.1 Continuous Phase Transitions

Phase transitions are classified in terms of how thermodynamics quantities behave close to the critical value of an external parameter, say ϵ, which divides two phases (or more) with different macroscopic properties. This critical value of the external parameter is known as critical point, and it represents any quantity whose variation triggers a phase transition at $\epsilon_c = 0$, such as temperature or magnetic field intensity.[2] For example, discontinuities in thermodynamic quantities lead to first-order phase transitions, whose main hallmark consists in the *coexistence* of the two phases of matter at the critical point and a non-zero latent heat [1, 14, 26]. However, while these transitions are certainly of importance (solid-liquid phase transitions), we will direct our attention to another kind, namely, continuous phase transitions. The latter display a diverging correlation at the critical point, and thus, the system adopts a unique, yet critical, state. As a consequence, the features of the system vary in a continuous manner as the critical point is traversed. In the following we provide a more detailed explanation of this type of phase transitions.

Let us briefly recall the basic procedure to determine thermodynamic quantities at thermal equilibrium of a system governed by a Hamiltonian H. In this respect, the so-called partition function Z plays a key role [1],

$$Z = \sum_{\sigma_n} e^{-\beta H(\sigma_n)} \tag{1.1}$$

where σ_n represents the nth possible microstate with energy $H(\sigma_n)$ and $\beta^{-1} \equiv k_B T$ the thermal energy, where k_B corresponds to the Boltzmann constant. In this manner, the probability of finding the a macroscopic state in the nth microstate reads $e^{-\beta H(\sigma_n)}/Z$, and thus, macroscopic quantities are obtained as $\langle O \rangle = \sum_n O_n e^{-\beta H(\sigma_n)}/Z$, where O_n stands for the value of O in the nth microstate. The thermodynamic functions follow from Z, as for example the Helmholtz free energy

$$h = -\frac{1}{\beta} \ln Z. \tag{1.2}$$

However, these thermodynamic functions may undergo singularities at particular values of an external parameter, denoted by ϵ, with $\epsilon_c = 0$ the critical point. This very fact seems to contradict the analytic behavior not only of Z, but also of these thermodynamic functions. However, singularities take place only in the thermody-

---

[2]For temperature, for example, this parameter would become $\epsilon = T/T_c - 1$ where T_c stands for the critical temperature of the phase transition.

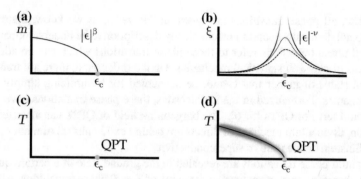

Fig. 1.1 Schematic illustration of the main traits of a continuous phase transition, namely, order parameter m (**a**) and correlation length ξ (**b**) as a function of the external parameter ϵ. The order parameter vanishes in the disorder and increases as $|\epsilon|^\beta$ for $\epsilon < \epsilon_c = 0$, while the correlation length diverges as $\xi \sim |\epsilon|^{-\nu}$. In **b** the impact of finite-size systems is illustrated by dashed lines. The panels **c** and **d** show the possible situation between QPT and classical or thermal phase transitions. In **c** the system exhibits only a quantum critical point at ϵ_c and there is no phase transition at finite temperature, as it is the case of a one-dimensional Ising model (Sects. 1.2.2 and 1.2.3). In **d** the system portrays thermal phase transitions along the line, which terminates at $T = 0$ with a QPT

namic limit, $N, L \to \infty$ keeping the density finite. This limit grants Eq. (1.1) the possibility to develop singularities since the limit function of a sequence of analytic functions need not be analytic.

Continuous phase transitions are characterized by the emergence of a spontaneous ordering for values of the external parameter $\epsilon < \epsilon_c = 0$. In contrast to first-order phase transitions, there is no abrupt jump, but rather a continuous variation on the properties of the system. The underlying mechanism resides in a diverging correlation across the system, $\xi \propto |\epsilon|^{-\nu}$, where ξ represents the correlation length. Besides ξ, it is possible to quantify the amount of order by the so-called order parameter m. This quantity vanishes in the entire disordered phase ($\epsilon > 0$), while it increases as $m \propto |\epsilon|^\beta$ for $\epsilon < 0$. We will comment in Sect. 1.1.1 the relevance of the exponents ν and β (called critical exponents), the impact of finite-size systems and the universality of these phase transitions. These features are sketched in Fig. 1.1a and b, where in the latter we highlight the impact of finite-size systems, that show no true singularity, but they rather exhibit a smooth behavior.

The contribution of L. Landau deserves special mention as he deeply improved our understanding of phase transitions and critical phenomena [26]. The model currently known as Ginzburg–Landau model provides a correct description, on phenomeno-logical and mean-field grounds, of the behavior of the order parameter in the vicinity of the critical point. As this model can be used to address critical behavior in several scenarios, it explicitly highlighted the universality of continuous phase transitions. However, we do not provide here a detailed explanation of this model as it will be analyzed in Chap. 2.

Finally, we remark that, since quantum mechanics provides the most accurate description of nature, everything should obey the laws of quantum mechanics. In this sense, all phase transitions are quantum. However, as we know, many problems do not demand a quantum mechanically description to be understood, and thus classical phase transitions refer to those phase transitions which can be addressed by means of classical statistical mechanics. On the order hand, there are transitions between states of matter that cannot be accounted for by utilizing simply classical mechanics. Formalized in the last decades, these phase transitions have gained the natural term of QPTs [5, 6]. The burgeoning field of QPTs has attracted great attention, as quantum mechanical effects are behind exotic phases of matter such as Bose–Einstein condensate or superconductivity.

Quantum phase transitions are revealed in the ground state of a certain quantum system, therefore, they often receive the name of $T = 0$ phase transitions. Although there are first-order QPTs, we shall focus on their continuous version. Let $H = H_0 + gH_1$ be the Hamiltonian of the system where H_0 and H_1 do not commute, $[H_0, H_1] \neq 0$, showing different symmetries. Loosely speaking, we may identify a critical point g_c at which there is an avoided crossing and from which ground-state properties differ. In the thermodynamic limit, the avoided crossing becomes a non-analytic point at which the energy gap Δ vanishes, thus, g_c is a quantum critical point. Moreover, the theory of classical continuous phase transitions can be translated to this context, such as the emergence of long-range correlations ξ and order parameters m. Yet, although QPTs emerge at $T = 0$, they provide a good description of thermodynamic quantities at non-zero temperatures T and for wide range of $|g - g_c|$ [5]. The discussion on critical exponents, scaling laws and universality provided in the following lines also applies to QPTs upon the identification $\epsilon = g/g_c - 1$. Finally, we point out that there are two possible scenarios, as we will discuss later in the Ising model, that are illustrated in Fig. 1.1c and d. For the former there is only a phase transition at strictly zero temperature (e.g. one-dimensional Ising model, see Sects. 1.2.2 and 1.2.3), while for the latter there is a critical line at which classical phase transitions take place; such line terminates at $T = 0$, which leads to a QPT (e.g. two-dimensional Ising model, see Sect. 1.2.2). For further elements on the phenomenology of QPTs see [5, 6].

1.1.1 Critical Exponents

A continuous phase transition is characterized by a set of positive numbers named critical exponents, which indicate how certain macroscopic quantities behave close to its critical point $\epsilon_c = 0$ [9, 10]. We will denote critical behavior when these quantities show a power-law dependence as $\epsilon \to 0$ with a particular exponent, say κ, i.e., $|\epsilon|^\kappa$, and from this it follows that κ is a critical exponent. Certainly, these critical exponents are of primary importance in the understanding of continuous phase transitions as well as their classification in the so-called universality classes (see Sect. 1.1.3).

In general, certain quantities close to a continuous phase transition obey $\mathcal{A} = \mathcal{A}_0|\epsilon|^\kappa$ where $\mathcal{A}_0$ is simply a constant value depending on the microscopic details of the system. Then, a continuous phase transition is characterized by a set of critical exponents, which can be gathered as[3]

$$c_v \sim |\epsilon|^{-\alpha}, \quad m \sim |\epsilon|^{\beta}, \quad \chi \sim |\epsilon|^{-\gamma}, \quad m \sim |g|^{1/\delta}, \quad \xi \sim |\epsilon|^{-\nu}, \text{ and } \tau \sim |\epsilon|^{-z\nu},$$
$$(1.3)$$

which correspond to the heat capacity c_v, order parameter m, susceptibility $\chi = (\partial m / \partial g)|_{g=0}$, correlation length ξ and relaxation time τ. Note that the relation $m \sim |g|^{1/\delta}$ takes place at the critical point $\epsilon_c = 0$, where g denotes an external source field, such as magnetic field in a standard Ising model (see Sect. 1.2). In addition, we stress that β must not be confused with the inverse of the temperature—a common but unfortunate notation, which is kept also in this thesis. The previous expressions indicate that an infinitely extended system exhibits an infinitely correlation length at the critical point, a clear symptom of ordering, and that it will take infinitely long time to relax. In this regard, the two-point correlation function $G(\mathbf{r}_1, \mathbf{r}_2)$ provides a quantification of how correlations spread across a system, ultimately leading to the definition of a correlation length ξ. In particular, with $\mathbf{m}(\mathbf{r})$ the local order parameter,

$$G(\mathbf{r}_1, \mathbf{r}_2) = \langle \mathbf{m}(\mathbf{r}_1) \cdot \mathbf{m}(\mathbf{r}_2) \rangle - \langle \mathbf{m}(\mathbf{r}_1) \rangle \langle \mathbf{m}(\mathbf{r}_2) \rangle \quad (1.4)$$

shows an exponential decay $G(\mathbf{r}_1, \mathbf{r}_2) \sim e^{-|\mathbf{r}_1 - \mathbf{r}_2|/\xi}$ with a typical length scale ξ, called correlation length, typically defined as $\xi^2 = \sum_{\mathbf{r}} |\mathbf{r}|^2 G(\mathbf{r}, 0) / \sum_{\mathbf{r}} G(\mathbf{r}, 0)$. In the vicinity of the critical point, ξ diverges in the aforementioned fashion, leading to a power-law decaying correlation $G(\mathbf{r}_1, \mathbf{r}_2) \sim |\mathbf{r}_1 - \mathbf{r}_2|^{2-d-\eta}$ where η is yet another critical exponent. However, among all these critical exponents, ν and z will play a crucial role in the remaining of the thesis.

In contrast to the rest of critical exponents that are solely determined by static properties, the dynamical critical exponent z depends on the dynamics. In the particular case of isolated quantum systems, a relevant energy Δ in the Hamiltonian governing the system provides this relaxation time, $\tau = \Delta^{-1}$, that vanishes at the critical point as $\Delta \sim |\epsilon|^{z\nu}$ [5]. Note that, since QPTs are originated in the ground state of a quantum system, Δ typically stands for the energy gap between ground and first excited state. In Table 1.1 we provide the critical exponents of two paradigmatic universality classes, namely, the two-dimensional Ising model and the mean field (Ginzburg–Landau model discussed in Chap. 2). Finally, it is worth noting that critical exponents are related among them by the so-called scaling laws, that may also include the spatial dimension of the system [9, 10, 27].

---

[3]There is in principle no underlying reason to assume that these exponents take the same value at both sides of the critical point, however, the inspection of phase transitions in most systems teaches us that this is actually the case. Nevertheless, since $m = 0$ in the disordered phase, it is evident that β is, by definition, an exception. It is worth mentioning however that there are special situations where $\kappa_+ \neq \kappa_-$.

Table 1.1 Critical exponents of the two-dimensional (classical) Ising model and those of the standard mean-field description. The heat capacity exhibits a logarithmic divergence in the Ising, and a finite jump in mean-field phase transitions, respectively [27]. Note that the dynamical critical exponent z is not given. As we will see in detail in Chap. 2 in a Ginzburg–Landau model, this exponent depends on the dynamics

	α	β	γ	δ	η	ν
2d Ising	0 (log)	1/8	7/4	15	1/4	1
Mean field	0 (jump)	1/2	1	3	0	1/2

1.1.2   Finite-Size Scaling Hypothesis

Phase transitions occur only in the thermodynamic limit, $N, L \to \infty$. It is only upon this limit when these striking power-law dependencies are valid (see Eq. 1.3), while for any system consisting of a finite number of components they will eventually break down as no true singularity can take place. This naturally leads us to ponder on how critical features emerge as the system size increases. The beauty of phase transitions resides in that the behavior in the limiting case allows us to understand how these properties are recovered by approaching the thermodynamic limit, i.e., by increasing its components, and vice versa.

A system near a critical point exhibits a remarkable feature, which was first pointed out by B. Widom on phenomenological grounds and later formalized by M. E. Fisher and M. N. Barber on the so-called finite-size scaling hypothesis [28, 29]. Finally, these arguments were better explained with the advent of renormalization group (RG) techniques, suggested by L. P. Kadanoff and firmly developed by Wilson [30]. Briefly, RG methods exploit the scale invariance or self-similarity that the system exhibits at the critical point. In particular, the system shows a long-range correlation, featuring fluctuations within a length scale ξ, and therefore, microscopic degrees of freedom at smaller scales are irrelevant and can be neglected. This process is commonly known as coarse-grain. One then needs to rescale spatial coordinates and renormalize parameters of the system. However, we do not rely on RG arguments [30] but rather introduce the finite-size scaling theory in a more heuristic fashion (the reader is referred to [9, 10] for detailed explanation of RG). In the vicinity of the critical point the system looks the same within length ξ. This self-similar condition motivates the identification of $x = L/\xi$ as a scaling variable, dropping the microscopic details of the system. Thus, if $\mathcal{A} = \mathcal{A}_0|\epsilon|^\kappa$ in the thermodynamic limit, and because any thermodynamic quantity must be well behaved for finite-size systems $L < \infty$,

$$\mathcal{A}(L) = \mathcal{A}_0|\epsilon|^\kappa F_{\mathcal{A}}\left(|\epsilon|^\nu L\right) \tag{1.5}$$

where we have introduced the finite-size scaling function $F_{\mathcal{A}}(x)$ satisfying the asymptotic conditions $\lim_{x\to\infty} F_{\mathcal{A}}(x) = 1$ and $\lim_{x\to 0} F_{\mathcal{A}}(x) \sim x^{-\kappa/\nu}$, which restores the expected behavior in the $L \to \infty$ and prevents singular behavior for any finite L. We note that finite-size scaling functions are constructed on grounds of the leading order scaling, i.e., neglecting higher-order corrections. Hence, the larger the system size the more accurate the finite-size scaling hypothesis.

From Eq. (1.5) one can draw two striking consequences. First, at the critical point $\epsilon = 0$ ($x = 0$) a simple scaling relation is accomplished

$$\mathcal{A}(L) \sim L^{-\kappa/\nu}. \tag{1.6}$$

Therefore, as L grows the quantities scale as $L^{-\kappa/\nu}$ at the critical point. In particular, ξ scales linearly with L, $\xi \sim L$, while the relaxation time grows as $\tau \sim L^z$. Second, Eq. (1.5) implies that $\mathcal{A}|\epsilon|^{-\kappa}$ becomes independent of the specific values of ϵ and L alone. Instead, $\mathcal{A}|\epsilon|^{-\kappa}$ adopts a universal form in terms of $x = |\epsilon|^\nu L$: data of $\mathcal{A}$ obtained for different ϵ and L values *collapse* into a single curve, disclosing the functional form of the finite-size scaling function $F_\mathcal{A}(x)$. We stress that $F_\mathcal{A}(x)$ has in general a different shape for $\epsilon < 0$ and $\epsilon > 0$. However, in order to lighten the notation this is not explicitly written.

1.1.3 Universality

The universality of continuous phase transitions is perhaps the most powerful tool we possess to understand critical behaviors. As we have seen, phase transitions are characterized by critical exponents, spatial dimension and their finite-size scaling functions. The reader may have already noticed that these properties do not depend on the microscopic details. This quite remarkable fact establishes a correspondence between systems, that are a priori professedly distinct. Indeed, attending merely to critical exponents, dimension and scaling functions, one can observe that their physics is equivalent up to constant factors accounting for irrelevant microscopic details [9, 10]. Therefore, systems that share these properties are said to fall into the same *universality class*. This fact marked a milestone and constitutes one of the central pillars in the theory of critical phenomena [31].

Therefore, either classical or quantum, phase transitions can be sorted in a series of *universality classes*. Systems undergoing phase transitions that belong to the same universality class behave similarly. This grants a powerful tool for the understanding of the great diversity of phases and transitions between them. For example, the paradigmatic Ising ferromagnetic phase transition, that we will present below, falls into the same universality class as the liquid-gas phase transition, while a two-component Heisenberg model, typically known as classical XY model or quantum rotor, shares the same physics as that of superfluid-normal transition in, say, liquid helium. Hence, by analyzing the Ising ferromagnetic phase transition one learns about the liquid-gas phase transition. This adds importance to the study of systems that are easier to handle, since their examination teaches us phenomena from much more complicated models that fall into the same universality class.

1.2   Ising Model

The Ising model is perhaps the most emblematic and popular system in the realm
of phase transitions and critical phenomena, with applications to diverse areas of
research. Although it was originally suggested by W. Lenz as a simplified model to
examine ferromagnetism, the model was named after his student E. Ising for historical
reasons [32], who in 1925 solved the one-dimensional case [33] but failed to predict
its behavior in higher dimensions. It was then L. Onsager who in 1944 demonstrated
that a two-dimensional system does undergo a phase transition [12], bringing the
Ising model to the scientific spotlight and furnishing it with renewed interest. E. Ising
believed the model he was dealing with was too simple to be physically relevant. Quite
the opposite, although simple, the Ising model is currently regarded as a cornerstone
in statistical mechanics. Moreover, thanks to its simple formulation it can be applied
to several frameworks, ranging from condensed matter to social sciences.

The Ising model describes the interaction of N spins arranged in a d-dimensional
lattice, whose state can be either up ($\sigma_i = +1$) or down ($\sigma_i = -1$) and interact, in
the standard case, only between nearest neighbors. We will analyze a long-range
interaction in Chap. 6. The Hamiltonian of such a system can be written as

$$H_{\text{Ising}} = -J \sum_{\langle ij \rangle} \sigma_i \sigma_j - g \sum_i \sigma_i \qquad (1.7)$$

where the sum $\langle ij \rangle$ runs over neighboring lattices sites. In addition, the parameter J
accounts for the type of spin-spin interaction ($J > 0$ ferromagnetic or $J < 0$ anti-
ferromagnetic) and g represents an external magnetic field. Throughout this Section,
where we introduce and illustrate the phenomenology of continuous phase transitions
in an Ising model, we will focus on ferromagnetic couplings, but the same arguments
can be applied to $J < 0$ since the models are mathematically identical.[4] In Fig. 1.2
we plot a schematic representation of the model, for $J > 0$, illustrating the possible
ordered phases emerging at sufficiently low temperatures.

1.2.1   Mean-Field Theory

As in any other branch of physics, mean-field theories provide a helpful tool to gain
insight into a complex system of many interacting particles. In the realm of phase
transitions, mean-field approximations represent a usual and suitable strategy for
gaining analytic understanding, typically otherwise unattainable. This is however
not the case for the Ising model in one or two dimensions, which admits an exact
solution. Nevertheless, we provide here a brief discussion on how to apply a mean-

---

[4]Note that an anti-ferromagnetic Ising model with zero external field can be mapped onto its
ferromagnetic counterpart by redefining spin variables as $\tilde{\sigma}_i = (-1)^i \sigma_i$.

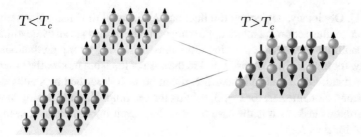

Fig. 1.2 Schematic illustration of the phases of a two-dimensional Ising model. Below the critical point, $T < T_c$, two equivalent ordered states arise in which all the spins are aligned, while for $T > T_c$ the spontaneous ordering is lost. See main text for details

field theory on the Ising model, discussing its benefits and flaws, while detailed discussions can be found in any standard textbook on statistical mechanics (see [1] for example). As aforementioned, Ginzburg–Landau model of phase transitions is a mean-field theory, which we will analyze in Chap. 2.

The essential assumption of the mean-field theory resides in neglecting fluctuations. In particular, for a d-dimensional Ising model, Eq. (1.7), the spin value can be written as $\sigma_i = \langle \sigma_i \rangle + \tilde{\sigma}_i$, with $\langle \sigma_i \rangle = m = 1/N \sum_i \langle \sigma_i \rangle$ the mean magnetization and $\tilde{\sigma}_i$ a fluctuation. Setting $\tilde{\sigma}_i = 0$, the local Hamiltonian of a spin ith interacting with γ nearest-neighbors reads

$$H_{\sigma_i} = -J \sum_{\langle j \rangle} \sigma_i \sigma_j - g\sigma_i \stackrel{\text{MF}}{=} -\sigma_i(\gamma Jm + g), \qquad (1.8)$$

and thus, we have effectively reduced the problem of N interacting spins to a single spin subject to an effective magnetic field $\gamma Jm + g$. Hence, as $\langle \sigma_i \rangle = m$, the magnetization can be obtained from

$$m = \langle \sigma_i \rangle = \frac{\text{Tr}\, \sigma_i e^{\beta\sigma_i(\gamma Jm+g)}}{\text{Tr}\, e^{\beta\sigma_i(\gamma Jm+g)}} = \tanh\left(\beta(\gamma Jm + g)\right) \qquad (1.9)$$

The previous self-consistent expression has always a trivial solution for $g = 0$, namely $m = 0$, while the interesting symmetry-breaking solutions ($m \neq 0$) take place when $\beta_c \gamma J > 1$, which leads to the critical temperature $k_B T_c = \gamma J$. Moreover, the critical exponents of the mean-field phase transition can be worked out [27], and are collected in Table 1.1, equivalent to those of the Ginzburg–Landau model.

Interestingly, for $d = 1$ the mean-field approach erroneously predicts the existence of a phase transition, as E. Ising proved in [33]. For a two-dimensional cubic lattice, $\gamma = 4$, there is indeed a phase transition at $k_B T_c 2J / \ln(1 + \sqrt{2}) \approx 2.269 J$ as shown by Onsager [12], which differs from this crude mean-field approximation, $k_B T_c = 4J$. Note however that better estimates can be attained by means of more refined mean-field procedures, as the typical Bragg–Williams or Bethe–Peierls theories [1]. Obviously, neglecting the fluctuations in small dimensional systems with short-range interaction is a major approximation, which leads to an overestimation of the capabilities of a system to order, ultimately conveying wrong predictions. Interestingly, for higher dimensions, $d > 4$ in the case of the Ising model, the system falls into the mean-field class and therefore, it can be well described by standard mean-field theory. Nonetheless, for $d \leq 3$, solving the seemingly harmless Ising model can be an arduous task, as it is the case for $d = 2$, or even impossible as it seems to be the case for $d = 3$.

1.2.2 Exact Solutions: Outline

The relatively easy solution of the one-dimensional Ising model, carried out by Ising [33], contrasts with the one for two-dimensional Ising model, carried out by Onsager [12]. In the following we briefly outline these solutions, which demonstrate the absence of a phase transition in one dimension at finite temperature. For details about the derivation we refer the interested reader to [1, 12, 33–35]. A solution to the three-dimensional Ising model is to date not known, and thus, its properties are typically obtained by means of numerical simulations.

1.2.2.1 One-Dimensional Model

For a one-dimensional system the partition function Z reads

$$Z = \sum_{\sigma} e^{-\beta H_{\text{Ising}}} = \sum_{\sigma} e^{\beta[J \sum_i \sigma_i \sigma_{i+1} + \frac{g}{2} \sum_i (\sigma_i + \sigma_{i+1})]} \tag{1.10}$$

$$= \sum_{\sigma_1} \sum_{\sigma_2} \cdots \sum_{\sigma_N} \langle \sigma_1 | T | \sigma_2 \rangle \langle \sigma_2 | T | \sigma_3 \rangle \ldots \langle \sigma_N | T | \sigma_1 \rangle \tag{1.11}$$

$$= \sum_{\sigma_1} \langle \sigma_1 | T^N | \sigma_1 \rangle = \text{Tr } T^N = \lambda_1^N + \lambda_2^N \tag{1.12}$$

where a periodic system, $\sigma_{N+1} = \sigma_1$, has been considered for convenience, and the two-spin matrix T, whose entries are given by $\langle \sigma_i | T | \sigma_j \rangle = e^{\beta[J\sigma_i\sigma_j + g(\sigma_i + \sigma_j)/2]}$, has two eigenvalues $\lambda_{1,2}$ which read

$$\lambda_{1,2} = e^{\beta J} \left[\cosh(\beta g) \pm e^{\beta(J-g)} \sqrt{e^{-4\beta J} + \sinh^2(\beta g)} \right]. \tag{1.13}$$

Since we are interested in the thermodynamic limit $N \to \infty$, and because $\lambda_1 > \lambda_2$, the partition function simplifies to $Z = \lambda_1^N$, from where the average magnetization can be readily obtained

$$m = -\frac{1}{N}\frac{1}{\beta}\frac{\partial \ln Z}{\partial g} = \frac{e^{\beta(2J-g)}(e^{2\beta g}-1)}{2\sqrt{1+e^{4\beta J}\sinh^2(\beta g)}}. \qquad (1.14)$$

As soon as $g = 0$, the magnetization vanishes, and thus, there is no critical temperature below which spontaneous ordering occurs. There is no classical phase transition in a one-dimensional Ising model. Note however that, as we will show in Sect. 1.2.3, it is possible to attain criticality in a one-dimensional Ising model, but it happens at $T = 0$ and it is a QPT instead, thus showing a phase diagram as illustrated in Fig. 1.1c.

1.2.2.2 Two-Dimensional Model

The solution two-dimensional Ising model turns out to be rather lengthy and mathematically involved. For that reason, we refer the reader to the original work by L. Onsager in 1944 [12], as well as to the work by Yang [34], or to the standard textbooks [1, 35], while we quote here the main results.

The partition function of the two-dimensional Ising model, Eq. (1.7), with zero field can be written again as $Z = \text{Tr } T^n$ where the number of spins is $N = n^2$. However, finding the largest eigenvalue of the transfer matrix T, now of $2^n \times 2^n$ dimension, is a laborious task. The Helmholtz free energy per spin becomes

$$h = -\lim_{N\to\infty}\frac{1}{N}\frac{1}{\beta}\ln Z = -\lim_{n\to\infty}\frac{1}{n}\frac{1}{\beta}\ln \lambda_1 \qquad (1.15)$$

$$= -\frac{1}{\beta}\ln[2\cosh(2\beta J)] - \frac{1}{2\pi\beta}\int_0^\pi d\phi \, \ln\left[\frac{1}{2}+\frac{1}{2}\sqrt{1-K^2\sin^2\phi}\right] \qquad (1.16)$$

where $K = 2\left(\cosh(2\beta J)\coth(2\beta J)\right)^{-1}$. From there, one can obtain the internal energy per spin u

$$u = -\lim_{N\to\infty}\frac{1}{N}\frac{\partial \ln Z}{\partial \beta} = \frac{\partial(\beta h)}{\partial \beta} \qquad (1.17)$$

and subsequently, the heat capacity $c_v = \frac{\partial u}{\partial T}$, which reveals a logarithmic divergence at a particular critical temperature. However, the real order-disorder transition—the emergence of a spontaneous symmetry breaking for $T < T_c$—is quantified in terms of the magnetization $m = \langle\frac{1}{N}\sum_i \sigma_i\rangle$, which was worked out by Yang in [34], and reads

$$m = \left[1 - \sinh(2\beta J)^{-4}\right]^{1/8} \qquad (1.18)$$

The critical temperature can be then obtained when $m = 0$, and corresponds to $T_c = 2J\,\text{arcsinh}^{-1}(1) = 2J\ln^{-1}(1+\sqrt{2})$. Moreover, from these results one extracts the critical exponents, as $\beta = 1/8$ from Eq. (1.18), collated in Table 1.1.

1.2.3 Transverse-Field Quantum Ising Model

The one-dimensional quantum Ising model with a transverse field, is perhaps the most paradigmatic many-body system undergoing a QPT, whose fame certainly stems from possessing an exactly solution [5, 35]. Moreover, as we have seen, such a one-dimensional system lacks of a thermal phase transition, being the QPT the only relevant critical phenomena of this system, and thus its phase diagram corresponds to the one depicted in Fig. 1.1c. The Hamiltonian can be written as

$$H_{\text{TFIM}} = -J \sum_j \left(\sigma_j^z \sigma_{j+1}^z + g\sigma_j^x \right), \tag{1.19}$$

where now each spin is described quantum mechanically through the Pauli matrices of the spin-$\frac{1}{2}$ algebra,

$$\sigma^x = \begin{pmatrix} 0 & 1 \\ 1 & 0 \end{pmatrix}, \quad \sigma^y = \begin{pmatrix} 0 & -i \\ i & 0 \end{pmatrix}, \quad \text{and} \quad \sigma^z = \begin{pmatrix} 1 & 0 \\ 0 & -1 \end{pmatrix}, \tag{1.20}$$

which satisfy $[\sigma_i, \sigma_j] = 2i\epsilon_{ijk}\sigma_k$. We remark that the interaction is limited to nearest neighbors, while its long-ranged counterpart will be examined in Chap. 6. Here we shall consider periodic boundary conditions, $\sigma_{N+1}^{x,y,z} = \sigma_1^{x,y,z}$. As commented, H_{TFIM} admits an exact solution, which is accomplished by mapping spins onto spinless fermions, represented by the usual creation and annihilation operators, $c_i^\dagger$ and c_i, respectively, satisfying $\{c_i, c_j^\dagger\} = \delta_{i,j}$. Indeed, $\sigma_j^z = 1 - 2c_j^\dagger c_j$, and thus, spin up corresponds to an empty fermionic site, and vice versa. Moreover, raising a spin simply leads to the identification $\sigma_j^+ = c_j^\dagger$. Although one may think that the previous map can be directly introduced in H_{TFIM}, one must notice that while spin operators acting on different sites commute, fermionic operators anticommute. The previous map is valid for a single spin, but it fails when dealing with more sites. The solution to this problem was found by Jordan and Wigner [36], who proposed the following transformation

$$\sigma_j^+ = \prod_{i<j}(1 - 2c_i^\dagger c_i)c_j \tag{1.21}$$

$$\sigma_j^- = \prod_{i<j}(1 - 2c_i^\dagger c_i)c_j^\dagger. \tag{1.22}$$

under which commutation and anticommutation relations are indeed fulfilled. In order to solve H_{TFIM} we however take $\sigma_z \rightarrow \sigma_x$ and $\sigma_x \rightarrow -\sigma_z$, that is

$$\sigma_j^x = (1 - 2c_j^\dagger c_j) \tag{1.23}$$

$$\sigma_j^z = -(c_j + c_j^\dagger) \prod_{i<j}(1 - 2c_i^\dagger c_i). \tag{1.24}$$

Moreover, after performing the previous map to H_{TFIM} and Fourier transforming the fermionic operators, $c_k = N^{-1/2} \sum_j c_j^{-ikr_j}$, the quantum Ising model becomes

$$H_{\text{TFIM}} = J \sum_k \left(2\left[g - \cos(ka)\right] c_k^\dagger c_k + i \sin(ka) \left[c_{-k}^\dagger c_k^\dagger + c_{-k}c_k \right] - g \right), \tag{1.25}$$

where a corresponds to the spacing between neighboring spins. The previous Hamiltonian can be diagonalized employing Bogoliubov transformation, that transforms c_k into a new fermionic operator γ_k, $\left\{ \gamma_k, \gamma_{k'}^\dagger \right\} = \delta_{k,k'}$, such that its number is conserved,

$$\gamma_k = u_k c_k - i v_k c_{-k}^\dagger, \tag{1.26}$$

with $u_{-k} = u_k$ and $v_{-k} = -v_k$ real coefficients that satisfy $u_k^2 + v_k^2 = 1$. Their specific value is determined demanding only terms $\gamma_k^\dagger \gamma_k$ in the transformed H_{TFIM}. Choosing $u_k = \cos(\theta_k/2)$ and $v_k = \sin(\theta_k/2)$ with $\tan(\theta_k) = \sin(ka)/(g - \cos(ka))$, we achieve

$$H_{\text{TFIM}} = \sum_k \epsilon_k \left(\gamma_k^\dagger \gamma_k - \frac{1}{2} \right), \quad \text{with} \quad \epsilon_k = 2J \left(1 + g^2 - 2g\cos(ka)\right)^{1/2}. \tag{1.27}$$

The ground state corresponds to a vacuum state $|0\rangle$, such that $\gamma_k |0\rangle = 0 \ \forall k$, while excited states are obtained occupying single-particle states, $\gamma_{k_1}^\dagger \gamma_{k_2}^\dagger \cdots \gamma_{k_n}^\dagger |0\rangle$. The minimum relevant energy gap takes place for $ka = 2n\pi$ with $n = 0, 1, \ldots$, $\Delta = 2J|g - g_c|$ where $g_c = 1$ locates the critical point. Therefore, $z\nu = 1$ for the one-dimensional transverse field quantum Ising model. A closer inspection reveals that $\nu = 1$, $z = 1$ and $\beta = 1/8$, as the classical two-dimensional Ising model (see Table 1.1) [5]. This fact actually brings us to the quantum-classical map, that exploits the similarity between $e^{-\beta H}$ and the time-evolution propagator e^{-itH} upon the identification of an imaginary time. Briefly, a QPT in d dimensions is related to a classical phase transition in $(d + z)$ dimensions (see Refs. [5, 6] for a detailed discussion).

1.3 Nonequilibrium Dynamics and the Kibble–Zurek Mechanism

The properties of a system at thermal equilibrium is of primary interest for understanding collective phenomena, different phases of matter and their corresponding transition between them, as we have briefly explained in the previous lines. However, most interesting physics occurs when a system finds itself out of equilibrium. Nonequilibrium phenomena are omnipresent at every level of physics, playing an essential role in nature [14]. However, and despite their great relevance, nonequilibrium scenarios are, in general, much less understood than their equilibrium or static counterparts.

Among the vast territory of nonequilibrium phenomena, it is perhaps worth highlighting different relevant topics such as thermalization (or the lack of), transport properties, defect formation and the more traditional out-of-equilibrium thermodynamics dealing with entropy production and irreversibility [14]. In this thesis we are interested in the nonequilibrium aspects of phase transitions, or, in other words, in how the system reacts when it is driven across or close to a critical point. This particular issue has attracted attention during the last decades, being an active field of research. However, the inspection of critical dynamics has encouraged novel questions and problems that still remain to be understood [37]. Even restricting ourselves to the seemingly specific topic of dynamics of phase transitions, there is enough broad phenomenology whose mention largely exceeds the purpose of this thesis (see [9, 15–17]). For that reason, we specify the particular problem we will analyze when dealing with nonequilibrium dynamics.

In particular, we will consider mostly linear ramps across or towards the critical point at which a continuous phase transition takes place, that is, we are mainly interested in the dynamics of symmetry breaking, an important problem of statistical mechanics with application to several scenarios. For that reason, we rely on a protocol

$$\epsilon(t) = \epsilon_0 + (\epsilon_1 - \epsilon_0)\frac{t}{\tau_Q}, \qquad 0 \le t \le \tau_Q, \qquad (1.28)$$

with $\epsilon_{0,1}$ being the initial and final values of the external parameter, and a quench rate $\dot\epsilon(t) = (\epsilon_1 - \epsilon_0)/\tau_Q$. The system is driven through the critical point into the ordered phase, which as a consequence of the divergence of the relaxation time $\tau \sim |\epsilon|^{-z\nu}$, unavoidably departs from equilibrium and the choice of a new symmetry-breaking state must be done locally. Hence, different spatial regions adopt a different symmetry, and the system is divided in a series of *domains*. Within a domain, the system is ordered, namely, the order parameter adopts a homogeneous value $\mathbf{m}_n$ in these spatial regions, i.e., $\mathbf{m}(\mathbf{r}_i) = \mathbf{m}_n \ \forall \mathbf{r}_i \in D_n$ with $n = 1, 2 \dots N_d$ and N_d the number of different domains. These domains are separated from others with different order $\mathbf{m}_n \neq \mathbf{m}_m$, by an interface or defects, which are topologically stable objects. Needless to say, these defects increase the energy of the system, and thus they are excitations promoted by the divergences at the critical point. Intuitively, the slower the critical point is traversed, the more adiabatic the protocol and less number of defects is formed, ideally leading to $N_d = 1$ in a purely adiabatic ramp.

This dynamical problem was first suggested by T. W. B. Kibble as a possible mechanism to explain topological defect formation in a cosmological context, where causality prevents the acquisition of common symmetry-breaking states in regions not connected through the light cone [38–40]. It was then W. H. Zurek who harvested the ideas of Kibble and applied them to condensed matter problems, addressing defect formation as a fundamental consequence of symmetry breaking at finite rate [41, 42]. Moreover, thanks to the universality of continuous phase transitions, the number of defects is predicted to follow a universal power-law scaling in terms of the quench rate $\dot{\epsilon}(t)$. This theory is currently known as KZ mechanism, and has become a milestone in the realm of dynamics of phase transitions [20]. One of the seminal works [43] clearly summarizes the aim of the KZ mechanism in its suggestive title: *"Density of kinks after a quench: when symmetry breaks, how big are the pieces?"*. The KZ mechanism has been acknowledged to be relevant across several systems [43–47], extended to inhomogeneous systems [48–50] and successfully verified in a variety of experiments [51–54]. Moreover, KZ mechanism can be also applied to QPTs, as indicated in [55–58]. Nevertheless, the experimental confirmation of the quantum KZ (QKZ) is evidently more challenging, and it has been only recently when QKZ scaling laws have been observed [59, 60].

The KZ mechanism, in its standard formulation, can be summarized as follows. Let a system undergoing a continuous phase transition be characterized by its diverging correlation length ξ and relaxation time τ,

$$\xi = \xi_0 \, |\epsilon|^{-\nu}, \quad \text{and} \quad \tau = \tau_0 \, |\epsilon|^{-z\nu}, \tag{1.29}$$

respectively, where ϵ corresponds to the external varied parameter, following the protocol given in Eq. (1.28), and ξ_0 and τ_0 are simply constant values. At $\epsilon_c = 0$ there is a continuous phase transition at $g = g_c$. Then, because the system is driven towards the critical point, there is a competition between two time scales. Far away from the critical point the relaxation time τ is sufficiently short such that the systems follows the changes imposed by the external $\epsilon(t)$, thus behaving adiabatically. However, as $\epsilon(t) \to 0$, the critical slowing down, $\tau \to \infty$, implies that the system will not have time to react—it remains *frozen* according to the KZ terminology—and consequently will depart from equilibrium. Since $\tau \to \infty$, no matter how slow the protocol is performed, the system will always abandon equilibrium. KZ mechanism establishes a well-defined location of the *freeze-out* instant, which separates adiabatic from impulse dynamics and can be estimated when the time remaining to the critical point matches with the relaxation time (see Fig. 1.3)

$$\tau_0 |\hat{\epsilon}|^{-z\nu} = \tau_Q |\hat{\epsilon}| \quad \Rightarrow \quad |\hat{\epsilon}| = \left(\frac{\tau_Q}{\tau_0} \right)^{-1/(z\nu+1)}, \tag{1.30}$$

and also provides a *freeze-out* time, $\hat{t} \sim \tau_Q^{z\nu/(z\nu+1)}$. Then, the system remains frozen while traversing the critical point, and thus the correlation length at $\hat{\epsilon}$ determines the average size of domains selecting a common symmetry-breaking state. Hence, $\hat{\xi} = \xi_0 |\hat{\epsilon}|^{-\nu} \sim \tau^{\nu/(z\nu+1)}$. In the ordered phase, the density of defects (or independently chosen domains) can be approximated by

$$n_d \approx \left(\frac{L}{\hat{\xi}} \right)^d \sim \tau_Q^{-d\nu/(z\nu+1)} \tag{1.31}$$

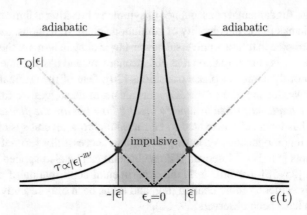

Fig. 1.3 Schematic representation of the KZ mechanism. The diverging relaxation time τ at the critical point $\epsilon_c = 0$ prevents the system to react to external changes in a finite time. First, $\epsilon(t)$ varies in a time scale much longer than the relaxation time, and hence, the dynamics results adiabatic. To the contrary, as $\epsilon(t)$ approaches the critical point, τ becomes much larger than the time in which $\epsilon(t)$ varies. Therefore, close to ϵ_c, the dynamics turns into impulsive (shaded region), after which, adiabatic regime is again recovered. The distance to ϵ_c at which the *freeze-out* instant takes place is estimated when both time scales match (dark points). This yields $|\hat{\epsilon}| \sim \tau_Q^{-1/(z\nu+1)}$ on which subsequent scaling predictions are based

where d corresponds to the spatial dimension of the system. Thus, the density of defects crucially depends on the quench time τ_Q in a universal fashion. It is worth noting that KZ arguments rely on nearly adiabatic ramps, that is, when non-adiabatic effects are mainly caused by the critical point. Hence, the previous scaling is expected to break down as $\tau_Q \to 0$, i.e., as the quench is performed too fast. In addition, we must take into account that the relevance of Eq. (1.31) resides in the scaling, as it is usually the case in critical phenomena. Indeed, KZ predictions typically overestimate the actual number while correctly describing the power-law exponent.

The arguments supporting the KZ mechanism grabbed the attention in the scientific community since they seemed to suggest that defect formation is decided before the critical point is actually traversed. Certainly, while KZ arguments provide a good estimate of the observed power-law scaling, they must be interpreted carefully. First, KZ mechanism does not anticipate how many defect will form (normally it is overestimated), it provides solely a reliable power-law exponent. Second, it is definitely important what happens at the other side of the critical point [61]. Third, this simple KZ scenario does not suffice in general, as one must take into account diverse factors which may affect the outcome, such as the sound velocity, i.e., the velocity at which information can propagate along the system, which might set a tighter bound to domain formation than KZ prediction. This is of particular relevance in systems with large inhomogeneities, leading to a modification of the previously derived power-law scaling relations [48, 50, 62]. Moreover, KZ predictions may fail when addressing passages in systems with a more intricate phase diagram, and when traversing multicritical points as shown in [63].

Finally, it is worth mentioning that the same nonequilibrium scaling relations can be obtained by recasting the equation of motion by rescaling coordinates such that the dependence on the quench rate is removed [64, 65]. This will be exemplified in Chap. 2. Moreover, as we will see explicitly, KZ scaling can be encompassed in a broader class of universality, applying finite-size scaling theory to a nonequilibrium context, as we show in Chaps. 2 and 4.

1.4 Structure and Contents

During this thesis, we have covered and faced different topics with a common denominator, phase transitions and their associated dynamics. Yet, distinct techniques will be used depending on the particular system and scenario, introduced in due course. Here follows a description of each of the chapters presented in this thesis.

In Chap. 2 we tackle a paradigmatic model of phase transitions, namely, a Ginzburg–Landau model. The inspection of this classical phase transition in a nonequilibrium situation, in contact to a thermal bath, establishes the grounds for subsequent analysis, such as KZ scaling laws and nonequilibrium finite-size scaling functions. For that, we rely on a Fokker–Planck approach, that aims to determine probability distributions in a deterministic manner [66]. In the two paradigmatic dynamical regimes, namely, overdamped and underdamped, we find a good agreement with reported scaling laws. Moreover, in Chap. 2 we also examine a Coulomb crystal system which can be experimentally realized in ion traps. In its one-dimensional version, this model undergoes a continuous phase transition from linear to zigzag configuration, and it represents a suitable and realistic platform where defect formation can be examined, as recently demonstrated in [52, 53]. The advantage of the developed method, based on a Fokker–Planck description, resides in the capability of determining the probability distributions of the state at any time, and thus it may have potential applications in the avenue of examining critical dynamics from a thermodynamic perspective. The main results presented in Chap. 2 have been published in Ref. [25].

In Chap. 3 we introduce the QRM, which will accompany us through Chaps. 4 and 5. Consisting of a single spin and a single bosonic mode, this model portrays one of the most fundamental systems in quantum physics. In Chap. 3 we will discuss why the QRM is a suitable model to study critical phenomena, and demonstrate that the QRM undergoes a QPT in a limit that differs from the conventional thermodynamic limit. Indeed, we attain a QPT without scaling up the number of system constituents, that is, quantum critical behavior can be attained even in a system comprising a single spin and a bosonic mode. We dubbed finite-component system phase transition this novel manner of achieving critical behaviors. Moreover, despite the absence of a proper thermodynamic limit, we are able to apply finite-size scaling theory, which reveals the universality class to which it belongs. Briefly, we show that this particular QPT and that of the Dicke model, the so-called superradiant QPT, fall into the same universality class. We close Chap. 3 investigating the impact of criticality on excited states, for which a semiclassical approach of the QRM becomes advisable. Part of the materials presented in Chap. 2 have been published in Refs. [21, 22, 24].

We extend the scrutiny of the QRM to the nonequilibrium realm, in the spirit of the KZ mechanism, which is the subject of study of Chap. 4. Because the QRM adopts a simple form in the limit in which its QPT takes place, we have access to the dynamics when the system is quenched towards the critical point at finite rate. In this regard, it is worth mentioning that, prior to our work, KZ arguments were believed to fail or to be not applicable to zero-dimensional systems [67, 68]. Quite the contrary, we show that KZ scaling laws are precisely retrieved, thus KZ mechanism successfully explains the dynamical behavior of the system when adiabaticity breaks down. However, since the QRM does not comprise a spatial dimension, it is certainly not a standard system where QKZ or KZ physics applies, and therefore, in order to avoid any misunderstanding, we stress that we apply KZ arguments that correctly describe the observed scaling. Furthermore, we go beyond the analysis of the strict limit in which the QPT takes place, and analyze how these universal power-law relations break down as one moves away from criticality or when considering finite-size systems. This latter issue brings us to examine nonequilibrium finite-size scaling functions, which encompass KZ scaling as a limiting case. Remarkably, this extends the concept of universality to a nonequilibrium scenario, which again coincides with that of the universality class in which Dicke model falls. Moreover, we comment that, because any realistic realization of the QRM is inevitably constrained to finite-size systems, their inspection is of considerable importance. Some results of this chapter have been published in Refs. [21, 24].

In Chap. 5 we propose a platform where the QPT of the QRM can be probed. Indeed, a single trapped-ion experiment with coherent interaction with the motional degrees of freedom can be engineered such that its physics is effectively dictated by a QRM. In this manner, the physics of a QRM can be realized, and therefore, that of its universality class. Although the realization of a QRM by means of a single ion subject to classical radiation sources is already well established [69], the extreme parameter regime demanded to observe the QPT in the QRM cannot be trivially achieved. In Chap. 5 we first introduce the basics of a trapped ion setup, to then propose a scheme in which larger parameter regimes can be explored, based on a standing-wave disposition of irradiation sources. Moreover, although the QRM may be correctly realized in an isolated trapped-ion Hamiltonian, the unavoidable presence of experimental imperfections may spoil the targeted QRM and thus these effects must be taken into account to elucidate whether the dynamics of its QPT can be observed. Pure dephasing noise affecting the internal levels of the ion, encoding the qubit, appears typically as the most relevant decoherence process, but it is still possible to retrieve nonequilibrium scaling functions of QRM. The second part of Chap. 5, Sect. 5.2, focuses on the application of continuous dynamical techniques to cope with this source of noise, while at the same time allows to explore certain interesting features of the QRM. The results presented in this chapter have been published in Refs. [23, 24].

In Chap. 6 we address the QKZ mechanism. As discussed in Sect. 1.3, the quantum counterpart of the KZ mechanism considers the problem of traversing a QPT at finite rate, where quantum excitations are promoted in the vicinity of the critical point due to the loss of adiabaticity, and thus universal power-law dependencies on the quench rate are expected. In particular, motivated by the great experimental progress in ion-trap technologies [70], in Chap. 6 we consider a one-dimensional long-range Ising model with transverse field as a testbed for a potential exploration of QKZ physics in a truly quantum many-body system. We present a brief introduction of the basic properties of such a system depending on the range of the interactions and on the ferromagnetic or anti-ferromagnetic nature of the couplings. Indeed, the range of the interaction largely affects the critical properties of the system. Yet, for sufficiently short-range interaction, there is a QPT of the same universality class of that of its nearest-neighbors counterpart. Nevertheless, due to the exponential growth of the Hilbert space, an exploration of large system sizes becomes soon impracticable, thus demanding approximate methods for their computation, such as density matrix renormalization group. These techniques are extremely helpful to address the properties of quantum many-body systems, however, we leave them for future work. Yet, for a reasonable large number of spins, comparable to the experimental capabilities, an exact procedure still allows to test the dynamics of the system, which suggests that KZ scaling could be readily attained within current technology.

Finally, although we have provided a summary and possible prospects of the work presented in each of the aforementioned chapters, in Chap. 7 we close with a detailed summary of the main topics covered throughout this thesis. There, we place the main outcomes in a broader context and discuss interesting directions for future research related to the presented work.

References

1. K. Huang, *Statistical Mechanics* (Wiley, New York, 1987)
2. P.W. Anderson, More is different. Science **177**, 393 (1972). https://doi.org/10.1126/science.177.4047.393
3. C. Cohen-Tannoudji, B. Diu, F. Laloë, *Quantum Mechanics* (Wiley-VCH, Berlin Heidelberg, 2005)
4. A. Messiah, *Quantum Mechanics* (Dover Publications, New York, 1961)
5. S. Sachdev, *Quantum Phase Transitions*, 2nd edn. (Cambridge University Press, Cambridge, UK, 2011)
6. M. Vojta, Quantum phase transitions. Rep. Prog. Phys. **66**, 2069 (2003), http://stacks.iop.org/0034-4885/66/i=12/a=R01
7. R.P. Feynman, Simulating physics with computers. Int. J. Theor. Phys. **21**, 467 (1982). https://doi.org/10.1007/BF02650179
8. E. Schrödinger, Are there quantum jumps? Part II. Brit. J. Phil. Sci. **3**, 233 (1952), http://www.jstor.org/stable/685266
9. H.E. Stanley, *Introduction to Phase Transitions and Critical Phenomena* (Clarendon Press, Oxford, 1971)

10. J. Cardy, *Scaling and Renormalization in Statistical Physics* (Cambridge University Press, Cambridge, UK, 1996)
11. G. Jaeger, The Ehrenfest classification of phase transitions: introduction and evolution. Arch. Hist. Exact Sci. **53**, 51 (1998). https://doi.org/10.1007/s004070050021
12. L. Onsager, Crystal statistics. I. A two-dimensional model with an order-disorder transition. Phys. Rev. **65**, 117 (1944). https://doi.org/10.1103/PhysRev.65.117
13. J.M. Kosterlitz, D.J. Thouless, Ordering, metastability and phase transitions in two-dimensional systems. J. Phys. C **6**, 1181 (1973), http://stacks.iop.org/0022-3719/6/i=7/a=010
14. R. Balescu, *Statistical Dynamics: Matter Out of Equilibrium* (Imperial College Press, London, 1997)
15. P.C. Hohenberg, B.I. Halperin, Theory of dynamic critical phenomena. Rev. Mod. Phys. **49**, 435 (1977). https://doi.org/10.1103/RevModPhys.49.435
16. A. Polkovnikov, K. Sengupta, A. Silva, M. Vengalattore, Colloquium: nonequilibrium dynamics of closed interacting quantum systems. Rev. Mod. Phys. **83**, 863 (2011). https://doi.org/10.1103/RevModPhys.83.863
17. U.C. Taeuber, *Critical Dynamics: A Field Theory Approach to Equilibrium and Nonequilibrium Scaling Behavior* (Cambridge University Press, Cambridge, UK, 2014)
18. M.A. Nielsen, I.L. Chuang, *Quantum Computation and Quantum Information* (Cambridge University Press, Cambridge, England, 2000)
19. A.K. Chandra, A. Das, and B.K.C. (eds.), *Quantum Quenching, Annealing and Computation* (Springer, Berlin Heidelberg, 2010)
20. A. del Campo, W.H. Zurek, Universality of phase transition dynamics: topological defects from symmetry breaking. Int. J. Mod. Phys. A **29**, 1430018 (2014). https://doi.org/10.1142/S0217751X1430018X
21. M.-J. Hwang, R. Puebla, M.B. Plenio, Quantum phase transition and universal dynamics in the Rabi model. Phys. Rev. Lett. **115**, 180404 (2015). https://doi.org/10.1103/PhysRevLett.115.180404
22. R. Puebla, M.-J. Hwang, M.B. Plenio, Excited-state quantum phase transition in the Rabi model. Phys. Rev. A **94**, 023835 (2016a). https://doi.org/10.1103/PhysRevA.94.023835
23. R. Puebla, J. Casanova, M.B. Plenio, A robust scheme for the implementation of the quantum Rabi model in trapped ions. New J. Phys. **18**, 113039 (2016b), http://stacks.iop.org/1367-2630/18/i=11/a=113039
24. R. Puebla, M.-J. Hwang, J. Casanova, M.B. Plenio, Probing the dynamics of a superradiant quantum phase transition with a single trapped ion. Phys. Rev. Lett. **118**, 073001 (2017a). https://doi.org/10.1103/PhysRevLett.118.073001
25. R. Puebla, R. Nigmatullin, T.E. Mehlstäubler, M.B. Plenio, Fokker-Planck formalism approach to Kibble-Zurek scaling laws and nonequilibrium dynamics. Phys. Rev. B **95**, 134104 (2017b). https://doi.org/10.1103/PhysRevB.95.134104
26. L.D. Landau, E.M. Lifshitz, *Statistical Physics*, 3rd edn. (Butterworth-Heinemann, Oxford, 1980)
27. N. Goldenfeld, *Lectures on Phase Transitions and the Renormalization Group* (Addison-Wesley, 1992)
28. J.G. Brankov, *Introduction to Finite-Size Scaling* (Leuven University Press, Leuven, 1996)
29. M.E. Fisher, M.N. Barber, Scaling theory for finite-size effects in the critical region. Phys. Rev. Lett. **28**, 1516 (1972). https://doi.org/10.1103/PhysRevLett.28.1516
30. K.G. Wilson, Renormalization group and critical phenomena. I. Renormalization group and the Kadanoff scaling picture. Phys. Rev. B **4**, 3174 (1971). https://doi.org/10.1103/PhysRevB.4.3174
31. H.E. Stanley, Scaling, universality, and renormalization: three pillars of modern critical phenomena. Rev. Mod. Phys. **71**, S358 (1999). https://doi.org/10.1103/RevModPhys.71.S358
32. S.G. Brush, History of the Lenz-Ising model. Rev. Mod. Phys. **39**, 883 (1967). https://doi.org/10.1103/RevModPhys.39.883
33. E. Ising, Beitrag zur Theorie des Ferromagnetismus. Z. Phys. **31**, 253 (1925). https://doi.org/10.1007/BF02980577

34. C.N. Yang, The spontaneous magnetization of a two-dimensional Ising model. Phys. Rev. **85**, 808 (1952). https://doi.org/10.1103/PhysRev.85.808
35. R.J. Baxter, *Exactly Solved Models in Statistical Mechanics* (Academic Press, London, 1989)
36. P. Jordan, E. Wigner, Über das Paulische Äquivalenzverbot. Z. Phys. **47**, 631 (1928). https://doi.org/10.1007/BF01331938
37. J. Eisert, M. Friesdorf, C. Gogolin, Quantum many-body systems out of equilibrium. Nat. Phys. **11**, 124 (2015). https://doi.org/10.1038/nphys3215
38. T.W.B. Kibble, Topology of cosmic domains and strings. J. Phys. A Math. Gen. **9**, 1387 (1976), http://stacks.iop.org/0305-4470/9/i=8/a=029
39. T.W.B. Kibble, Some implications of a cosmological phase transition. Phys. Rep. **67**, 183 (1980). https://doi.org/10.1016/0370-1573(80)90091-5
40. T.W.B. Kibble, Phase-transition dynamics in the lab and the universe. Phys. Today **60**, 47 (2007). https://doi.org/10.1063/1.2784684
41. W.H. Zurek, Cosmological experiments in superfluid helium? Nature **317**, 505 (1985). https://doi.org/10.1038/317505a0
42. W. Zurek, Cosmological experiments in condensed matter systems. Phys. Rep. **276**, 177 (1996). https://doi.org/10.1016/S0370-1573(96)00009-9
43. P. Laguna, W.H. Zurek, Density of kinks after a quench: when symmetry breaks, how big are the pieces? Phys. Rev. Lett. **78**, 2519 (1997). https://doi.org/10.1103/PhysRevLett.78.2519
44. P. Laguna, W.H. Zurek, Critical dynamics of symmetry breaking: quenches, dissipation, and cosmology. Phys. Rev. D **58**, 085021 (1998). https://doi.org/10.1103/PhysRevD.58.085021
45. W.H. Zurek, Topological relics of symmetry breaking: winding numbers and scaling tilts from random vortex-antivortex pairs. J. Phys. Condens. Matter **25**, 404209 (2013), http://stacks.iop.org/0953-8984/25/i=40/a=404209
46. W.H. Zurek, Causality in condensates: gray solitons as relics of BEC formation. Phys. Rev. Lett. **102**, 105702 (2009). https://doi.org/10.1103/PhysRevLett.102.105702
47. R. Nigmatullin, A. del Campo, G. De Chiara, G. Morigi, M.B. Plenio, A. Retzker, Formation of helical ion chains. Phys. Rev. B **93**, 014106 (2016). https://doi.org/10.1103/PhysRevB.93.014106
48. A. del Campo, G. De Chiara, G. Morigi, M.B. Plenio, A. Retzker, Structural defects in ion chains by quenching the external potential: the inhomogeneous Kibble-Zurek mechanism. Phys. Rev. Lett. **105**, 075701 (2010). https://doi.org/10.1103/PhysRevLett.105.075701
49. G. De Chiara, A. del Campo, G. Morigi, M.B. Plenio, A. Retzker, Spontaneous nucleation of structural defects in inhomogeneous ion chains. New J. Phys. **12**, 115003 (2010), http://stacks.iop.org/1367-2630/12/i=11/a=115003
50. A. del Campo, A. Retzker, M.B. Plenio, The inhomogeneous Kibble-Zurek mechanism: vortex nucleation during Bose-Einstein condensation. New J. Phys. **13**, 083022 (2011), http://stacks.iop.org/1367-2630/13/i=8/a=083022
51. S.-Z. Lin, X. Wang, Y. Kamiya, G.-W. Chern, F. Fan, D. Fan, B. Casas, Y. Liu, V. Kiryukhin, W.H. Zurek, C.D. Batista, S.-W. Cheong, Topological defects as relics of emergent continuous symmetry and Higgs condensation of disorder in ferroelectrics. Nat. Phys. **10**, 970 (2014). https://doi.org/10.1038/nphys3142
52. K. Pyka, J. Keller, H.L. Partner, R. Nigmatullin, T. Burgermeister, D.M. Meier, K. Kuhlmann, A. Retzker, M.B. Plenio, W.H. Zurek, A. del Campo, T.E. Mehlstäubler, Topological defect formation and spontaneous symmetry breaking in ion Coulomb crystals. Nat. Commun. **4**, 2291 (2013). https://doi.org/10.1038/ncomms3291
53. S. Ulm, J. Roßnagel, G. Jacob, C. Degünther, S.T. Dawkins, U.G. Poschinger, R. Nigmatullin, A. Retzker, M.B. Plenio, F. Schmidt-Kaler, K. Singer, Observation of the Kibble-Zurek scaling law for defect formation in ion crystals. Nat. Commun. **4**, 2290 (2013). https://doi.org/10.1038/ncomms3290
54. N. Navon, A.L. Gaunt, R.P. Smith, Z. Hadzibabic, Critical dynamics of spontaneous symmetry breaking in a homogeneous Bose gas. Science **347**, 167 (2015). https://doi.org/10.1126/science.1258676

55. B. Damski, The simplest quantum model supporting the Kibble-Zurek mechanism of topological defect production: Landau-Zener transitions from a new perspective. Phys. Rev. Lett. **95**, 035701 (2005). https://doi.org/10.1103/PhysRevLett.95.035701
56. W.H. Zurek, U. Dorner, P. Zoller, Dynamics of a quantum phase transition. Phys. Rev. Lett. **95**, 105701 (2005). https://doi.org/10.1103/PhysRevLett.95.105701
57. J. Dziarmaga, Dynamics of a quantum phase transition: exact solution of the quantum Ising model. Phys. Rev. Lett. **95**, 245701 (2005). https://doi.org/10.1103/PhysRevLett.95.245701
58. A. Polkovnikov, Universal adiabatic dynamics in the vicinity of a quantum critical point. Phys. Rev. B **72**, 161201 (2005). https://doi.org/10.1103/PhysRevB.72.161201
59. M. Anquez, B.A. Robbins, H.M. Bharath, M. Boguslawski, T.M. Hoang, M.S. Chapman, Quantum Kibble-Zurek mechanism in a spin-1 Bose-Einstein condensate. Phys. Rev. Lett. **116**, 155301 (2016). https://doi.org/10.1103/PhysRevLett.116.155301
60. L.W. Clark, L. Feng, C. Chin, Universal space-time scaling symmetry in the dynamics of bosons across a quantum phase transition. Science **354**, 606 (2016). https://doi.org/10.1126/science.aaf9657
61. N.D. Antunes, P. Gandra, R.J. Rivers, Is domain formation decided before or after the transition? Phys. Rev. D **73**, 125003 (2006). https://doi.org/10.1103/PhysRevD.73.125003
62. J. Dziarmaga, P. Laguna, W.H. Zurek, Symmetry breaking with a slant: topological defects after an inhomogeneous quench. Phys. Rev. Lett. **82**, 4749 (1999). https://doi.org/10.1103/PhysRevLett.82.4749
63. S. Deng, G. Ortiz, L. Viola, Anomalous nonergodic scaling in adiabatic multicritical quantum quenches. Phys. Rev. B **80**, 241109 (2009). https://doi.org/10.1103/PhysRevB.80.241109
64. G. Nikoghosyan, R. Nigmatullin, M.B. Plenio, Universality in the dynamics of second-order phase transitions. Phys. Rev. Lett. **116**, 080601 (2016). https://doi.org/10.1103/PhysRevLett.116.080601
65. M. Kolodrubetz, B.K. Clark, D.A. Huse, Nonequilibrium dynamic critical scaling of the quantum Ising chain. Phys. Rev. Lett. **109**, 015701 (2012). https://doi.org/10.1103/PhysRevLett.109.015701
66. H. Risken, *The Fokker-Planck Equation: Methods of Solution and Applications*, 2nd edn. (Springer, New York, 1984)
67. T. Caneva, R. Fazio, G.E. Santoro, Adiabatic quantum dynamics of the Lipkin-Meshkov-Glick model. Phys. Rev. B **78**, 104426 (2008). https://doi.org/10.1103/PhysRevB.78.104426
68. O.L. Acevedo, L. Quiroga, F.J. Rodríguez, N.F. Johnson, New dynamical scaling universality for quantum networks across adiabatic quantum phase transitions. Phys. Rev. Lett. **112**, 030403 (2014). https://doi.org/10.1103/PhysRevLett.112.030403
69. J.S. Pedernales, I. Lizuain, S. Felicetti, G. Romero, L. Lamata, E. Solano, Quantum Rabi model with trapped ions. Sci. Rep. **5**, 15472 (2015). https://doi.org/10.1038/srep15472
70. D. Leibfried, R. Blatt, C. Monroe, D. Wineland, Quantum dynamics of single trapped ions. Rev. Mod. Phys. **75**, 281 (2003). https://doi.org/10.1103/RevModPhys.75.281

Chapter 2
Structural Phase Transitions

Phase transitions appear in any scenario analyzed in physics, describing transition of distinct phases of matter, which ranges from daily life experience to more exotics realms as quantum physics or cosmology [1–3]. Yet, despite the disparate systems in which they manifest and the fundamentally distinct physics involved, essential features are shared among them. Indeed, the observation of these common properties across different fields strongly suggested and motivated a search for a deeper underlying theory of phase transitions. As explained in the Introduction, the first successful attempt to establish such a common ground is known nowadays as the Ehrenfest classification (see [4] for an explanation and history of this classification). This classification differentiates phase transitions according to their *order* [5], namely, the order of the derivative respect to the free external parameter (typically temperature) in which the free energy becomes discontinuous. However, although the examined phase transition in this Chapter can be well described in terms of the Ehrenfest classification, second-order, we will refer to it as a continuous phase transition, according to the modern classification (see Introduction). Since this thesis aims to explore different physics related to continuous phase transitions, we shall commence this journey with the analysis of a classical phase transition and later move into the quantum realm.

The present Chapter is devoted to the study of the paradigmatic model of a continuous classical phase transitions, the Ginzburg-Landau model. We will briefly motivate its origin and its success in addressing variety of critical systems, as well as the different critical phenomena associated with this simple model [6], presented in Sect. 2.1. Moreover, as phase transitions occur only in the thermodynamic limit, an inspection of finite-size effects is pertinent. Then, we will tackle a realistic linear to zigzag phase transition in an ion Coulomb crystal [7, 8], and its corresponding nonequilibrium behavior when the symmetry breaking transition is traversed at a finite rate. This finite-rate symmetry breaking process promotes the formation of nonequilibrium excitations that can stabilize, leading to topological defects, and it is known as Kibble-Zurek (KZ) mechanism [9–11] (see Introduction, Sect. 1.3),

© Springer Nature Switzerland AG 2018
R. Puebla, *Equilibrium and Nonequilibrium Aspects of Phase Transitions in Quantum Physics*, Springer Theses, https://doi.org/10.1007/978-3-030-00653-2_2

which has been investigated in several experiments (see [12] for a review). However, while standard KZ arguments are sufficient to successfully explain results in spatially homogeneous systems, some experimental systems are inevitably affected significantly by inhomogeneities and finite-size effects, such as Bose-Einstein condensates [13–15], and ion Coulomb crystals [16, 17], which lead to a modification of KZ scaling laws. Therefore, the measured power-law scaling may not agree with the prediction in the thermodynamic limit. In these cases numerical simulations are particularly valuable to gain insights into nonequilibrium dynamics, which typically rely on the computation of several stochastic trajectories in order to attain an accurate estimate of statistical quantities, as the average number of defects. The calculation of statistical estimates from individual tracking of stochastic trajectories is commonly known as *Langevin* approach. Yet, there is another equivalent approach that aims to compute, in a deterministic manner, the time evolution of the probability distributions of the whole system. This description is based on the so-called *Fokker-Planck* equations [18]. It is however worth noting that, since Fokker-Planck equations determine the probability over the whole configuration space, the resulting differential equations may be involved and computationally demanding, and in certain cases, the evaluation of individual stochastic trajectories might be a more suitable procedure.

The current Chapter pursues a manifold goal. First, we examine nonequilibrium aspects of the Ginzburg-Landau theory, in the spirit of the KZ problem, and in different parameter regimes that feature different physics, namely overdamped and underdamped dynamics. Second, we make use of the Fokker-Planck treatment, discussing its advantages and under what circumstances it becomes more suitable than the standard Langevin approach. Finally, we apply the developed framework to the realistic linear to zigzag phase transition in a Coulomb ion crystal. The results collected in this Chapter have been published in [19].

2.1 Ginzburg-Landau Theory

The Ginzburg-Landau (GL) theory of phase transitions has become the paradigmatic model to describe, in a phenomenological manner, critical phenomena in a broad variety of systems. In particular, although we refer to this theory as GL, it was L. Landau who first established the grounds for a general framework of continuous phase transitions exploiting symmetry arguments and in terms of a free energy functional $\mathcal{F}$ [5]

$$\mathcal{F}[\phi] = \int d\mathbf{r} \left\{ c_0 + c_1\phi(\mathbf{r}) + c_2\phi^2(\mathbf{r}) + c_3\phi^3(\mathbf{r}) + c_4\phi^4(\mathbf{r}) + d_2(\nabla\phi(\mathbf{r}))^2 + \ldots \right\}. \quad (2.1)$$

There $\phi(\mathbf{r})$ stands for the order parameter, $\mathbf{r}$ defines the d-dimensional vector position, and the set of parameters $\{c_n, d_n\}$ depend on the microscopic details of the particular system. The free energy $\mathcal{F}[\phi]$ is constructed such that it correctly describes the system when $\phi(\mathbf{r})$ and its gradients are small, as it consists just in a series expansion in the vicinity of the phase transition. In this manner, the minimization of $\mathcal{F}[\phi]$ with respect $\phi(\mathbf{r})$ permits to locate the critical point, if any, as well as the behavior of the order parameter and the order of the phase transition. In particular, a nth order phase transition features a discontinuity in $\partial^n \mathcal{F}/\partial\varepsilon^n$ at a certain critical value ε_c, where ε represents an external parameter whose variation triggers the phase transition. Heuristically, we can exemplify the arguments as follows. If the system possesses different symmetry with $\phi(\mathbf{r}) = 0$ and $\phi(\mathbf{r}) \neq 0$, the parameter c_1 must vanish $c_1 \equiv 0$. Hence, since $\phi(\mathbf{r}) = 0$ corresponds to a stable configuration at one side ($c_2 > 0$) and $\phi(\mathbf{r}) \neq 0$ to the other (only possible if $c_2 < 0$), then $c_2 = 0$ corresponds to the critical point. Moreover, if by symmetry of the system it holds that $\mathcal{F}[\phi] = \mathcal{F}[-\phi]$, it follows that $c_3 \equiv 0$, while $c_4 > 0$ is required for stability. It is worth noting however that more complex scenarios can be described depending on the specific $\{c_n, d_n\}$ and on distinct order parameters, such as in magnetism or superconductivity where $\phi(\mathbf{r})$ represents a vector field or a complex function, respectively. The great success of the GL theory lies therefore in its elegance and multidisciplinary nature—despite its simple formulation, it encompasses variety of distinct critical phenomena. In short, this theory manifests the universality across different phase transition problems [20]. Nevertheless, although Landau's theory represented a big step forward in the understanding of critical phenomena, we know nowadays that, while symmetries are crucial, the role of fluctuations is also very important (see Introduction, Sect. 1.2). Certainly, the latter feature does not appear in this simple theory. Later developments indeed showed that Landau's theory becomes a suitable description of phase transitions when mean field arguments work, as evidenced by the wrong predictions regarding exactly solvable models [21], as dealing with one- and two-dimensional Ising models (see Introduction). In terms of the Ginzburg criterion, systems above the upper critical dimension ($D = 4$) are well addressed by mean field treatment as fluctuations of the order parameter become negligible, although it may be also possible to find mean-field systems at lower dimensions ($d < D$) if they contain long-range interactions.

In particular, we focus on continuous phase transitions, which have been already discussed in the Introduction, and within GL theory can be described by the customary single- to double-well potential. Note that, the partition function of the system reads

$$Z(\beta) = \int D[\phi]\, e^{-\beta\mathcal{F}[\phi]} \qquad (2.2)$$

where $\beta^{-1} \equiv k_B T$ gives account of the thermal energy at temperature T, k_B the Boltzmann constant, and $\int D[\phi]$ represents the integration over all possible configurations of the order parameter, i.e., summing over all microscopic states. From the previous partition function, all thermodynamic quantities, including critical exponents follow [22]. Certainly, while equilibrium properties of phase transitions are of primarily interest, nonequilibrium scenarios encompass much richer physics, which is precisely the subject of the present Chapter. Although the general equilibrium properties of continuous phase transitions have been discussed in the Introduction, we will comment some of them in this part regarding the GL theory.

For the following developments, we shall consider a scalar one-dimensional order parameter $\phi(x, t)$, where x represents the spatial dimension and t the time instant. The free energy of the system can be written then as

$$\mathcal{F} = \frac{1}{2} \int dx \left[h^2 (\partial_x \phi)^2 + V(\phi) \right],$$ (2.3)

where the GL potential $V(\phi)$ reads

$$V(\phi) = \frac{\varepsilon}{2} \phi^2 + \frac{\lambda}{4} \phi^4,$$ (2.4)

with the parameters λ and h depending on the microscopic details of the particular system. The critical point takes place at $\varepsilon_c = 0$, which corresponds to a symmetry-breaking transition from $\phi = 0$ for $\varepsilon > 0$ to two energetically equivalent choices at $\phi = \pm\sqrt{-\varepsilon/\lambda}$ for $\varepsilon < 0$, as shown in Fig. 2.1. Therefore, the critical exponent associated to the order parameter becomes $\beta = 1/2$, since $\phi \sim |\varepsilon - \varepsilon_c|^\beta$ as explained in the Introduction.

2.1.1 Nonequilibrium Dynamics

Here, we will consider the GL model to study the nonequilibrium dynamics resulting from a finite-rate symmetry-breaking, forced by externally varying the parameter $\varepsilon(t)$ and in contact with a Markovian heat bath. In particular, we shall consider a linear functional dependence of ε on time, the simplest case, that is

$$\varepsilon(t) = \varepsilon_0 + \frac{t}{\tau_Q} (\varepsilon_1 - \varepsilon_0), \qquad 0 \leq t \leq \tau_Q.$$ (2.5)

where $\varepsilon(0) = \varepsilon_0 > 0$ and $\varepsilon(\tau_Q) = \varepsilon_1 < 0$. Therefore, the systems finds itself in the symmetric phase at the start of the quench protocol, and brought towards the critical point to end in the symmetry broken phase. The rate at which the critical point is traversed is $d\varepsilon(t)/dt = (\varepsilon_1 - \varepsilon_0)/\tau_Q$, and thus, it is determined by the quench time τ_Q once ε_0 and ε_1 are fixed.

The considered phenomenological equation governing the dynamics of the system corresponds to the model A of the classification done in [6],

$$\left(\frac{\partial^2}{\partial t^2} + \eta \frac{\partial}{\partial t} \right) \phi(x, t) = h^2 \frac{\partial^2}{\partial x^2} \phi(x, t) - \frac{\delta V(\phi)}{\delta \phi} + \zeta(x, t)$$ (2.6)

which is also known as time-dependent GL equation. There, the parameter η corresponds to the friction, and $\zeta(x, t)$ represents the stochastic force produced by a thermal bath. Both are related through the fluctuation-dissipation theorem (see Appendix A). It is worth noting that, while different model's dynamics described in [6] deal with particular conservation laws, the dynamics under Eq. (2.6) does not conserve the order parameter. Moreover, although these dynamical equations should be regarded simply as phenomenological descriptions, they can give account of realistic systems in the vicinity of critical points, and thus, serve as good models to study defect formation, the emergence of scaling laws in out-of-equilibrium stages and the relevance of finite-size effects.

A continuous phase transition is characterized by a diverging correlation length at the critical point, which follows from the two-point correlation function,

$$G(x_1, x_2, t) = \langle \phi(x_1, t)\phi(x_2, t)\rangle - \langle \phi(x_1, t)\rangle \langle \phi(x_2, t)\rangle \tag{2.7}$$

since $G(x_1, x_2, t) \propto e^{-|x_1 - x_2|/\xi}$, and where $\langle \ldots \rangle$ stands for stochastic average, i.e., average over many stochastic realizations under Eq. (2.6). In equilibrium, $\xi \propto |\varepsilon - \varepsilon_c|^{-\nu}$ with $\nu = 1/2$ the mean-field critical exponent. Indeed, the correlation length can be extracted from Eq. (2.7) as

$$\xi_L(t) = \frac{\sqrt{\int_0^{L/2} dx\, x^2 G(x, t)}}{\sqrt{2\int_0^{L/2} dx\, G(x, t)}} \tag{2.8}$$

where we have assumed a spatially homogeneous system, $G(x_1, x_2, t) \equiv G(x_1 - x_2, t)$ and finite extension, $x \in [0, L]$ such that it forms a ring. Note that the previous expression retrieves ξ when $G(x, t) \propto e^{-|x|/\xi}$ and taking the limit $L \to \infty$. Recall that KZ arguments crucially rely on the scaling of the correlation length in nonequilibrium systems close to the critical point. Therefore, it also advisable to analyze the number of defects promoted during the evolution, and that stabilize in the region $\varepsilon < 0$. In this particular case, defects are related to the number of different choices of the symmetry-breaking, whose number follows by simply counting how many times $\phi(x, t)$ crosses zero, as studied in the seminal papers [23, 24]. The density of defects can be then formulated as [25]

$$n_L(t) = \frac{1}{L}\left\langle \int_0^L dx\, \delta[\phi(x, t)] \right\rangle \tag{2.9}$$

Closely related to $n_L(t)$ we find the quantity $g_L(t)$, which we proposed as a good quantification of defect formation. When one defect is created, the field $\phi(x, t)$ interpolates rapidly, yet smoothly, between the two distinct chosen configurations, $\phi = \pm\sqrt{-\varepsilon/\lambda}$, and thus, $\phi(x, t)$ exhibits large spatial variations. On the contrary, a field with no defects has small spatial variations. Therefore, we introduce the following quantity

$$g_L(t) = L\frac{\int_0^L dx \ \langle (\partial_x \phi(x, t))^2 \rangle}{\int_0^L dx \ (\langle \phi(x, t) \rangle)^2}, \tag{2.10}$$

which gives account of the amount of different choices of symmetry-breaking regions. Besides the diverging correlation length $\xi \propto |\varepsilon - \varepsilon_c|^{-\nu}$, a continuous phase transition is also characterized by a diverging relaxation time, $\tau \propto |\varepsilon - \varepsilon_c|^{-z\nu}$, where z is the dynamical critical exponent [6]. If the fourth and higher order terms of ϕ in Eq. (2.4) are negligible then $\nu = 1/2$ (mean field exponent), while depending on the dynamical regime, $z = 2$ or $z = 1$, for overdamped or underdamped dynamics [9, 12, 23, 24, 26, 27]. The former regime is found when $|\eta\dot{\phi}| \gg |\ddot{\phi}|$, while the latter takes place in the opposite limit. In particular, overdamped dynamics exhibits a relaxation time $\tau \simeq |\phi/\dot{\phi}|$. Thus, changes in the potential lead to dynamics in a typical time scale $\tau \propto \eta/|\varepsilon|$, which readily determines the dynamical critical exponent $z = 2$ in the overdamped limit [23, 24, 27]. In contrast, in underdamped dynamics the term $\eta\dot{\phi}$ can be neglected in favor of $\ddot{\phi}$, and thus, $\tau \simeq |\phi/\ddot{\phi}|^{1/2}$. For that reason, $\tau \propto |\varepsilon|^{-1/2}$ which gives the critical exponent $z = 1$ in the underdamped regime [12, 24, 26, 27]. Note however that these regimes emerge as limiting cases of the more general situation in which the terms $\eta\dot{\phi}$ and $\ddot{\phi}$ compete. In particular, a fixed friction coefficient η determines the dynamical regime and thus the corresponding z, which together with ν fix the KZ scaling laws as the phase transition is traversed.

In order to derive scaling laws, resulting from traversing with finite rate a continuous phase transition, we resort to the KZ arguments, as given in the Introduction and briefly summarize here. Due to the diverging relaxation time near the critical point, the system *freezes*, using the KZ terminology. This *freeze-out* instant takes place at $\hat{t}$ from which the system remains insensitive to the external changes, and thus, it does not recover equilibrium properties during the quench [9, 12]. As we have seen in the Introduction, combining the divergence of correlation length and relaxation time, and crossing the critical point at a rate τ_Q^{-1}, entails $\hat{\xi} \sim \tau_Q^{\nu/(1+z\nu)}$, where $\hat{\xi}$ represents the correlation length of the system at the *freeze-out* instant. Hence, one can estimate the scaling of the density of defects as $n_L \sim L^d/\hat{\xi}^d \sim \tau_Q^{-d\nu/(1+z\nu)}$, where d is the dimension of the system.[1] Remarkably, these KZ scaling laws can be derived by transforming the equations of motion as shown in [28–31], without relying on the intuitive, yet physical, arguments of the transition between adiabatic and impulsive dynamics [9, 12].

2.1.1.1 Langevin Approach

The stochastic Eq. (2.6) determines the evolution of $\phi(x, t)$, which after considering the form of $V(\phi)$ leads to [6, 27, 32]

---

[1]Note that defects in this model have zero dimension as they simply separate one-dimensional spatial regions of $\phi(x, t)$ with different symmetry breaking choices.

Fig. 2.1 Schematic
representation of the
potential
$V(\phi) = \lambda\phi^4/4 + \varepsilon\phi^2/2$
(solid lines) and the resulting
harmonic approximation
$V(\phi) \approx \varepsilon\phi^2/2$ (dashed
lines) for three characteristic
scenarios, namely $\varepsilon > 0$
(light grey), $\varepsilon = \varepsilon_c = 0$
(grey) and $\varepsilon < 0$ (black)

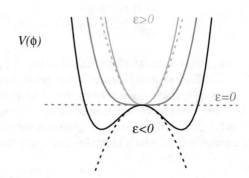

$$\left(\frac{\partial^2}{\partial t^2} + \eta\frac{\partial}{\partial t}\right)\phi(x,t) = h^2\frac{\partial^2}{\partial x^2}\phi(x,t) - \varepsilon\phi(x,t) - \lambda\phi^3(x,t) + \zeta(x,t) \quad (2.11)$$

where the stochastic force $\zeta(x,t)$ fulfills

$$\langle\zeta(x,t)\rangle = 0, \quad (2.12)$$

$$\langle\zeta(x,t)\zeta(x',t')\rangle = \frac{2\eta}{\beta}\delta(x-x')\delta(t-t'), \quad (2.13)$$

that is, the noise is considered white and holds the fluctuation-dissipation theorem [33] (see Appendix A). At the beginning of the quench, $\varepsilon \geq 0$, the field $\phi(x,0)$
is considered to be in thermal equilibrium, and thus, it has small amplitude such that
$|\lambda\phi^3| \ll |\varepsilon\phi|$. In this manner, to a good approximation, the higher order terms of ϕ in
$V(\phi)$ can be neglected, which results in an approximated harmonic or Gaussian GL
potential, $V(\phi) \approx \varepsilon\phi^2/2$, whose comparison with the ϕ^4 is plotted in Fig. 2.1. This
procedure may seem a rough simplification, however, it will be of great utility later
on, while at the same time, it correctly describes dynamics of realistic models [26,
27, 32], and agrees with well-known results including the nonlinear term ϕ^4 [23, 24]
It is important to remark that, within this harmonic approximation, the system cannot
equilibrate since there exists no stable configuration for $\varepsilon < 0$. Nevertheless, as we
aim to derive and describe properties of the system when it finds itself far from equilibrium, the lack of equilibration for $\varepsilon < 0$ is not a crucial limitation. Moreover, the
harmonic approximation still reproduces essential features of the ϕ^4 model. Roughly
speaking, within the harmonic approximation, although the amplitude of the field
increases without limit for $\varepsilon < 0$, the number of defects remains equal as that of the
ϕ^4. In addition, we are interested in the emergent scaling laws which are established
at or just after the critical point is traversed.

For convenience, we consider periodic boundary conditions, $\phi(0,t) = \phi(L,t)$.
In the Fourier space, $\phi(x,t)$ can be written as $\phi(x,t) = \sum_n \varphi_n(t)e^{ik_nx}$, being $k_n = 2\pi n/L$ the wave-vector associated to the nth mode $\varphi_n(t)$, which fulfills $\varphi_n(t) = \varphi^*_{-n}(t)$ since the field is real. Performing the harmonic approximation, the stochastic
equation of motion decouples for each mode as

$$\left(\frac{\partial^2}{\partial t^2} + \eta \frac{\partial}{\partial t}\right) \varphi_n(t) = \left(-k_n^2 h^2 - \varepsilon(t)\right) \varphi_n(t) + \zeta_n(t), \qquad (2.14)$$

where the noise now obeys $\langle \zeta_n(t) \rangle = 0$ and $\langle \zeta_n(t) \zeta_m(t') \rangle = 2\eta \beta^{-1} \delta_{nm} \delta(t - t')$.

Different observables are then attained after stochastic average, that is, after averaging over many noise realizations to provide a reliable statistical estimate. In this manner, the ensemble averaged value of a macroscopic quantity $\mathcal{A}$ can be denoted as $\langle \mathcal{A}[\phi(t), \dot{\phi}(t); t] \rangle$. Formally, this is achieved by integrating over all possible field configurations, which in the Fourier space can be written as

$$\langle \mathcal{A} \rangle = \int_{-\infty}^{+\infty} d\varphi \int_{-\infty}^{+\infty} d\dot{\varphi} \mathcal{A} \prod_n P_n(t, \varphi_n, \dot{\varphi}_n), \qquad (2.15)$$

where $d\varphi$ and $d\dot{\varphi}$ denote integration over all the different modes, and $P_n(t, \varphi_n, \dot{\varphi}_n)$ represents the time-dependent probability distribution for the nth mode in the phase space, that is, it quantifies the probability of obtaining the specific values φ_n and $\dot{\varphi}_n$ at time t, independently of the rest of the modes. Indeed, as they represent probability distributions they must fulfill

$$\int_{-\infty}^{+\infty} \int_{-\infty}^{+\infty} d\varphi_n d\dot{\varphi}_n \, P_n(t, \varphi_n, \dot{\varphi}_n) = 1. \qquad (2.16)$$

In the Langevin approach, one can obtain approximately $\mathcal{A}$, and thus, the corresponding probability distributions by repeatedly solving Eq. (2.14).

2.1.1.2   Fokker-Planck Approach

As aforementioned, the Fokker-Planck formalism sacrifices the knowledge of individual stochastic trajectories in favor of a deterministic computation of probability distributions. Thus, one may abandon the set of stochastic Langevin equations to achieve deterministic partial differential equations that dictate the evolution of the probability distributions $P_n(t, \varphi_n, \dot{\varphi}_n)$ [18]. However, for the sake of brevity, we leave the derivation of the Fokker-Planck equation to the Appendix A, and quote here the result of the counterpart of Eq. (2.14) that reads

$$\frac{\partial}{\partial t} P_n(t, \varphi_n, \dot{\varphi}_n) =$$
$$\left[-\frac{\partial}{\partial \varphi_n} \dot{\varphi}_n + \frac{\partial}{\partial \dot{\varphi}_n} \left(\eta \dot{\varphi}_n + \left(h^2 k_n^2 + \varepsilon(t)\right) \varphi_n\right) + \frac{\eta}{\beta} \frac{\partial^2}{\partial \dot{\varphi}_n^2}\right] P_n(t, \varphi_n, \dot{\varphi}_n), \quad (2.17)$$

and it is known as the Kramers Equation [18].

2.1.2 Overdamped Regime: Smoluchowski Equation

Before analyzing the general case of the dynamics dictated by Eq. (2.14) or its Fokker-Planck counterpart, Eq. (2.17), it is useful to consider the overdamped regime or pure relaxational dynamics [6]. This regime takes place when the friction coefficient dominates, that is, $|\eta\dot{\phi}| \gg |\ddot{\phi}|$. In such a limit, the equation of motion can be approximated as

$$\eta \frac{\partial}{\partial t} \phi(x, t) = \left(h^2 \frac{\partial^2}{\partial x^2} - \varepsilon(t) \right) \phi(x, t) + \zeta(x, t), \qquad (2.18)$$

or in the Fourier space as

$$\eta \frac{\partial}{\partial t} \varphi_n(t) = \left(-h^2 k_n^2 - \varepsilon(t) \right) \varphi_n(t) + \zeta_n(t), \qquad (2.19)$$

Note that in this regime the probability distributions depend on φ_n but not on $\dot{\varphi}_n$, and thus, we will simply write $P_n(\varphi_n, t)$. Furthermore, as $\varphi_n(t)$ is in general complex, it is useful to express $\phi(x, t)$ as

$$\phi(x, t) = \frac{\varphi_0^R(t)}{\sqrt{N_c}} + \sqrt{\frac{2}{N_c}} \sum_{n=1}^{(N_c-1)/2} \varphi_n^R(t) \cos(k_n x) + \varphi_n^I(t) \sin(k_n x), \qquad (2.20)$$

where we have introduced a momentum cut-off k_c that sets a maximum number of modes $(N_c - 1)/2$, and $\varphi_n^R(t) \equiv \mathrm{Re}(\varphi_n(t))$ and $\varphi_n^I(t) \equiv \mathrm{Im}(\varphi_n(t))$. We choose an odd number N_c, without loss of generality. The Fokker-Planck counterpart of Eq. (2.19) is known as Smoluchowski equation [18] (see Appendix A)

$$\eta \frac{\partial P_n(t, \varphi_n^{R,I})}{\partial t} = \frac{\partial}{\partial \varphi_n^{R,I}} \left[\frac{1}{\beta} \frac{\partial}{\partial \varphi_n^{R,I}} + \left(h^2 k_n^2 + \varepsilon(t) \right) \varphi_n^{R,I} \right] P_{o,n}(t, \varphi_n^{R,I}). \qquad (2.21)$$

Initially, at ε_0 the system is in thermal equilibrium, which is characterized by being a stationary state, $\partial_t P_n^{\mathrm{th}} = 0$. In particular, one finds that the following probability distribution fulfills this requirement,

$$P_n^{\mathrm{th}}(\varphi_n^{R,I}) = \frac{f_n^{\mathrm{th}}}{\sqrt{\pi}} e^{-(f_n^{\mathrm{th}} \varphi_n^{R,I})^2}, \qquad \text{with} \quad f_n^{\mathrm{th}} = \sqrt{\beta \left(h^2 k_n^2 + \varepsilon_0 \right)/2}. \qquad (2.22)$$

Certainly, these probability distributions correspond to the Boltzmann distribution at a given temperature $1/\beta$, as one would have demanded. Interestingly, due to the harmonic approximation, such a thermal state exists only as long as $\varepsilon_0 > \varepsilon_c = 0$. Moreover, we consider an Ansatz for the time-dependent probability distributions, $P_n(\varphi_n^{R,I}, t) = \frac{f_n(t)}{\sqrt{\pi}} e^{-(f_n(t)\varphi_n^{R,I})^2}$, that is, a Gaussian probability distribution, with a time-dependent variance that obeys a non-linear differential equation

$$\frac{\partial}{\partial t} f_n(t) = -\frac{2}{\eta \beta} f_n^3(t) + \frac{1}{\eta} \left(h^2 k_n^2 + \varepsilon(t) \right) f_n(t), \qquad (2.23)$$

obtained upon substituting the Ansatz in Eq. (2.21). The initial condition $f_n(0)$ is simply that of thermal equilibrium, i.e., $f_n(0) = f_n^{\text{th}}$. In this manner, we achieve the full knowledge of the probabilistic dynamics as it follows from a set of uncoupled and simple differential equations determining the variance of the probability distributions. Note however that, while this has been achieved after a number of assumptions and approximations, the procedure correctly captures nonequilibrium dynamics of the system.

Assuming the Gaussian form of the probability distributions, we can calculate the quantities of interests, namely $\xi_L(t)$ and $g_L(t)$. The two-point correlation function given by Eq. (2.7) simplifies to

$$G(x_1, x_2, t) = \frac{1}{2N_c f_0^2(t)} + \sum_{n=1}^{N_c-1} \frac{\cos(k_n(x_1 - x_2))}{N_c f_n^2(t)}. \qquad (2.24)$$

from where it is obvious that $G(x_1, x_2, t) = G(x_1 - x_2, t)$. In addition, recall that due to symmetry, $\langle \phi(x, t) \rangle \equiv 0$. Then, the correlation length follows from Eq. (2.8),

$$\xi_L(t) = \frac{L}{2\sqrt{6}} \sqrt{1 + 12 f_0^2(t) \sum_{n=1}^{(N_c-1)/2} \frac{(-1)^n}{f_n^2(t) n^2 \pi^2}}. \qquad (2.25)$$

as well as the expression for $g_L(t)$, Eq. (2.10),

$$g_L(t) = L \frac{\sum_{n=1}^{(N_c-1)/2} \frac{k_n^2}{f_n^2(t)}}{\frac{1}{2 f_0^2(t)} + \sum_{n=1}^{(N_c-1)/2} \frac{1}{f_n^2(t)}}. \qquad (2.26)$$

The evaluation of the previous expressions will allow us to test the emergence of the scaling laws. However, before entering in the finite-rate passages, we discuss to limiting cases, namely, thermal equilibrium and sudden quenches, $\tau_Q \to \infty$ and $\tau_Q \to 0$, respectively.

2.1.2.1   Thermal Equilibrium and Sudden Quenches

Thermal equilibrium at any ε is attained in the limiting case of an infinitely slow quench, i.e., $\tau_Q \to \infty$, and allows us to gain valuable insights. From a practical point of view, it is sufficient to replace $f_n(t)$ by f_n^{th} in the expression for two-point correlation function, correlation length ξ_L and g_L. In addition, we comment once again that thermal equilibrium is only possible for $\varepsilon > 0$ as a consequence of the harmonic approximation. Taking the continuum limit of Eq. (2.25), one gets

$$\xi_L^{\mathrm{th}}(\varepsilon > 0) = \sqrt{\frac{h^2}{\varepsilon} - \frac{hL}{2\varepsilon^{1/2}\sinh(\varepsilon^{1/2}L/2h)}}. \qquad (2.27)$$

The thermodynamic limit is of particular interest as it must reveal the diverging ξ_L, that is, $\xi_L \propto |\varepsilon - \varepsilon_c|^{-\nu}$ as $\varepsilon \to \varepsilon_c$. The thermodynamic limit consists in taking $L, N_c \to \infty$, while keeping the cut-off k_c finite. In this manner, $\xi_{L\to\infty}^{\mathrm{th}}(\varepsilon > 0) = h\varepsilon^{-1/2}$, and as expected for GL theory, the critical exponent results in $\nu = 1/2$, and the two-point correlation function becomes

$$G^{\mathrm{th}}(x) = \frac{e^{-\sqrt{\varepsilon}|x|/h}\pi}{h\beta\sqrt{\varepsilon}}. \qquad (2.28)$$

Moreover, for finite systems and at $\varepsilon = \varepsilon_c = 0$, ξ_L adopts a simple expression

$$\xi_L^{\mathrm{th}}(\varepsilon = \varepsilon_c) = \frac{L}{2\sqrt{6}}, \qquad (2.29)$$

which agrees with the expected finite-size scaling theory, $\xi_L(\varepsilon_c = 0) \propto L$ [6, 34], and gives account of the maximum correlation length within the harmonic approximation, which evidently does not exceed the system size, $\xi_L \lesssim L$. In Fig. 2.2 we plot the two-point correlation function, the correlation length together with its scaling close to the critical point. In a similar way, we can calculate the $g_L^{\mathrm{th}}(\varepsilon)$ in the thermodynamic limit as

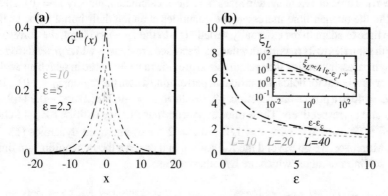

Fig. 2.2 Two-point correlation function $G^{\mathrm{th}}(x)$ (**a**) and correlation length ξ_L^{th} (**b**). The former corresponds to the thermodynamic limit calculation for different values of ε, namely, 10 (dotted light-grey line), 5 (dashed grey line) and 2.5 (dotted-dash black line). In **b**, ξ_L^{th} is plotted as a function of ε and for $L = 10$, 20 and 40, while in the inset its scaling as $\varepsilon \to \varepsilon_c = 0$ is shown, with the parameters $h = 5$ and $\beta = 1$. Note that ξ_L precisely follows the scaling law $\xi_L \propto |\varepsilon - \varepsilon_c|^{-\nu}$ with $\nu = 1/2$ (solid black line) until the saturation value is attained, $L/(2\sqrt{6})$

$$g_{L\to\infty}^{\mathrm{th}}(\varepsilon > 0)/L \approx \frac{\int_0^{k_c} dk \frac{k^2}{\beta(h^2k^2+\varepsilon)}}{\int_0^{k_c} dk \frac{1}{\beta(h^2k^2+\varepsilon)}} = \frac{\varepsilon^{1/2}k_c}{h\arctan(hk_c\varepsilon^{-1/2})} - \frac{\varepsilon}{h^2}, \qquad (2.30)$$

and therefore $g_{L\to\infty}^{\mathrm{th}}/L \sim \frac{2k_c}{h\pi}(\varepsilon - \varepsilon_c)^{1/2}$, as the critical point is approached, which discloses the critical exponent $1/2$.

In the sudden quench limit to the contrary, $\tau_Q \to 0$, the system has no time to equilibrate at intermediate ε values, and thus, the quantities do not vary their initial value, which is that of thermal equilibrium at ε_0.

2.1.2.2 Finite-Rate Quenches

After discussing thermal equilibrium and sudden quenches properties, we finally commence the study of finite-rate passages across the critical point. For that reason, we numerically solve the set of Eq. (2.23) that determine the nonequilibrium evolution of an initially thermal state with temperature β, traversing the phase transition by the time-varying $\varepsilon(t)$. The control parameter is varied from $\varepsilon(0) = \varepsilon_0$ to $\varepsilon(\tau_Q) = \varepsilon_1$ in a time τ_Q, as given in Eq. (2.5). For the simulations we consider $\eta = 10$, $h = 5$ and $\beta = 1$, the initial value $\varepsilon_0 = 100$ and $\varepsilon_1 = -10$. In addition, we set the momentum cutoff $k_c = 5\pi$ which fixes the number of modes for a particular system size L.

In Fig. 2.3 we show the main results for the nonequilibrium overdamped dynamics. In the first panel the scaling of the correlation length ξ_L and g_L of the resulting nonequilibrium state at the critical point is represented as a function of the quench time τ_Q. Three different system sizes have been calculated, namely $L = 10, 20$ and 40. As the quench time increases, the quantities abandon their initial value as they have time to adapt to the externally varied $\varepsilon(t)$. In the $\tau_Q \to \infty$ limit, the correlation length saturates to its maximum value, as discussed previously, while g_L tends to zero. In both cases, increasing the system size L provides a wider range in which an evident power-law behavior holds. In particular, performing a fit in the region $\tau_Q \in [10^0, 10^3]$ for $L = 40$ to a power-law function, τ_Q^α, results in $\alpha = -0.248(2)$ and $-0.244(2)$ for $1/\xi_L$ and g_L, respectively. These fitted exponents indeed agree with the KZ prediction, as $\tau_Q^{-\nu/(1+z\nu)} = \tau_Q^{-1/4}$ with $\nu = 1/2$ and $z = 2$ for overdamped dynamics [23, 24, 26]. Moreover, we go beyond the KZ scaling and analyze the nonequilibrium finite-size scaling functions, which are hypothesized as

$$1/\xi_L \sim \tau_Q^{-\nu/(1+z\nu)} f_\xi(\tau_Q L^{-1/\nu-z}), \quad \text{and} \quad g_L \sim \tau_Q^{-\nu/(1+z\nu)} f_g(\tau_Q L^{-1/\nu-z}) \quad (2.31)$$

with $f_\xi(x)$ and $f_g(x)$ nonequilibrium scaling functions fulfilling $f_\xi(x \ll 1) \sim 1$ or simply a constant, and $f_\xi(x \gg 1) \sim x^{\nu/(1+z\nu)}$. Note that while the former condition gives account of the KZ scaling, the latter ensures that for adiabatic quenches the finite-size scaling is recovered [32]. The previous expressions are tested in Fig. 2.3b, where we observe a collapse of the data points when plotting L/ξ_L and Lg_L against $L^{-4}\tau_Q$ for different L and τ_Q values. Finally, we can also test the universality of the equation of motions in the same spirit as discussed in [28], which is accomplished by finding an appropriate transformation of physical parameters, such as space and time, that removes the dependence on τ_Q from the equations of motion. In the case of overdamped dynamics, the transformation consists in rescaling the spatial and time coordinates as $x \rightarrow x\tau_Q^{-1/4}$ and $t \rightarrow (t - t_c)\tau_Q^{-1/2}$, where $t_c = \tau_Q \frac{\varepsilon_0}{\varepsilon_0 - \varepsilon_1}$ corresponds to the time instant at which $\varepsilon(t_c) = \varepsilon_c$. This is briefly shown in the following lines. After shifting the time $t' = t - t_c$, the equation of motion reads

$$\eta \frac{\partial}{\partial t'}\phi(x, t') = \left(h^2 \frac{\partial^2}{\partial x^2} - \frac{\varepsilon_1 - \varepsilon_0}{\tau_Q}t' \right) \phi(x, t') + \zeta(x, t'), \qquad (2.32)$$

and thus, introducing $t'' = \theta_t t'$ and $x' = \theta_x x$, it follows

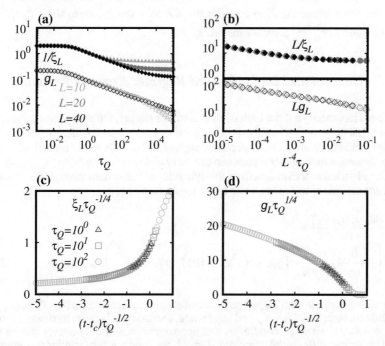

Fig. 2.3 Results for the overdamped dynamics. In **a** we plot the scaling of the inverse of the correlation length, $1/\xi_L$ (full points) and g_L (open points) at the critical point, ε_c, for three different system sizes, $L = 10$ (light grey), 20 (grey) and 40 (black). At intermediate quench times, $\tau_Q \in [10^0, 10^3]$, a fit results in $1/\xi_L \propto \tau_Q^{-0.248(2)}$ and $g_L \propto \tau_Q^{-0.244(2)}$ for $L = 40$. In **b** we show the nonequilibrium finite-size scaling $1/\xi_L \sim \tau_Q^{-\nu/(1+z\nu)} f_\xi(\tau_Q L^{-1/\nu-z})$ and $g_L \sim \tau_Q^{-\nu/(1+z\nu)} f_g(\tau_Q L^{-1/\nu-z})$ with $\nu = 1/2$ and $z = 2$ for overdamped dynamics. In **c** and **d** the collapse of the data into a single curve unveils the universality of the dynamics regardless of τ_Q for ξ_L and g_L, respectively. The parameters used in the simulations were $\beta = 1$, $\eta = 10$, $\varepsilon_0 = 100$, $\varepsilon_1 = -10$ and $h = 5$. See main text for further details regarding the transformations and discussion of the results

$$\eta \frac{\partial}{\partial t''} \phi(x', t'') = \left(h^2 \frac{\partial^2}{\partial (x')^2} \frac{\theta_x^2}{\theta_t} - \frac{\varepsilon_1 - \varepsilon_0}{\tau_Q \theta_t^2} t'' \right) \phi(x', t'') + \theta_t^{-1} \zeta(x', t''). \quad (2.33)$$

We now impose $\tau_Q \theta_t^2 = 1$ and $\theta_x^2 = \theta_t$, that lead to $\theta_t = \tau_Q^{-1/2}$ and $\theta_x = \tau_Q^{-1/4}$. Hence, the dynamics is expected to be universal in terms of (x', t''), that are achieved under the aforementioned transformation, $x' = x \tau_Q^{-1/4}$ and $t'' = (t - t_c) \tau_Q^{-1/2}$. Note that the stochastic term is enlarged by a factor $\tau_Q^{1/2}$, which indicates that the universality will be eventually lost for very slow quenches, $\tau_Q \rightarrow \infty$. However, τ_Q-independent dynamics still holds for small temperatures and a wide region of quench rates (see [28] for a detailed discussion on this topic). In short, performing this transformation the dynamics becomes universal in the sense that they do not depend on τ_Q and hence, $\xi_L \tau_Q^{-1/4}$ and $g_L \tau_Q^{1/4}$ versus $(t - t_c) \tau_Q^{-1/2}$ must show the same functional dependence regardless of the specific τ_Q value, which is precisely plotted in Fig. 2.3c and d. Note that this latter transformation conveys the KZ as a special case, since $\xi_L \tau_Q^{-1/4} \sim$ constant and $g_L \tau_Q^{1/4} \sim$ constant for $t = t_c$.

2.1.3 General and Underdamped Regime: Kramers Equation

So far we have analyzed the high friction limit, or simply, the overdamped dynamics in the GL theory, which besides being a good example where different scaling laws and the Fokker-Planck formalism can be applied, it has truly experimental relevance. Nevertheless, a more general scenario can be exploited starting from Eq. (2.17). The Fokker-Planck formalism describes the dynamics of the system through the so-called Kramers equation, which we write again for convenience,

$$\frac{\partial P_n(t, \varphi_n, \dot{\varphi}_n)}{\partial t} =$$
$$\left[-\frac{\partial}{\partial \varphi_n} \dot{\varphi}_n + \frac{\partial}{\partial \dot{\varphi}_n} \left(\eta \dot{\varphi}_n + \left(h^2 k_n^2 + \varepsilon(t) \right) \varphi_n \right) + \frac{\eta}{\beta} \frac{\partial^2}{\partial \dot{\varphi}_n^2} \right] P_n(t, \varphi_n, \dot{\varphi}_n). \quad (2.34)$$

The main difference with respect to the overdamped case resides in that $P_n(t, \varphi_n, \dot{\varphi}_n)$ depends on both variables, φ_n and $\dot{\varphi}_n$. Hence, the analysis is more intricate although still, due to the assumed conditions and approximations, the dynamics can be captured by simple differential equations. Indeed, we follow a very similar procedure as in Sect. 2.1.2. Thermal equilibrium conditions are deduced from $\partial_t P_n^{\text{th}} = 0$, whose solution in terms of $\varphi_n^R \equiv \text{Re}(\varphi_n)$ and $\varphi_n^I \equiv \text{Im}(\varphi_n)$ reads

$$P_n^{\text{th}} = \frac{\sqrt{(A_n^{\text{th}} B_n^{\text{th}})^2 - (C_n^{\text{th}})^2}}{2\pi} \text{Exp} \left[-\frac{1}{2} \left(\left(A_n^{\text{th}} \varphi_n^{R,I} \right)^2 + \left(B_n^{\text{th}} \dot{\varphi}_n^{R,I} \right)^2 - 2 C_n^{\text{th}} \varphi_n^{R,I} \dot{\varphi}_n^{R,I} \right) \right] \quad (2.35)$$

where the parameters of this two-dimensional Gaussian distribution are $A_n^{\text{th}} = \sqrt{\beta \left(h^2 k_n^2 + \varepsilon \right)}$, $B_n^{\text{th}} = \sqrt{\beta}$, and $C_n^{\text{th}} = 0$. The goal consists now in determining the evolution of these coefficients, which upon introducing the Gaussian Ansatz in

Eq. (2.34), it is straightforward to attain

$$\frac{\partial A_n(t)}{\partial t} = -\frac{C_n(t)}{A_n(t)}\left(\varepsilon(t) + h^2 k_n^2 + \frac{\eta C_n(t)}{\beta}\right), \tag{2.36}$$

$$\frac{\partial B_n(t)}{\partial t} = \eta B_n(t) - \frac{\eta}{\beta}B_n^3(t) + \frac{C_n(t)}{B_n(t)}, \tag{2.37}$$

$$\frac{\partial C_n(t)}{\partial t} = A_n^2(t) + \eta C_n(t) - B_n^2(t)\left(h^2 k_n^2 + \varepsilon(t) + \frac{2\eta C_n(t)}{\beta}\right). \tag{2.38}$$

As one would expect, the general case must coincide with the Smoluchowski treatment provided η is sufficiently large. It is then pertinent to define a reduced probability distribution integrating over $\dot{\varphi}_n$,

$$Q_n(t, \varphi_n^{R,I}) = \int_{-\infty}^{+\infty} d\dot{\varphi}_n^{R,I} \, P_n(t, \varphi_n^{R,I}, \dot{\varphi}_n^{R,I}) = \frac{F_n(t)}{\sqrt{\pi}} e^{-F_n^2(t)(\varphi_n^{R,I})^2}, \tag{2.39}$$

where the term $F_n(t)$ depends on the coefficients $A_n(t)$, $B_n(t)$ and $C_n(t)$ as

$$F_n(t) = \sqrt{\frac{1}{2B_n^2(t)}\left(A_n^2(t)B_n^2(t) - C_n^2(t)\right)}. \tag{2.40}$$

Note for example that F_n^{th} coincides with f_n^{th}, as they represent the same physical state (recall that the thermalization is guaranteed in both dynamical regimes, and it is unique for a given β). Hence, neither thermal equilibrium nor sudden quenches properties are discussed here, as they have been presented in Sect. 2.1.2.1. From a more practical point of view, calculating $F_n(t)$ enables the computation of $G(x, t)$, $\xi_L(t)$ and $g_L(t)$ by merely replacing $f_n(t)$ by the previous $F_n(t)$, as we will use in the following.

2.1.3.1 Finite-Rate Quenches

Solving numerically the three coupled differential equations per mode, Eqs. (2.36), (2.37) and (2.38), we study the low friction limit, i.e., underdamped dynamics, which is expected when $|\eta\dot{\phi}| \ll |\ddot{\phi}|$. We take the same parameters as for the overdamped dynamics $h = 5$, $\beta = 1$, $\varepsilon_0 = 100$ and $\varepsilon_1 = -10$, but $\eta = 0.1$ is chosen such that the scaling laws will resemble those of the asymptotic limit $\eta \to 0$.

The numerical results involving different quench times and system sizes are collected in Fig. 2.4. The inverse of the correlation length $1/\xi_L$ and g_L exhibit a power-law scaling, which becomes more evident and broader as L increases, and agrees with the KZ prediction. A fit for $L = 40$ in the range $\tau_Q \in [10^1, 10^3]$ provides $1/\xi_L \propto \tau_Q^{-0.34(1)}$ and $g_L \propto \tau_Q^{-0.36(1)}$, while the KZ mechanism predicts a $\tau_Q^{-\nu/(1+z\nu)}$ scaling with $\nu = 1/2$ and $z = 1$ for underdamped, and thus $\tau_Q^{-1/3}$. In Fig. 2.4b the nonequilibrium scaling functions are plotted for both quantities, supporting the existence of the functions $f_\xi(\tau_Q L^{-1/\nu - z})$ and $f_g(\tau_Q L^{-1/\nu - z})$, regardless of the specific

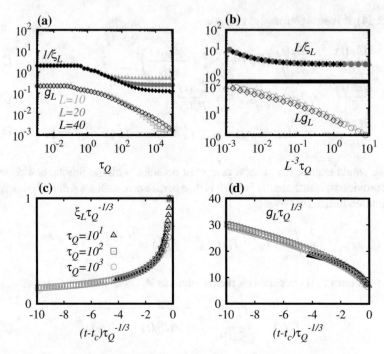

Fig. 2.4 Results for the underdamped dynamics. In **a** we plot the scaling of the inverse of the correlation length, $1/\xi_L$ (full points) and g_L (open points) at the critical point, ε_c, for three different system sizes, $L = 10$ (light grey), 20 (grey) and 40 (black). At intermediate quench times, $\tau_Q \in [10^0, 10^3]$, a fit results in $1/\xi_L \propto \tau_Q^{-0.34(1)}$ and $g_L \propto \tau_Q^{-0.36(1)}$ for $L = 40$. In **b** we show the nonequilibrium finite-size scaling $1/\xi_L \sim \tau_Q^{-\nu/(1+z\nu)} f_\xi(\tau_Q L^{-1/\nu-z})$ and $g_L \sim \tau_Q^{-\nu/(1+z\nu)} f_g(\tau_Q L^{-1/\nu-z})$ with $\nu = 1/2$ and $z = 1$ for overdamped dynamics. In **c** and **d** the data collapse unveil the universality of the dynamics regardless of τ_Q for ξ_L and g_L, respectively. The parameters used in the simulations were $\beta = 1$, $\eta = 0.1$, $\varepsilon_0 = 100$, $\varepsilon_1 = -10$ and $h = 5$. See main text for further details regarding the transformations and discussion of the results

values of L and τ_Q separately. Finally, for $L = 40$ we demonstrate the universality of the dynamics, as discussed in [28]. Recall that this universality resides in that the dynamics are self-similar in the parameter τ_Q, i.e., one can find a transformation such that τ_Q dependence is removed. As explained in the case of overdamped dynamics, Eq. (2.33), after shifting the time $t' = t - t_c$ with $t_c = \tau_Q \frac{\varepsilon_0}{\varepsilon_0 - \varepsilon_1}$ we introduce $x' = \theta_x x$ and $t'' = \theta_t t'$ such that the τ_Q dependence is removed from the equation of motion. In this case the transformation turns out to be $\theta_t = \theta_x = \tau_Q^{-1/3}$, that is, by transforming the spatial and time coordinates as $x \to x\tau_Q^{-1/3}$ and $t \to (t - t_c)\tau_Q^{-1/3}$, we find that the results for ξ_L and g_L collapse into a single curve when plotted against $(t - t_c)\tau_Q^{-1/3}$, as shown in Fig. 2.4c and d.

2.2 Coulomb Crystal in Ion Traps

In this section we abandon the GL theory, at least for the moment, to tackle a realistic setup, namely Coulomb crystals in ion traps, which has become an excellent platform where novel theoretical aspects can be tested and new regimes of matter can be explored thanks to the unprecedented control, cooling and isolation of ion traps [35, 36]. Indeed, Coulomb crystals in ion traps are assuredly one of the most accessible platforms where the exciting frontier between plasma physics, mesoscopic systems and quantum mechanics meet, and therefore, they appear as an excellent test-bed for statistical mechanics [36, 37]. Crystalline structures are formed in this setup as a consequence of Coulomb interaction, and thus, they differ from the customary solid-state crystals where atoms gather forming bonds overlapping wave-functions as a result of their electronic structure [38]. Although Coulomb crystallization of an electron gas was first predicted by E. Wigner in 1934 [39] and have been studied experimentally in certain systems such as a two-dimensional electron gas on the surface of liquid helium [40], the development of ion traps undoubtedly brought Coulomb crystals into the scientific spotlight, sparking renewed interest as well as offering novel opportunities for their systematically inspection. Therefore, we shall commence acknowledging the pioneer work of S. Earnshaw, who in 1842 proved that a set of charged particles cannot be trapped or confined solely by means of a time-independent electric field, and the memorable fabrication of the ion traps by the Physics Nobel laureates in 1989, H. G. Dehmelt and W. Paul. The former made use of a static electric field combined with an inhomogeneous magnetic field to attain a global confining potential, which is known as Penning trap [41]. Yet, another successful manner to confine charged particles was developed by W. Paul, which instead relies on time-dependent electromagnetic fields, and it is known as Paul or radio-frequency trap [42]. Since then, ion trap-based technologies have experienced a stunning advance, and has become one of the most promising platforms for quantum information processing [35, 43, 44]. It is worth commenting that, on top of these trap techniques, laser cooling is an essential technology to attain Coulomb crystallization [36] as well to exploit quantum effects [35].

Although the confinement is attained through different electromagnetic fields, the effective potential governing both physical systems turns out to be equivalent, up to corrections neglecting the so-called micromotion [45]. The full dynamics of the ions follows the well-known Mathieu equations, which determine the required stability conditions for the ion configurations depending on different parameters [35]. However, for the purposes of this thesis, micromotion effects are negligible and therefore we will assume that the system is well described by the effective potential V, while we refer to interested readers to the detailed analysis in [45],

$$V = \frac{1}{2} \sum_j m_j \left(\omega_x^2 x_j^2 + \omega_y^2 y_j^2 + \omega_z^2 z_j^2 \right) + \frac{1}{4\pi\epsilon_0} \sum_{j<k} \frac{Q_j Q_k}{|\mathbf{r}_j - \mathbf{r}_k|}. \qquad (2.41)$$

The first term captures the effect of the confinement provided by the electromagnetic fields, leading to effective harmonic potential with frequencies $\omega_{x,y,z}$, and where the position of the jth ion of mass m_j and charge Q_j has been denoted as $\mathbf{r}_j = (x_j, y_j, z_j)$, while the second term stands for Coulomb interaction. Note that, while different species of ions can be confined, we will consider in the following a unique type, i.e., $m_j \equiv m$ and $Q_j \equiv Q \; \forall j$.

At sufficiently low temperatures, the Coulomb repulsion becomes significant and the ions localize at their equilibrium position forming crystal-like patterns. This transition point can be quantified by the dimensionless coupling Γ, which gives account of the ratio between the nearest-neighbour Coulomb energy and the average thermal energy,

$$\Gamma = \frac{Q^2}{4\pi\epsilon_0 a_{\mathrm{WS}} k_B T} \tag{2.42}$$

where the Wigner-Seitz radius a_{WS} corresponds to the average distance between ions, and thus, it is related to the particle density $n = N/V$ as $(4/3)\pi n a_{\mathrm{WS}}^3 \equiv 1$. The ion plasma behaves as a gas when $\Gamma < 1$ and as liquid when $\Gamma > 2$. The crystallization however occurs at a much higher value, $\Gamma \approx 174$, as shown in [46]. Depending on the frequencies $\omega_{x,y,z}$, different crystal-like structures will be formed, as we illustrate in Fig. 2.5. Considering periodic boundary conditions in the x-axis, and ω_y sufficiently large such that the ions are confined in the xz-plane (a quasi two-dimensional system), a linear string or chain of ions is formed when ω_z is large, Fig. 2.5a, with zero transverse displacement $\langle z_j \rangle = 0$. If the transverse potential is reduced, that is, if ω_z becomes smaller than a critical value ω_z^c, a structural phase transition takes place, which brings the system into a zigzag structure with $\langle z_j \rangle \neq 0$ as it becomes energetically more favorable Fig. 2.5b. Reducing even more ω_z, one will find a three-row structure (Fig. 2.5c), then four rows and so on [7], which ultimately leads to a triangular lattice. Three-dimensional crystals (bcc) are also possible, and will naturally emerge if ω_y does not constrain the system to the xz-plane. Although the specific equilibrium positions $\mathbf{r}_j^0$ can be calculated minimizing the potential energy, $\partial_{\mathbf{r}_j} V |_{\mathbf{r}_j^0} = 0$, which for a linear string were tabulated by James [47], an estimate of

(a)   **(b)**   **(c)**

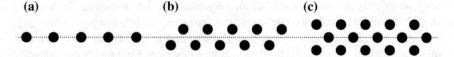

Fig. 2.5 Crystal structures emerging in a homogeneous ion trap, depending on the ion species and confinement potentials. For simplicity, ω_y is assumed to be very large such that the crystal is formed in the xz-plane, as well as x-axis periodic boundary conditions. A string or linear chain of ions is formed when $\omega_z > \omega_z^c$ with transverse displacement $\langle z_j \rangle = 0$ (dashed line) (**a**), while a zigzag structure emerges since it is energetically more favorable when the transverse potential becomes shallower, $\omega_z < \omega_z^c$, characterized by $\langle z_j \rangle \neq 0$ (**b**). Reducing even more the transverse potential, a three-row is formed (**c**)

the distance between ions, a, results from the condition in which the confinement harmonic force compensates the Coulomb repulsion, typically leading to few μm [36]. Then, assuming typical values $n = 3 \times 10^{14}$ m^{-3} ($a_{WS} \approx 10$ μm) and ion charge $Q = -e = 1.6 \times 10^{-19}$ C, a temperature lower than 10 mK would be required to form an ion crystal. This low temperature is achievable by means of Doppler cooling, a widely used and suitable laser cooling technique [48], and therefore, Coulomb crystals can be explored nowadays in trapped-ion platforms as demonstrated in several works [16, 17, 49–51] (see [36] for a review). Finally, before starting a more detailed analysis of the structural phase transition from linear to a zigzag chain [8, 27, 52], it is worth commenting that quantum effects become important when distinct particle wavefunctions overlap, that is, when the thermal de Broglie wavelength λ_B becomes comparable to the interparticle distance a_{ws}. Nevertheless, we will consider a situation in which $\lambda_B = \left(h^2/(3mk_BT)\right)^{1/2} \ll a_{ws}$ with h the Planck constant,[2] and thus, the system can be treated classically.

2.2.1 Ginzburg-Landau Map

The linear to zigzag phase transition was firstly investigated numerically in [53], and later also including theoretical calculations [7]. It was however years later when it was proven that the linear to zigzag phase transition is continuous [8], and also providing a GL description. As we have already introduced in Sect. 2.1, GL theory is a powerful tool that properly casts critical phenomena in a variety of systems, although fails in certain circumstances (see Sect. 2.1). As we present in the following lines, GL theory is suitable for the linear to zigzag phase transition.

The GL description of this physical problem starts with symmetry arguments. In this regard, we should notice that, as depicted in Fig. 2.5, while the linear phase features a unique stable equilibrium, $\langle z_j \rangle = 0$, the zigzag structure admits an equivalent conformation, namely, a *zagzig*, which is the z-reflection counterpart of the zigzag. That is, while one stable conformation in the zigzag phase can be written as $z_j = (-1)^j|z_D|$ (with z_D the transverse displacement amplitude), its symmetric counterpart reads $z_j = (-1)^{j+1}|z_D|$. In this manner, we can map the ion transverse coordinate to a field as $\phi = (-1)^j z_j$ in order to attain a smooth ϕ function. Therefore, we can already notice that, as the free energy within GL theory must preserve symmetry properties, $\mathcal{F}[\phi] = \mathcal{F}[-\phi]$, which prevents $\mathcal{F}$ of containing any odd power of ϕ. This is again the typical scenario of a symmetry breaking phase transition since any fluctuation in ϕ will break the delicate Z_2 symmetry.[3] For simplicity, we consider identical ion, periodic boundary condition in the x axis, and ω_y sufficiently

---

[2]The de Broglie wavelength reads $\lambda_B = h/p$. For a particle of mass m and in thermal equilibrium with temperature T, p takes the value $p^2 = 3mk_BT$, as a consequence of the equipartition theorem.

[3]Consider that the field $\phi(x)$ fluctuates around its mean value, ϕ_0, such that $\phi(x) = \phi_0 + \psi(x)$. Then, while in the linear phase ($\phi_0 = 0$) the system is symmetric by construction under fluctuation reflections, $\mathcal{F}[\psi(x)] = \mathcal{F}[\psi(x)]$, in the zigzag phase the symmetry is not preserved since $\mathcal{F}[\phi_0 + \psi(x)] \neq \mathcal{F}[\phi_0 - \psi(x)]$.

large to confine the ions to the xz-plane, although the procedure can be performed in a straightforward manner for a more general situation [32]. Then, instead of the general Eq. (2.41), the quasi two-dimensional system obeys the potential

$$V = \frac{1}{2} \sum_j m\omega_z^2 z_j^2 + \frac{Q^2}{4\pi\epsilon_0} \sum_{j<k} \frac{1}{\sqrt{(x_j - x_k)^2 + (z_j - z_k)^2}} \tag{2.43}$$

which, upon Taylor expanding it around the equilibrium positions with $\phi = (-1)^j z_j$, it can be shown that [8, 27]

$$\mathcal{F}[\phi(x)] = \frac{m}{2a} \int dx \left\{ \delta\phi^2(x) + h^2 \left(\frac{\partial\phi(x)}{\partial x}\right)^2 + \Lambda\phi^4(x) \right\}, \tag{2.44}$$

with the interspacing ion distance a, $h = \omega_0 a\sqrt{\ln 2}$, $\delta = \omega_z^2 - \omega_{z,c}^2$ and $\Lambda = 93\zeta_R(5)\omega_0^2/(32a^2)$ with $\zeta_R(s) = \sum_s^\infty n^{-s}$ the Riemann zeta function. The natural frequency of the system and the critical value of the linear to zigzag transition are given by $\omega_0 = Q/(4\pi\epsilon_0\sqrt{ma^3})$ and $\omega_c = \omega_0\sqrt{7\zeta_R(3)/2}$, respectively. Remarkably, Eq. (2.44) corresponds to a scalar ϕ^4 potential as discussed in Sect. 2.1. Briefly, the equilibrium configuration of the field ϕ_0 follows from $\partial_\phi V_{GL}(\phi) = 0$, where the potential is simply,

$$V_{GL}(\phi_0) = \frac{m}{2a}\delta\phi_0^2 + \frac{m\Lambda}{2a}\phi_0^4 \tag{2.45}$$

which corresponds to the aforementioned Eq. (2.4). The situation is then completely similar to the cases studied in the Sect. 2.1, and then, the linear to zigzag phase transition is tantamount to the paramagnetic to ferromagnetic phase transition, among others. In this manner, the order parameter ϕ_0 undergoes a smooth transition from $\phi_0 = 0$ for $\delta > 0$ ($\omega_z > \omega_{z,c}$) to $\phi_0 = \pm\sqrt{-\delta/2\Lambda}$ for $\delta < 0$ ($\omega_z < \omega_{z,c}$) (Fig. 2.6). More specifically, $\phi_0 \propto |\delta - \delta_c|^\beta$ (recall that δ corresponds to ε in the previous Section, and $\delta_c = 0$) with $\beta = 1/2$ the mean-field critical exponent associated to the order parameter, and as we have derived in Sect. 2.1, the correlation length $\xi \approx h/\sqrt{\delta}$ for $|\delta - \delta_c| \ll 1$, and thus, it diverges as $\xi \propto |\delta - \delta_c|^{-\nu}$ with $\nu = 1/2$. Finite-size systems do not exhibit these divergences but rather show a smooth behavior, as we have shown in Sect. 2.1.

2.2.2 Defect Formation and Kibble-Zurek Scaling

The problem of linear to zigzag phase transition can be mapped onto a GL theory with suitable parameters to a very good approximation. In this manner, it is straightforward to apply the same analysis as in Sect. 2.1 to study nonequilibrium dynamics across the critical point. However, we do not resort to the previously explained map, but apply

Fig. 2.6 Schematic representation of the GL potential $V_{GL}(\phi_0)$ derived in Eq. (2.45), for different situations and their associated crystal conformation, namely, for $\delta > 0$ a linear string and for $\delta < 0$ the two equivalent choices $\phi_0 = \pm\sqrt{-\delta/2\Lambda}$, which correspond to the two symmetric zigzag (zagzig) disposition

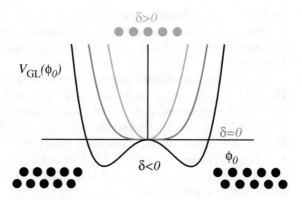

the procedure directly to the realistic Coulomb crystal problem. Before introducing the Fokker-Planck approach, we introduce the customary Langevin dynamics of the ions and how kinks or defects are formed when the zigzag phase is induced by varying the global confining potential [37, 54]. This situation, in the spirit of the KZ mechanism, has been experimentally addressed in ion traps [16, 17] as well as numerically in a number of works, where GL theory is commonly used [27, 32, 55]. Note however that here we consider periodic boundary conditions, which largely simplifies the problem as well as the derivation of KZ scaling laws (see [27, 55–57] for inhomogeneous systems).

The potential of the system can be written as

$$V = \frac{1}{2}\sum_j^N m\left(\omega_y^2 y_j^2 + \omega_z^2(t)z_j^2\right) + \frac{Q^2}{4\pi\epsilon_0}\sum_{j<k}^N \frac{1}{|\mathbf{r}_j - \mathbf{r}_k|}. \tag{2.46}$$

where the transverse potential frequency is varied linearly in time, from the linear to the zigzag phase,

$$\omega_z(t) = \begin{cases} \omega_{z,0} & \text{for } t < 0 \\ \omega_{z,0} + \frac{\omega_{z,1}-\omega_{z,0}}{\tau_Q}t & \text{for } 0 \leq t < \tau_Q \\ \omega_{z,1} & \text{for } t \geq \tau_Q, \end{cases} \tag{2.47}$$

with τ_Q the total time of the quench and $\omega_{z,0} > \omega_{z,c}$ and $\omega_{z,1} < \omega_{z,c}$. Hence, $\tau_Q \to \infty$ leads to an adiabatic passage through the phase transition. Note that, while we consider ω_y large enough to constrain the ions to the xz-plane, i.e., $y_j \approx 0$, the impact fluctuations and confinement potential in the y-axis have been taken into account in the molecular dynamics simulations.

Initially, the N ions are in thermal equilibrium in a linear chain configuration, $\omega_{z,0}$. Then, as a consequence of the finite-rate passage across the phase transition, the system is driven out of equilibrium entering into the zigzag phase, where symmetry-breaking states can be chosen only locally and thus, promoting the appearance of defects or kinks, as exemplified in Fig. 2.7. In the following we address the dynamics of the system by means of molecular dynamics simulations (solving the Langevin equations) and, under the harmonic approximation, using the Fokker-Planck approach.

The numerical simulations were performed with realistic parameters, namely, $a = 10$ μm, $N = 21$ ions, mass $m = 172$ amu and charge $Q = -e = 1.6 \times 10^{-19}$ C, which corresponds to $^{172}\text{Yb}^{+}$ ions. The temperature is set to $T = 5$ mK and considered to be constant during the whole process. The frequency $\omega_0 = 2\pi \times 135$ kHz and the critical frequency amounts to $\omega_{z,c} = 2\pi \times 278$ kHz. The initial and final frequencies are $\omega_{z,0} = 2\pi \times 477$ kHz and $\omega_{z,1} = 2\pi \times 159$ kHz. In addition, we consider $\omega_y = \omega_{z,0} = 2\pi \times 477$ kHz, sufficiently large to ensure a quasi two-dimensional Coulomb crystal.

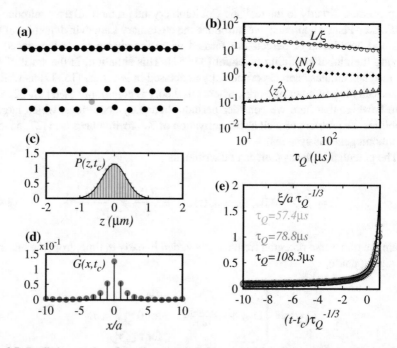

Fig. 2.7 Realistic crystal configurations are plotted in **a**, obtained solving the Langevin equations. At the top (bottom), the ions before (after) the ramp across the critical point for $N = 21$ ions with periodic boundary conditions. Note the presence of a defect in the zigzag phase, depicted in light grey (see main text for further details). The **b** panel shows the scaling of the correlation length L/ξ, number of defects $\langle N_d \rangle$ and mean-square displacement $\langle z^2 \rangle$ as a function of the quench time τ_Q. In **c**, we show the probability distribution for the transverse displacement z, while **d** corresponds to the two-point correlation function $G(x, t)$ at the critical point for a quench $\tau_Q = 41.7$ μs. In **e** the rescaled correlation length $\xi/a\tau_Q^{-1/3}$ is shown for different quench times which collapse into a single curve, for both approaches. All the results of the molecular dynamics simulations were obtained averaging 2000 stochastic trajectories, while the solid lines represent the results using the presented Fokker-Planck approach

2.2.2.1 Langevin Approach

The dynamics of the system is dictated by the following stochastic equations

$$m\frac{d^2x_j}{dt^2} + \eta\frac{dx_j}{dt} + \frac{\partial V}{\partial x_j} = \zeta_j^x(t), \tag{2.48}$$

$$m\frac{d^2y_j}{dt^2} + \eta\frac{dy_j}{dt} + \frac{\partial V}{\partial y_j} = \zeta_j^y(t), \tag{2.49}$$

$$m\frac{d^2z_j}{dt^2} + \eta\frac{dz_j}{dt} + \frac{\partial V}{\partial z_j} = \zeta_j^z(t). \tag{2.50}$$

There, η is the friction coefficient and $\zeta_j^{x,y,z}$ is the stochastic force that satisfies the fluctuation-dissipation theorem (see Appendix A),

$$\langle \zeta_j^\alpha(t) \rangle = 0, \tag{2.51}$$

$$\langle \zeta_j^\alpha(t)\zeta_k^\gamma(t') \rangle = 2\eta\beta^{-1}\delta_{\alpha\gamma}\delta_{jk}\delta(t-t'), \tag{2.52}$$

where $\alpha, \gamma \in \{x, y, z\}$ and $j, k \in \{1, 2, ..., N\}$.

The friction coefficient can be estimated as $\eta = 1.5 \times 10^{-21}$ kg s^{-1} [16], while the rest of the parameters have been given previously. The dynamics is then solved by Langevin impulse integration method as explained in [58], and choosing a timestep of 10 ns. Then, in order to ensure that the system initially is in thermal equilibrium, we evolve the system under $\omega_{z,0}$ during 100 μs before starting the quench protocol. The quench time τ_Q is varied from 10–400 μs. For each value of τ_Q, we perform 2000 simulations that provides an accurate estimation of statistical observables, as we show here for the two-point correlation function $G(x, t)$, correlation length $\xi(t)$, number of defects at the end of the quench $\langle N_d \rangle$, and the probability distribution of the transverse displacement $z(t)$, which due to translational symmetry does not depend on the ion position.

In Fig. 2.7 we have collected some important results. First, it is noticeable that, due to the periodic boundary conditions and thus, the different topology, an odd (even) number of ions leads to an odd (even) number of defects. As we have selected $N = 21$, the minimum amount of defects will be one (Fig. 2.7a), therefore $\langle N_d \rangle$ saturates to 1 as $\tau_Q \to \infty$. To the contrary, too fast passages will destroy the crystal as too much energy is suddenly given to the crystal, driven far from stability. Hence, we shall focus on moderate passages across the critical point. In the panel (b), we show the scaling of relevant quantities, namely, the average number of defects, correlation length L/ξ and mean-square transverse displacement $\langle z^2 \rangle$ at the critical point. A fit to a τ_Q^α gives $\alpha = -0.31(1)$ and $-0.30(1)$ in the range of $\tau_Q \in [40, 200]$ μs for L/ξ and $\langle N_d \rangle$ using Langevin approach. These fitted exponents are indeed in agreement with the KZ predictions for underdamped dynamics. Probability distributions can be retrieved through histograms of the possible outcomes, as shown in Fig. 2.7c for the probability distribution of the transverse displacement, $P(z, t)$. In (d), the two-point correlation function $G(x, t)$ is plotted, both for $\tau_Q = 47.7$ μs. Finally, in (e) the collapse of the correlation length applying the transformation is verified. Remarkably, while the Langevin results include the full non-linear terms, the Fokker-Planck description neglecting non-linear terms turns out to be a good approximation, which we discuss in the following.

2.2.2.2 Harmonic Approximation and Fokker-Planck Approach

Since the linear to zigzag phase transition can be mapped onto a GL theory, we can proceed as in Sect. 2.1, that is, dropping non-linear terms and relying on a Fokker-Planck approach to attain probability distributions in a deterministic manner. We emphasize however that we do not resort to the GL mapping, and thus, we apply the introduced method to the ion crystal. As in the GL case, it is convenient to move to the Fourier space, where the stochastic Langevin equations decouple for each mode provided non-linear terms are neglected, which is a good approximation in the vicinity of the critical point. The Fokker-Planck equations, or more specifically, the resulting Kramers equation dictates the dynamics of the system, from where statistical observables follow.

The equilibrium configuration of the ions for $\omega_z > \omega_{z,c}$ is given by $\mathbf{r}_j^{(0)} = (x_j^{(0)}, 0, 0)$, where for convenience we take $x_i > x_j$ for $i > j$. As a consequence of the translationally invariant system, equilibrium inter-particle distances are constant, $a = x_{j+1}^{(0)} - x_j^{(0)}$, in contrast to the inhomogeneous traps where, due to the harmonic confinement potential, the distance a becomes smaller in the center than on the edges [37]. Considering small fluctuations around the equilibrium position, $q_j = x_j - x_j^{(0)}$, we obtain a linear approximation of the equations of motion upon Taylor expanding the potential. Indeed, up to second order, transverse and axial motion decouple [27], and the transverse displacement z_j evolves according to

$$m\ddot{z}_j + \eta\dot{z}_j + m\omega_t^2 z_j - \frac{1}{2}\sum_{i\neq j}\mathcal{K}_{i,j}\left(z_j - z_i\right) = \zeta_j^z(t) \qquad (2.53)$$

where $\mathcal{K}_{i,j} \equiv -\partial^2 V/\partial x_j \partial x_i\big|_{x_j^{(0)}}$ is given by

$$\mathcal{K}_{i,j} = \frac{Q^2}{2\pi\epsilon_0}\frac{1}{\left|x_i^{(0)} - x_j^{(0)}\right|^3}. \qquad (2.54)$$

Finally, as the Eq. (2.53) describes the motion of coupled oscillators, it adopts a simpler form in terms of the normal modes, Ψ_n

$$z_j = \frac{1}{\sqrt{N}}\Psi_0^+ + \sqrt{\frac{2}{N}}\sum_{n=1}^{(N-1)/2}\left(\Psi_n^+ \cos(k_n ja) + \Psi_n^- \sin(k_n ja)\right), \qquad (2.55)$$

where $k_n = 2\pi n/Na$ and the sign $+(-)$ indicates the parity under $k_n \to -k_n$. Introducing Eq. (2.55) in the expression (2.53) gives the Langevin equation under the harmonic approximation (i.e., neglecting non-linear interaction terms) in the Fourier space,

$$m\ddot{\Psi}_n^\pm + \eta\dot{\Psi}_n^\pm + m\omega_n^2(t)\Psi_n^\pm = \zeta_n^\pm(t), \tag{2.56}$$

where $\omega_n(t)$ defines the frequency of the normal modes,

$$\omega_n^2(t) = \omega_t^2(t) - 2\left(\frac{Q^2}{4\pi\epsilon_0 ma^3}\right)\sum_{j=1}^{N}\frac{1}{j^3}\sin^2\left(\frac{k_n ja}{2}\right), \tag{2.57}$$

and $\zeta_n^\pm(t)$ represents the stochastic force in the normal mode space, which again fulfills $\langle\zeta_n^p(t)\zeta_m^q(t')\rangle = 2\eta/\beta\delta_{pq}\delta_{nm}\delta(t-t')$. As we have seen in Sect. 2.1, Eq. (2.56) adopts the form of a Kramers equation within the Fokker-Planck approach, which now reads

$$\frac{\partial P_n(t, \Psi_n, \dot{\Psi}_n)}{\partial t} =$$
$$\left[-\frac{\partial}{\partial\Psi_n}\dot{\Psi}_n + \frac{\eta}{2\beta m^2}\frac{\partial^2}{\partial\dot{\Psi}_n^2} + \frac{\partial}{\partial\dot{\Psi}_n}\left(\frac{\eta}{m}\dot{\Psi}_n + \omega_n^2(t)\Psi_n\right)\right]P_n(t, \Psi_n, \dot{\Psi}_n), \tag{2.58}$$

and in this manner, having the probability distributions for each mode allows to compute any statistical quantity. We solve the previous equation resorting to the same procedure explained in Sect. 2.1, i.e., assuming Gaussian probability distributions. Therefore, we can obtain the two-point correlation functions, g_L and correlation length ξ in a straightforward manner. Moreover, from Eq. (2.55) and since $\Psi_n^\pm$ are statistically independent and Gaussian distributed, we can calculate the probability over the transverse displacement, $P(z, t)$, which adopts also a Gaussian form and independent of j,

$$P(z, t) = \frac{1}{\sqrt{2\pi\sigma^2(t)}}e^{-z^2/(2\sigma^2(t))}, \tag{2.59}$$

with a time-dependent variance

$$\sigma^2(t) = \frac{1}{N}\left(\frac{1}{2F_0^2(t)} + \sum_{n=1}^{(N-1)/2}\frac{1}{F_n^2(t)}\right), \tag{2.60}$$

where $F_n(t)$ is obtained from Eq. (2.58) in the same way as explained in Sect. 2.1.3. Despite the approximate procedure, the Fokker-Planck results still reproduce the essential features of the nonequilibrium states, as shown in Fig. 2.7.

2.3   Conclusion and Outlook

In this chapter we have examined the celebrated GL theory of continuous phase transitions. Although the equilibrium properties of this model have been discussed, more emphasis have been devoted to analyze nonequilibrium passages across the critical point, in the spirit of the KZ problem. Indeed, this constitutes the simplest scenario in which KZ arguments can be tested, as shown by numerical simulations in [23, 24] where the ideas developed in the seminal works by W. H. Zurek and T. W. B. Kibble where confirmed [9, 10]. A finite-rate ramp traversing the critical point forces the system to break the symmetry only locally, and thus, promoting the formation of defects. In this chapter, we have studied this nonequilibrium scenario in two distinct dynamical regimes, namely overdamped and underdamped dynamics, resorting to a Fokker-Planck description and neglecting non-linear terms in the GL potential. This apparent naive procedure turns out to be a good approximation in the vicinity of the critical point. Moreover, Coulomb crystals in ion traps appear as a suitable and realistic platform to explore KZ physics [27, 32, 37, 55], as shown by the recent experimental works [16, 17]. In particular, we focus on quasi two-dimensional crystals undergoing a linear to zigzag phase transition, which can be well described by the GL theory [8, 27]. Therefore, the results presented in the GL can be readily applied to this realistic platform. Nevertheless, we do not resort to this map and study by means of the Fokker-Planck procedure (upon neglecting non-linear interaction terms) and molecular dynamics simulations. We discuss the range of validity of the performed approximation as well as the advantages it entails with respect to the customary Langevin approach. Among them, we find the deterministic computation of probability distributions and the possibility to apply the developed framework to systems that are not well described by a GL model.

It is worth recalling and keeping in mind that the presented method to obtain an advantageous Fokker-Planck approach relies on certain approximations. The most remarkable certainly resides in the lack of thermal equilibrium in the symmetry broken phase, as higher order terms are neglected. While these non-linear terms are crucial to address and achieve thermalization in this phase, here we are interested solely in nonequilibrium features, far from equilibrium, and therefore they need not to be accounted for our purposes. In this respect, developing a Fokker-Planck theory which allows to obtain valuable insights while including non-linear interaction terms may be an interesting direction for future work.

Moreover, the results presented here had the clear goal of addressing the nonequilibrium finite-size scaling functions and KZ scaling. However, there are several attractive directions which remain to be disclosed and investigated. For example, the developed framework may allow to gain theoretical insights in how correlations are built in certain systems and in diverse situations, not necessarily involving phase transitions. Indeed, the determination of entropy rates or information-inspired quantities, such as Shannon entropy, is left for future work. This naturally brings us to the question of whether the analyzed scaling laws emerge in typical thermodynamic quantities, such as work done and entropy production, and their inspection is attracting an increasing attention from different scientific communities. In this regard, it is worth noting the recent works combining ideas from quantum or mesoscopic thermodynamics and nonequilibrium critical phenomena [59–61]. The natural confluence of these two fields will assuredly uncover intriguing physics.

References

1. H.E. Stanley, *Introduction to Phase Transitions and Critical Phenomena* (Clarendon Press, Oxford, 1971)
2. K. Huang, *Statistical Mechanics* (Wiley, New York, 1987)
3. N. Goldenfeld, *Lectures on Phase Transitions and the Renormalization Group* (Addison-Wesley, 1992)
4. G. Jaeger, The Ehrenfest classification of phase transitions: introduction and evolution. Arch. Hist. Exact Sci. **53**, 51 (1998). https://doi.org/10.1007/s004070050021
5. L.D. Landau, E.M. Lifshitz, *Statistical Physics*, 3rd edn. (Butterworth-Heinemann, Oxford, 1980)
6. P.C. Hohenberg, B.I. Halperin, Theory of dynamic critical phenomena. Rev. Mod. Phys. **49**, 435 (1977). https://doi.org/10.1103/RevModPhys.49.435
7. G. Piacente, I.V. Schweigert, J.J. Betouras, F.M. Peeters, Generic properties of a quasi-one-dimensional classical Wigner crystal. Phys. Rev. B **69**, 045324 (2004). https://doi.org/10.1103/PhysRevB.69.045324
8. S. Fishman, G. De Chiara, T. Calarco, G. Morigi, Structural phase transitions in low-dimensional ion crystals. Phys. Rev. B **77**, 064111 (2008). https://doi.org/10.1103/PhysRevB.77.064111
9. T.W.B. Kibble, Topology of cosmic domains and strings. J. Phys. A: Math. Gen. **9**, 1387 (1976). http://stacks.iop.org/0305-4470/9/i=8/a=029
10. W.H. Zurek, Cosmological experiments in superfluid helium? Nature **317**, 505 (1985). https://doi.org/10.1038/317505a0
11. T.W.B. Kibble, Phase-transition dynamics in the lab and the universe. Phys. Today **60**, 47 (2007). https://doi.org/10.1063/1.2784684
12. A. del Campo, W.H. Zurek, Universality of phase transition dynamics: topological defects from symmetry breaking. Int. J. Mod. Phys. A **29**, 1430018 (2014). https://doi.org/10.1142/S0217751X1430018X
13. L.E. Sadler, J.M. Higbie, S.R. Leslie, M. Vengalattore, D.M. Stamper-Kurn, Spontaneous symmetry breaking in a quenched ferromagnetic spinor Bose-Einstein condensate. Nature **443**, 312 (2006). https://doi.org/10.1038/nature05094
14. C.N. Weiler, T.W. Neely, D.R. Scherer, A.S. Bradley, M.J. Davis, B.P. Anderson, Spontaneous vortices in the formation of Bose-Einstein condensates. Nature **455**, 948 (2008). https://doi.org/10.1038/nature07334
15. G. Lamporesi, S. Donadello, S. Serafini, F. Dalfovo, G. Ferrari, Spontaneous creation of Kibble-Zurek solitons in a Bose-Einstein condensate. Nat. Phys. **9**, 656 (2013). https://doi.org/10.1038/nphys2734
16. K. Pyka, J. Keller, H.L. Partner, R. Nigmatullin, T. Burgermeister, D.M. Meier, K. Kuhlmann, A. Retzker, M.B. Plenio, W.H. Zurek, A. del Campo, T.E. Mehlstäubler, Topological defect formation and spontaneous symmetry breaking in ion Coulomb crystals. Nat. Commun. **4**, 2291 (2013). https://doi.org/10.1038/ncomms3291
17. S. Ulm, J. Roßnagel, G. Jacob, C. Degünther, S.T. Dawkins, U.G. Poschinger, R. Nigmatullin, A. Retzker, M.B. Plenio, F. Schmidt-Kaler, K. Singer, Observation of the Kibble-Zurek scaling law for defect formation in ion crystals. Nat. Commun. **4**, 2290 (2013). https://doi.org/10.1038/ncomms3290

18. H. Risken, *The Fokker-Planck Equation: Methods of Solution and Applications*, 2nd edn. (Springer, New York, 1984)
19. R. Puebla, R. Nigmatullin, T.E. Mehlstäubler, M.B. Plenio, Fokker-Planck formalism approach to Kibble-Zurek scaling laws and nonequilibrium dynamics. Phys. Rev. B **95**, 134104 (2017). https://doi.org/10.1103/PhysRevB.95.134104
20. L.P. Kadanoff, More is the same; phase transitions and mean field theories. J. Stat. Phys. **137**, 777 (2009). https://doi.org/10.1007/s10955-009-9814-1
21. R.J. Baxter, *Exactly Solved Models in Statistical Mechanics* (Academic Press, London, 1989)
22. R. Balescu, *Statistical Dynamics: Matter Out of Equilibrium* (Imperial College Press, London, 1997)
23. P. Laguna, W.H. Zurek, Density of kinks after a quench: When symmetry breaks, how big are the pieces? Phys. Rev. Lett. **78**, 2519 (1997). https://doi.org/10.1103/PhysRevLett.78.2519
24. P. Laguna, W.H. Zurek, Critical dynamics of symmetry breaking: quenches, dissipation, and cosmology. Phys. Rev. D **58**, 085021 (1998). https://doi.org/10.1103/PhysRevD.58.085021
25. F. Liu, G.F. Mazenko, Defect-defect correlation in the dynamics of first-order phase transitions. Phys. Rev. B **46**, 5963 (1992). https://doi.org/10.1103/PhysRevB.46.5963
26. E. Moro, G. Lythe, Dynamics of defect formation. Phys. Rev. E **59**, R1303(R) (1999). https://doi.org/10.1103/PhysRevE.59.R1303
27. G. De Chiara, A. del Campo, G. Morigi, M.B. Plenio, A. Retzker, Spontaneous nucleation of structural defects in inhomogeneous ion chains. New J. Phys. **12**, 115003 (2010). http://stacks.iop.org/1367-2630/12/i=11/a=115003
28. G. Nikoghosyan, R. Nigmatullin, M.B. Plenio, Universality in the dynamics of second-order phase transitions. Phys. Rev. Lett. **116**, 080601 (2016). https://doi.org/10.1103/PhysRevLett.116.080601
29. S. Deng, G. Ortiz, L. Viola, Dynamical non-ergodic scaling in continuous finite-order quantum phase transitions. Europhys. Lett. **84**, 67008 (2008). http://stacks.iop.org/0295-5075/84/i=6/a=67008
30. M. Kolodrubetz, B.K. Clark, D.A. Huse, Nonequilibrium dynamic critical scaling of the quantum Ising chain. Phys. Rev. Lett. **109**, 015701 (2012). https://doi.org/10.1103/PhysRevLett.109.015701
31. A. Chandran, A. Erez, S.S. Gubser, S.L. Sondhi, Kibble-Zurek problem: universality and the scaling limit. Phys. Rev. B **86**, 064304 (2012). https://doi.org/10.1103/PhysRevB.86.064304
32. R. Nigmatullin, A. del Campo, G. De Chiara, G. Morigi, M.B. Plenio, A. Retzker, Formation of helical ion chains. Phys. Rev. B **93**, 014106 (2016). https://doi.org/10.1103/PhysRevB.93.014106
33. R. Kubo, The fluctuation-dissipation theorem. Rep. Prog. Phys. **29**, 255 (1966). http://stacks.iop.org/0034-4885/29/i=1/a=306
34. M.E. Fisher, M.N. Barber, Scaling theory for finite-size effects in the critical region. Phys. Rev. Lett. **28**, 1516 (1972). https://doi.org/10.1103/PhysRevLett.28.1516
35. D. Leibfried, R. Blatt, C. Monroe, D. Wineland, Quantum dynamics of single trapped ions. Rev. Mod. Phys. **75**, 281 (2003). https://doi.org/10.1103/RevModPhys.75.281
36. R.C. Thompson, Ion Coulomb crystals. Cont. Phys. **56**, 63 (2015). https://doi.org/10.1080/00107514.2014.989715
37. A. Retzker, R.C. Thompson, D.M. Segal, M.B. Plenio, Double well potentials and quantum phase transitions in ion traps. Phys. Rev. Lett. **101**, 260504 (2008). https://doi.org/10.1103/PhysRevLett.101.260504
38. N.W. Ashcroft, N.D. Mermin, *Solid State Physics* (Saunders College, Philadelphia, 1976)
39. E. Wigner, On the interaction of electrons in metals. Phys. Rev. **46**, 1002 (1934). https://doi.org/10.1103/PhysRev.46.1002
40. R. Crandall, R. Williams, Crystallization of electrons on the surface of liquid helium. Phys. Lett. A **34**, 404 (1971). https://doi.org/10.1016/0375-9601(71)90938-8
41. L.S. Brown, G. Gabrielse, Geonium theory: physics of a single electron or ion in a Penning trap. Rev. Mod. Phys. **58**, 233 (1986). https://doi.org/10.1103/RevModPhys.58.233

42. W. Paul, Electromagnetic traps for charged and neutral particles. Rev. Mod. Phys. **62**, 531 (1990). https://doi.org/10.1103/RevModPhys.62.531
43. H. Häffner, C.F. Roos, R. Blatt, Quantum computing with trapped ions. Phys. Rep. **469**, 155 (2008). https://doi.org/10.1016/j.physrep.2008.09.003
44. M.A. Nielsen, I.L. Chuang, *Quantum Computation and Quantum Information* (Cambridge University Press, Cambridge, 2000)
45. D.H.E. Dubin, T.M. O'Neil, Trapped nonneutral plasmas, liquids, and crystals (the thermal equilibrium states). Rev. Mod. Phys. **71**, 87 (1999). https://doi.org/10.1103/RevModPhys.71.87
46. M.D. Jones, D.M. Ceperley, Crystallization of the one-component plasma at finite temperature. Phys. Rev. Lett. **76**, 4572 (1996). https://doi.org/10.1103/PhysRevLett.76.4572
47. D. James, Quantum dynamics of cold trapped ions with application to quantum computation. App. Phys. B **66**, 181 (1998). https://doi.org/10.1007/s003400050373
48. D.J. Wineland, C. Monroe, W.M. Itano, D. Leibfried, B.E. King, D.M. Meekhof, Experimental issues in coherent quantum-state manipulation of trapped atomic ions. J. Res. Natl. Inst. Stand. Technol. **103**, 259 (1998). https://doi.org/10.6028/jres.103.019
49. T.B. Mitchell, J.J. Bollinger, D.H.E. Dubin, X.-P. Huang, W.M. Itano, R.H. Baughman, Direct observations of structural phase transitions in planar crystallized ion plasmas. Science **282**, 1290 (1998). https://doi.org/10.1126/science.282.5392.1290
50. W.M. Itano, J.J. Bollinger, J.N. Tan, B. Jelenković, X.-P. Huang, D.J. Wineland, Bragg diffraction from crystallized ion plasmas. Science **279**, 686 (1998). https://doi.org/10.1126/science.279.5351.686
51. L.L. Yan, W. Wan, L. Chen, F. Zhou, S.J. Gong, X. Tong, M. Feng, Exploring structural phase transitions of ion crystals **6**, 21547 (2016). https://doi.org/10.1038/srep21547
52. R. Nigmatullin, Formation and dynamics of structural defects in ion chains. Ph.D. Dissertation, Imperial College London, 2014
53. J.P. Schiffer, Phase transitions in anisotropically confined ionic crystals. Phys. Rev. Lett. **70**, 818 (1993). https://doi.org/10.1103/PhysRevLett.70.818
54. H.L. Partner, R. Nigmatullin, T. Burgermeister, J. Keller, K. Pyka, M.B. Plenio, A. Retzker, W.H. Zurek, A. del Campo, T.E. Mehlstäubler, Structural phase transitions and topological defects in ion Coulomb crystals. Phys. B **460**, 114 (2015). https://doi.org/10.1016/j.physb.2014.11.051
55. A. del Campo, G. De Chiara, G. Morigi, M.B. Plenio, A. Retzker, Structural defects in ion chains by quenching the external potential: the inhomogeneous Kibble-Zurek mechanism. Phys. Rev. Lett. **105**, 075701 (2010). https://doi.org/10.1103/PhysRevLett.105.075701
56. H. Saito, Y. Kawaguchi, M. Ueda, Kibble-Zurek mechanism in a trapped ferromagnetic Bose-Einstein condensate. J. Phys. Condens. Matter **25**, 404212 (2013). http://stacks.iop.org/0953-8984/25/i=40/a=404212
57. A. del Campo, A. Retzker, M.B. Plenio, The inhomogeneous Kibble-Zurek mechanism: vortex nucleation during Bose-Einstein condensation. New J. Phys. **13**, 083022 (2011). http://stacks.iop.org/1367-2630/13/i=8/a=083022
58. R.D. Skeel, J.A. Izaguirre, An impulse integrator for Langevin dynamics. Mol. Phys. **100**, 3885 (2002). https://doi.org/10.1080/0026897021000018321
59. F. Cosco, M. Borrelli, P. Silvi, S. Maniscalco, G. De Chiara, Nonequilibrium quantum thermodynamics in Coulomb crystals. Phys. Rev. A **95**, 063615 (2017). https://doi.org/10.1103/PhysRevA.95.063615
60. S. Deffner, Kibble-Zurek scaling of the irreversible entropy production. Phys. Rev. E **96**, 052125 (2017). https://doi.org/10.1103/PhysRevE.96.052125
61. B.-B. Wei, M.B. Plenio, Relations between dissipated work in non-equilibrium process and the family of Rényi divergences. New J. Phys. **19**, 023002 (2017). http://stacks.iop.org/1367-2630/19/i=2/a=023002

Chapter 3
Quantum Rabi Model: Equilibrium

The quantum Rabi model (QRM) is among the most fundamental models in quantum physics. Succinctly, it describes a two-level system interacting with a single bosonic mode, which besides of being a paradigmatic textbook example [1], has a relevance describing realistic phenomena across different fields, such as in quantum optics and quantum information [2]. This ubiquitous model is named after Isidor Isaac Rabi, whose pioneer work on the interaction of a two-level system with a classical radiation field [3, 4] established the basis of the nuclear magnetic resonance [5], for which he was awarded the Nobel prize in Physics in 1944. However, the first quantized version of the Rabi model is known as Jaynes–Cummings model [6], which represents a simplified version of the more general QRM, as we will discuss later on. Remarkably, while the Jaynes–Cummings model dates from 1963, these fundamental light-matter interaction models continue to be subject to an intense research both experimentally and theoretically [7]. Moreover, with the advent of quantum technologies, the QRM has recently experienced a renewed interest as it involves the key ingredients in quantum information processing, namely, a qubit (typically modeled as a two-level system) and a bosonic degree of freedom (harmonic oscillator). The unprecedented degree of precision and isolation, required to coherently control these fundamental quantum entities, is nowadays possible in different platforms, such as in circuit QED [8] or in trapped-ion settings [9]. It is worth emphasizing that in a trapped-ion experiment, the bosonic mode represents the quantized motion of the ion, i.e., phonons, while in circuit QED the bosonic degree of freedom stands for the photons of the field mode within the cavity. However, despite the distinct nature of the platforms, they share the basic, yet appealing physics well described by the Jaynes–Cummings model [10], namely, Rabi oscillations or collapses and revivals of quantum state populations [11], and it has been experimentally realized for the

© Springer Nature Switzerland AG 2018   55
R. Puebla, *Equilibrium and Nonequilibrium Aspects of Phase Transitions*
in Quantum Physics, Springer Theses, https://doi.org/10.1007/978-3-030-00653-2_3

preparation of Schrödinger cat states [12] among others. Furthermore, thanks to the great experimental progress in the recent years, the interaction strength can be now boosted and brought into a regime in which the simple Jaynes–Cummings model is no longer valid, and therefore the general QRM is a more suitable description of these interacting systems. These advances have not only promoted a renewed theoretical inspection of the QRM but also have opened the door for experimental exploration of the rich physics enclosed by light-matter interaction in novel coupling regimes.

In this chapter we will firstly introduce the QRM, paying attention to the different coupling regimes and with special emphasis on the motivation to analyze such a model from the point of view of phase transitions and critical phenomena. Then, in the Sect. 3.2 we present the mathematical derivation to attain an effective model that becomes exact in a suitable limit of the parameters. In that particular limit, the low-energy effective solution reveals the presence of a quantum phase transition (QPT) at a certain critical value of the interaction strength [13–15]. Certainly, QPTs are expected occur in the thermodynamic limit of infinitely many constituents [16–18], and therefore, the existence of a QPT in the QRM, which only involves a qubit and a bosonic mode, challenges the established notion of phase transitions. We will stress the relevance of this result and explain the underlying reason on how such a model undergoes a QPT without resorting to the thermodynamic limit, i.e., without scaling up the number of system constituents. The analysis presented in the first part of this chapter will motivate further inquiries presented in the following Chaps. 4 and 5, regarding nonequilibrium critical dynamics and a feasible experimental implementation to probe this QPT, respectively.

The rest of this chapter is devoted to examine the equilibrium fingerprints of the QPT, which bear similarities with that of the Dicke model [19] and Lipkin–Meshkov–Glick model [20]. Furthermore, it is well established that the existence of a phase transition entails a number of scaling relations in relevant quantities of the system as the thermodynamic limit is approached, which is typically known as finite-size scaling (see Introduction) [18, 21, 22]. Even though the QRM undergoes a QPT in a unusual limit, the finite-size scaling theory can be still applied. However, conventional finite-size effects translate to the QRM as finite-frequency values of the parameters (recall that the size of the QRM remains always finite), and therefore, we refer to this theory as finite-*frequency* theory to stress the difference. We verify the correctness of the derived critical and finite-frequency scaling exponents by means of numerical exact diagonalization of the QRM, corroborating the validity of the effective low-energy Hamiltonian. Finally, as we explain in detail in Sect. 3.3, the QRM also exhibits critical behavior in some excited states, at a particular excitation energy. This phenomenon is known as excited-state quantum phase transition (ESQPT), whose main hallmark resides in a singularity in the density of states [23–29]. In short, this chapter is entirely devoted to the inspection of the critical properties of the QRM at the eigenstate level, referred here as equilibrium and published in [30, 31], while the study of nonequilibrium dynamics involving the QPT is presented in Chap. 4.

3.1 Quantum Rabi Model

As we have already mentioned, the QRM describes the interaction of a two-level system, or simply, a qubit, and a single bosonic mode. A schematic representation of the physical scenario described by the QRM, together with the possible exchange processes between the qubit and the bosonic mode, can be found in Fig. 3.1. The Hamiltonian of the QRM can be written as

$$H_{QRM} = \omega_0 a^\dagger a + \frac{\Omega}{2}\sigma_z - \lambda\sigma_x(a + a^\dagger) \tag{3.1}$$

where we have introduced the usual bosonic annihilation and creation operators, a and $a^\dagger$, respectively, fulfilling $\left[a, a^\dagger\right] = 1$, that give account of the bosonic mode of frequency ω_0. The two-level system possesses a frequency splitting Ω and it is described by means of the Pauli $\frac{1}{2}$-spin matrices, $\vec{\sigma} = \left(\sigma_x, \sigma_y, \sigma_z\right)$,

$$\sigma_x = \begin{pmatrix} 0 & 1 \\ 1 & 0 \end{pmatrix}, \quad \sigma_y = \begin{pmatrix} 0 & -i \\ i & 0 \end{pmatrix}, \quad \text{and} \quad \sigma_z = \begin{pmatrix} 1 & 0 \\ 0 & -1 \end{pmatrix}, \tag{3.2}$$

which obey $\left[\sigma_i, \sigma_j\right] = 2i\epsilon_{ijk}\sigma_k$. Both subsystems interact through the latter term on the previous Hamiltonian, $\sigma_x\left(a + a^\dagger\right)$, whose interaction strength is given by λ, also known as coupling constant. This model shows a Z_2 parity symmetry, as the operator $\Pi = e^{i\pi(a^\dagger a + \sigma_+\sigma_-)} = e^{i\pi N_{exc}}$ commutes with H_{QRM}, $\left[H_{QRM}, \Pi\right] = 0$, with $\sigma_x = \sigma_+ + \sigma_-$ and $N_{exc} = a^\dagger a + \sigma_+\sigma_-$, the total number of excitations. Therefore, the Hilbert space can be split according to the parity, that is, in two parity subspaces, and so the eigenstates of H_{QRM} can be labeled depending on the parity of N_{exc}. The previous Hamiltonian can be simplified for small coupling constant by neglecting the so-called counter-rotating terms, namely, $\sigma_+ a^\dagger$ and $\sigma_- a$, and hence, $H_{QRM} \approx H_{JCM}$,

$$H_{JCM} = \omega_0 a^\dagger a + \frac{\Omega}{2}\sigma_z - \lambda(\sigma_+ a + \sigma_- a^\dagger). \tag{3.3}$$

In particular, this approximation can be safely performed when the coupling constant is small compared to $\Omega + \omega_0$, i.e., $\lambda \ll |\Omega + \omega_0|$, which is the result of applying a rotating-wave approximation (RWA), as explained in Appendix B. The resulting Hamiltonian is known as Jaynes–Cummings model (JCM) [6], which is exactly solvable [1, 32, 33] and it has served to explain a number of experiments in the quantum realm of light-matter interaction [10]. It is worth mentioning that the H_{JCM} conserves the total number of excitations since $[H_{JCM}, N_{exc}] = 0$. Yet, this approximation can lead to significant deviations from the actual physics dictated by the general H_{QRM} when the coupling constant becomes a large fraction of the bosonic frequency, the so-called ultrastrong coupling regime (USC) $0.1 \lesssim \lambda/\omega_0 \lesssim 1$. However, in the perturbative USC, i.e. $\lambda/\omega_0 \sim 0.1$, it is still possible to find an approximate solution of the QRM as follows [34, 35]

(a)   **(b)**

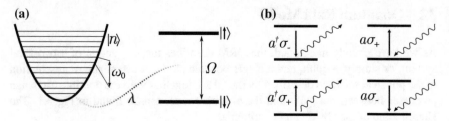

Fig. 3.1 Schematic representation of the quantum Rabi Model (QRM). In **a** we show the two interacting subsystem, namely, a two-level system or qubit, represented by the levels $|\downarrow\rangle$ and $|\uparrow\rangle$ separated by a frequency Ω, that is coupled to a single-mode bosonic field of frequency ω_0 with a strength λ (coupling constant). The bosonic field is represented by the usual annihilation and creation operators a and $a^\dagger$, which act on the Fock space $|n\rangle$ with $n = 0, 1, 2, \ldots$. In this thesis we will pay attention to the scenario $\omega_0 \ll \lambda \ll \Omega$, and in particular to the limit $\Omega/\omega_0 \to \infty$ and $\lambda/\omega_0 \to \infty$, while keeping $\lambda/\sqrt{\omega_0 \Omega}$ constant. In **b** the four possible exchange processes between the qubit and the bosonic mode are shown, that correspond to $\sigma_x(a + a^\dagger)$. At the top, the exchange consists in emitting one boson when the qubit undergoes the transition $|\uparrow\rangle \to |\downarrow\rangle$, and vice versa, as a consequence of the rotating terms. To the contrary, the counter-rotating terms produce the exchanges that consists in emitting a boson when the qubit population raises, $|\downarrow\rangle \to |\uparrow\rangle$, and vice versa

$$e^{-S_{\text{BS}}} H_{\text{QRM}} e^{S_{\text{BS}}} \approx H_{\text{BS}} = \omega_{\text{BS}} \sigma_z a^\dagger a + \frac{\omega_{\text{BS}}}{2} \sigma_z - \frac{\omega_{\text{BS}}}{2} + H_{\text{JCM}} \qquad (3.4)$$

which is valid up to second order in $\lambda/(\omega_0 + \Omega)$ and where the anti-Hermitian operator $S_{\text{BS}} = \lambda/(\omega_0 + \Omega) \ (a\sigma_- - a^\dagger\sigma_+) + \lambda^2/(2\omega_0(\omega_0 + \Omega))(a^2 - a^{\dagger 2})\sigma_z$ and $\omega_{\text{BS}} = \lambda^2/(\omega_0 + \Omega)$. This transformation reveals a Bloch–Siegert shift on the qubit frequency, which has been measured in experiments performed in circuit QED [36, 37]. In addition, it is worth mentioning that H_{BS} reduces to the Jaynes–Cummings when $\lambda \ll |\omega_0 + \Omega|$, as expected from the rough RWA.

Furthermore, a larger interaction strength between these two subsystems will potentially lead to a departure from the simple physics dictated by H_{JCM} or H_{BS} [35]. As a paradigmatic example, we find the deep strong coupling regime, $\lambda/\omega_0 \gtrsim 1$, in which an unexpected structured behavior of collapses and revivals emerges [38]. Remarkably, two recent experiments performed in circuit QED were able to explore the non-perturbative USC as the coupling constant was boosted to $\lambda/\omega_0 \sim 1$ [39, 40], showing the potential opportunities of these platforms to further explore the intriguing strong light-matter interaction. In this thesis we will examine the QRM in a parameter regime where neither the RWA nor the Bloch–Siegert approximation are applicable, and thus, counter-rotating terms are of great importance.

Even though the QRM may seem a simple model, its complexity should not be underestimated. From a mathematical point of view, during the last decades considerable efforts have been devoted to solve the QRM. However, it has been only recently when the existence of the Z_2 parity symmetry has been acknowledged to be sufficient to demonstrate the integrability of the QRM, Eq. (3.1), carried out by Braak in [41] (see [42] for a viewpoint). Therefore, according to Braak's criterion, the QRM is integrable and exactly solvable, which has posed similar questions on related models [43]. Nevertheless, even though the provided solution is of great value, it lacks of a closed form and therefore it is not directly applicable to the analysis performed in this thesis.

Finally, it is worth emphasizing again that the QRM comprises just two constituent particles. Phase transitions are however expected to take place in the limit of infinitely many particles, i.e., in the thermodynamic limit. In this regard, there are two paths to attain the thermodynamic limit of the QRM. One consists in increasing the number of qubits coupled to the bosonic mode, a multi-qubit version of the QRM, known as Dicke model [19] that we will comment later. The other path deals with a single qubit but coupled to several bosonic modes, typically a continuum, which primarily appeared to describe the inherent quantum dissipation of quantum systems, and it is known as spin-boson model [44, 45]. Indeed, both models exhibit a phase transition in their corresponding thermodynamic limit [14, 46–50]. Note that the phase transition of the Dicke model is known as *superradiant* as the system acquires a large number of bosonic excitations at one side of the critical point. Assuredly, the QRM is by definition far from being in the thermodynamic limit, and then, it raises the question of whether such a finite-component system may be able to display criticality in a certain parameter regime as studied in Refs. [51–55]. These studies indeed reported certain fingerprints of a phase transition, as for example the emergence of a pitchfork bifurcation, however, the used semiclassical approximation hindered the observation of a much richer physics, as revealed by a fully quantum mechanical description presented throughout this chapter. Although the QRM lacks a conventional thermodynamic limit, the fact that its Hilbert space is infinitely large can be exploited to realize a QPT even with a finite number of constituents, as we show in the following.

3.2 Superradiant Quantum Phase Transition

In this section we demonstrate that the QRM undergoes a QPT in the limit of $\Omega/\omega_0 \to \infty$ and $\lambda/\omega_0 \to \infty$, while keeping the ratio $\lambda/\sqrt{\omega_0\Omega}$ finite. This entails a parameter hierarchy, $\omega_0 \ll \lambda \ll \Omega$. For that purpose, we present in Sect. 3.2.1 the mathematical derivation used to obtain the low-energy effective Hamiltonians that become exact in the aforementioned limit, and reveal the presence of a QPT at a certain critical value of the coupling constant. Moreover, the exact effective Hamiltonian in this limit allows us to derive eigenstates, energy spectrum and expectation value of observables. In addition, we will pay special attention to scaling properties of these quantities close to the QPT, which will serve us as a test to confirm the validity of our theory by comparing them with the exact numerical diagonalization of the QRM. However, for the sake of readability, while we show the basic ingredients of the derivation in the following lines, we refer the interested readers to the Appendix C for further details regarding the mathematical aspects of the derivation.

3.2.1   Low-Energy Effective Hamiltonians

We start writing the Hamiltonian of the QRM as $H_{\mathrm{QRM}} = H_0 - \lambda V$ where

$$H_0 = \omega_0 a^\dagger a + \frac{\Omega}{2}\sigma_z, \text{ and } V = (a + a^\dagger)\sigma_x. \qquad (3.5)$$

The unperturbed Hamiltonian H_0 has decoupled spin subspaces $\mathcal{H}_\downarrow$ and $\mathcal{H}_\uparrow$, such that the entire Hilbert space $\mathcal{H} = \mathcal{H}_\uparrow \oplus \mathcal{H}_\downarrow$. For $\Omega/\omega_0 \gg 1$, the low-lying eigenstates of H_0 are confined in $\mathcal{H}_\downarrow$, and they are just that of a simple harmonic oscillator. The interaction Hamiltonian V, however, introduces coupling between the two spin subspaces so that virtual transitions between them would modify the nature of the low-lying eigenstates and energy eigenvalues. The goal is then to find a unitary transformation U which makes the transformed Hamiltonian, $U^\dagger H_{\mathrm{QRM}} U$, free of coupling terms between $\mathcal{H}_\downarrow$ and $\mathcal{H}_\uparrow$. This procedure is in the same spirit as the Schrieffer–Wolff (SW) transformation [56, 57].

We consider a unitary transformation $U = e^S$ where the generator S is anti-Hermitian and block-off-diagonal with respect to spin subspaces, i.e., $U^\dagger U = \mathbb{1}$ since $S^\dagger = -S$. Then, we can write the transformed Hamiltonian as

$$H' = e^{-S} H_{\mathrm{QRM}} e^S = \sum_{k=0}^{\infty} \frac{1}{k!}[H_{\mathrm{QRM}}, S]^{(k)} \qquad (3.6)$$

where $[H, S]^{(k)} \equiv [[H, S]^{(k-1)}, S]$ represents the k-order of nested commutators with $[H, S]^{(0)} \equiv H$. The transformed Hamiltonian can be now divided into diagonal and off-diagonal part using the fact that S is block-off-diagonal and V is block-diagonal, and denote them H'_d and H'_{od}, respectively. They are

$$H'_d = \sum_{k=0}^{\infty} \frac{1}{2k!}[H_0, S]^{(2k)} - \sum_{k=0}^{\infty} \frac{1}{(2k+1)!}[\lambda V, S]^{(2k+1)}, \qquad (3.7)$$

$$H'_{od} = \sum_{k=0}^{\infty} \frac{1}{(2k+1)!}[H_0, S]^{(2k+1)} - \sum_{k=0}^{\infty} \frac{1}{(2k)!}[\lambda V, S]^{(2k)}. \qquad (3.8)$$

We show explicitly the derivation for the strict $\Omega/\omega_0 \to \infty$ and $\lambda/\omega_0 \to \infty$ limit, while the dimensionless coupling constant $g = 2\lambda/\sqrt{\omega_0 \Omega}$ remains finite. In this manner, it only required that the lowest order terms in λ of the block-off-diagonal H'_{od} cancel out. However, for $\Omega/\omega_0 \gg 1$, one can still obtain an effective Hamiltonian but more orders need to be considered (see Appendix C). Here we only focus in the $\Omega/\omega_0 \to \infty$ limit, and thus, it is sufficient to require

$$[H_0, S] = \lambda V, \qquad (3.9)$$

which is fulfilled by the following generator

$$S = \frac{\lambda}{\Omega}(a + a^\dagger)(\sigma_+ - \sigma_-) = i\frac{g}{2}\sqrt{\frac{\omega_0}{\Omega}}(a + a^\dagger)\sigma_y, \tag{3.10}$$

and then, after introducing this anti-Hermitian operator in Eq. (3.6), we obtain

$$H' = \omega_0 a^\dagger a + \frac{\Omega}{2}\sigma_z + \frac{\omega_0 g^2}{4}(a + a^\dagger)^2\sigma_z + \frac{\omega_0 g^3}{6}\sqrt{\frac{\omega_0}{\Omega}}(a + a^\dagger)^3\sigma_x + \mathcal{O}\left(\frac{\omega_0}{\Omega}\right). \tag{3.11}$$

By construction, H' is block-diagonal up to the second order of the dimensionless coupling constant g. Note that the third order term in g and higher order terms are accompanied by a factor $(\frac{\omega_0}{\Omega})^\alpha$ with $\alpha \geq \frac{1}{2}$. Therefore, in the $\Omega/\omega_0 \to \infty$ limit, all the higher order terms have coefficients that vanish, leaving only non-zero coefficients up to second order of g and free of coupling between spin subspaces, that is,

$$H' = H'_d = \omega_0 a^\dagger a + \frac{\Omega}{2}\sigma_z + \frac{\omega_0 g^2}{4}(a + a^\dagger)^2\sigma_z. \tag{3.12}$$

This step deserves special attention. Note that since Eq. (3.11) contains unbounded operators, a and $a^\dagger$, it is not straightforward to prove that Eq. (3.11) converges to the previous expression, Eq. (3.12), even in the strict $\Omega/\omega_0 \to \infty$ limit. Indeed, a rigorous proof of convergence would be required, although it may be an arduous task in itself [58]. Instead, we propose a method to certify the validity of the effective model in the $\Omega/\omega_0 \to \infty$ limit based on numerically exact diagonalization of the QRM for finite Ω/ω_0, as well as utilizing the scaling theory near the critical point [21, 22] as we will show in Sect. 3.2.3.

Since the above transformed Hamiltonian is diagonal in the spin basis, projecting it onto the spin down subspace is an exact procedure,

$$H_{np} = \langle\downarrow| H'_d |\downarrow\rangle = \omega_0 a^\dagger a - \frac{\omega_0 g^2}{4}(a + a^\dagger)^2 - \frac{\Omega}{2}. \tag{3.13}$$

Furthermore, since H_{np} is quadratic it can be diagonalized by a standard procedure (see Appendix C for further details). Using the squeezing operator, $S[x] = e^{x/2(a^{\dagger 2} - a^2)}$, we obtain

$$S^\dagger[r_{np}(g)]H_{np}S[r_{np}(g)] = \epsilon_{np}(g)a^\dagger a + E_{G,np}(g), \tag{3.14}$$

with $\epsilon_{np}(g) = \omega_0\sqrt{1 - g^2}$ the excitation energy and $E_{G,np}(g) = (\epsilon_{np}(g) - \omega_0)/2 - \Omega/2$ the ground-state energy, and $r_{np}(g) = -\frac{1}{4}\ln(1 - g^2)$ the squeezing parameter. Remarkably, this effective Hamiltonian is only valid for $0 \leq g \leq 1$, as it is clear from the fact that ϵ_{np} becomes complex, and therefore, H_{np} is not longer Hermitian. Needless to say, this pinpoints that the previous procedure cannot be applied to H_{QRM} for $g > 1$. Although this will become more evident later, we can identify the failure of the derivation in that the bosonic mode acquires a non-zero coherence, and thus $\langle a \rangle \neq 0$, and with $\langle a^\dagger a \rangle$ of the order of Ω/ω_0. As a consequence, the suppressed higher-order terms in Eq. (3.11) are not longer negligible and must be taken intro

account. For that reason, in order to obtain an effective low-energy Hamiltonian for
$g > 1$, we first displace the field by $\alpha \in \mathbb{R}$, with $\mathcal{D}[x] = e^{xa^\dagger - x^* a}$,

$$\tilde{H}_{\text{QRM}}(\alpha) =$$
$$\mathcal{D}^\dagger[\alpha] H_{\text{QRM}} \mathcal{D}[\alpha] = \omega_0(a^\dagger + \alpha)(a + \alpha) + \frac{\Omega}{2}\sigma_z - \lambda(a + a^\dagger)\sigma_x - 2\lambda\alpha\sigma_x, \quad (3.15)$$

and then define a new spin basis, $|\tilde{\uparrow}\rangle = \cos\theta\,|\uparrow\rangle + \sin\theta\,|\downarrow\rangle$ and $|\tilde{\downarrow}\rangle = -\sin\theta\,|\uparrow\rangle +$
$\cos\theta\,|\downarrow\rangle$, with $\tan(2\theta) = -2\lambda\alpha/\Omega$ and $\tilde{\Omega} = \sqrt{\Omega^2 + 16\lambda^2\alpha^2}$ the spin frequency
splitting. Note that $\vec{\tau} = (\tau_x, \tau_y, \tau_z)$ are the Pauli matrices associated with the rotated
spin states $|\tilde{\uparrow}\rangle$ and $|\tilde{\downarrow}\rangle$. Then,

$$\tilde{H}_{\text{QRM}}(\alpha) =$$
$$\omega_0 a^\dagger a + \frac{\tilde{\Omega}}{2}\tau_z - \lambda\cos 2\theta(a + a^\dagger)\tau_x + \omega_0\alpha^2 + (\omega_0\alpha - \lambda\sin 2\theta\tau_z)(a + a^\dagger), \quad (3.16)$$

we set α such that the last term of the previous expression vanishes upon projection
onto $\mathcal{H}_{\tilde{\downarrow}}$, i.e., $\omega_0\alpha + \lambda\sin 2\theta = 0$. Two solutions are found, $\pm\alpha_g$,

$$\alpha_g = \sqrt{\frac{\Omega}{4g^2\omega_0}}\sqrt{g^4 - 1}, \quad (3.17)$$

which provide two equivalent spin states, namely, $|\downarrow^\pm\rangle$ depending on the sign of $\pm\alpha_g$.
The choice of this displacement will lead to an undesired term on $\mathcal{H}_{\tilde{\downarrow}}$ of $\pm 2\alpha_g\omega_0(a + a^\dagger)|\tilde{\uparrow}\rangle\langle\tilde{\uparrow}|$, however, such a perturbation scales as $(\Omega/\omega_0)^{-1/2}$. Therefore, this term
can be neglected as long as we are interested in the physics within the subspace $\mathcal{H}_{\tilde{\downarrow}}$
and $\Omega/\omega_0 \gg 1$, and thus

$$\tilde{H}_{\text{QRM}}(\pm\alpha_g) = \omega_0 a^\dagger a + \frac{\tilde{\Omega}}{2}\tau_z - \tilde{\lambda}(a + a^\dagger)\tau_x + \omega_0\alpha_g^2, \quad (3.18)$$

has the same form of the original H_{QRM} but with different parameters, namely,
$\tilde{\lambda} = \sqrt{\Omega\omega_0}/2g$ and $\tilde{\Omega} = g^2\Omega$. Therefore, applying now the same procedure as done
for $g < 1$ to Eq. (3.18) we achieve

$$H_{\text{sp}} = \langle\tilde{\downarrow}|\tilde{H}'_d|\tilde{\downarrow}\rangle = \omega_0 a^\dagger a - \frac{\omega_0}{4g^4}(a + a^\dagger)^2 - \frac{\Omega}{4}(g^2 + g^{-2}), \quad (3.19)$$

which is now valid only for $g > 1$. Again, and because its quadratic form, is possible
to diagonalize it (see Appendix C for further details)

$$S^\dagger[r_{\text{sp}}(g)]H_{\text{sp}}S[r_{\text{sp}}(g)] = \epsilon_{\text{sp}}(g)a^\dagger a + E_{\text{G,sp}}(g) \quad (3.20)$$

with an excitation energy $\epsilon_{sp}(g) = \omega_0\sqrt{1 - g^{-4}}$, ground-state energy $E_{G,sp}(g) = (\epsilon_{sp}(g) - \omega_0)/2 - \Omega(g^2 + g^{-2})/4$ and $r_{sp}(g) = -1/4\ln(1 - g^{-4})$. It is worth stressing that H_{sp} does not depend on the specific sign of α_g, and thus there are two equivalent low-energy Hamiltonians for $g > 1$.

Therefore, we have found the effective low-energy Hamiltonian of the QRM in the $\Omega/\omega_0 \to \infty$. In addition, as they can be diagonalized, we know their eigenstates,

$$\left|\varphi_{np}^n(g)\right\rangle = S[r_{np}(g)]\,|n\rangle\,|\downarrow\rangle \qquad 0 \leq g \leq 1 \qquad (3.21)$$

$$\left|\varphi_{sp}^n(g)\right\rangle_\pm = D[\pm\alpha_g(g)]S[r_{sg}(g)]\,|n\rangle\,\left|\downarrow^\pm\right\rangle \qquad g \geq 1 \qquad (3.22)$$

where $\left|\downarrow^\pm\right\rangle = \mp\sqrt{(1 - g^{-2})/2}\,|\uparrow\rangle + \sqrt{(1 + g^{-2})/2}\,|\downarrow\rangle$. Indeed, there are two equivalent ground states for $g > 1$, $\left|\varphi_{sp}^0(g)\right\rangle_\pm$, that is, they take *exactly* the same energy and do not respect the Z_2 symmetry of the QRM, $\Pi\left|\varphi_{sp}^0(g)\right\rangle_\pm \neq \pm\left|\varphi_{sp}^0(g)\right\rangle_\pm$. This indicates that a spontaneous Z_2 symmetry breaking takes place in the superradiant phase. Note that spontaneous symmetry breaking is an essential feature of phase transitions, and the QPT of the QRM is not an exception. Moreover, the superradiant nature of the QPT refers to the fact that the ground state features a macroscopic population of the bosonic mode. In fact, $\langle a^\dagger a\rangle \propto \Omega/\omega_0 \to \infty$. Note that this is also the case for the superradiant QPT in the Dicke model, where $\langle a^\dagger a\rangle \propto N \to \infty$ being N the number of qubits [19, 46–49].

Hitherto, we have only focused on deriving the effective low-energy Hamiltonians in the $\Omega/\omega_0 \to \infty$ limit. However, applying a more general procedure we can achieve higher-order corrections to these Hamiltonians for large but still finite Ω/ω_0, as shown in detail in Appendix C. In particular, one can show that the first-order correction Hamiltonian for $0 \leq g \leq 1$ takes the following form

$$H_{np}^\Omega = \omega_0 a^\dagger a - \frac{\omega_0 g^2}{4}(a + a^\dagger)^2 - \frac{\Omega}{2} + \frac{\omega_0^2 g^4}{16\Omega}(a + a^\dagger)^4 + \frac{\omega_0^2 g^2}{4\Omega}. \qquad (3.23)$$

Although the effective low-energy Hamiltonians, H_{np} and H_{sp}, allow us to obtain relevant information about the QPT taking place in the QRM, the impact of a finite ratio Ω/ω_0 should not be diminished. Indeed, H_{np}^Ω is of great relevance both to validate our effective theory and to provide measurable consequences that confirm the emergence of a QPT as the ratio Ω/ω_0 increases. Even though an exact diagonalization of this Hamiltonian is now hindered by the quartic term $(a + a^\dagger)^4$, relying on a variational method still allows us to gain valuable insight regarding the QRM for $\Omega/\omega_0 \gg 1$. As we will see in Sect. 3.2.3, the variational estimate of the ground state will be of considerable importance to analytically determine scaling exponents.

3.2.2 Signatures of the Quantum Phase Transition

As shown previously, the low-energy physics of the QRM in the $\Omega/\omega_0 \to \infty$ limit is described by an effective Hamiltonian, which crucially depends on the dimensionless coupling constant g. At the critical value $g = 1$, the QRM becomes gapless, among another remarkable features as we present in the following, such as spin and bosonic mode observables or the entanglement between both subsystems, emphasizing also their scaling behavior close to the critical point.

3.2.2.1 Energy Gap and Ground-State Energy

One of the main hallmarks of a QPT resides in the closing of the energy gap at the critical point [13, 15]. Indeed, the energy gap Δ is expected to follow a universal law close to the critical point as

$$\Delta(g) \sim |g - g_c|^{z\nu}, \tag{3.24}$$

where z and ν are critical exponents of the QPT, as we have explained in the Chap. 1. From the previous developments, it is straightforward to obtain the energy gap of the QRM, directly given by $\epsilon_{np}(g)$ and $\epsilon_{sp}(g)$

$$\Delta(g) = \begin{cases} \omega_0\sqrt{1-g^2} & 0 \le g \le 1 \\ \omega_0\sqrt{1-g^{-4}} & g > 1. \end{cases} \tag{3.25}$$

For g sufficiently close to $g_c = 1$, it follows that $\Delta(g) \sim |g - g_c|^{1/2}$, that is, $z\nu = 1/2$. Note that this value is a typical feature of mean-field phase transitions, which could have been anticipated for the QRM due to the nature of the model.

The ground-state energy also shows singular behavior when undergoing a QPT. In particular, the second derivative of ground-state energy with respect to the control parameter becomes discontinuous at the critical point.[1] This can be seen in the QRM, since the ground-state energy $E_G(g)$ reads

$$E_G(g) = \begin{cases} \frac{1}{2}\left(\omega_0\sqrt{1-g^2} - \omega_0 - \Omega\right) & 0 \le g \le 1 \\ \frac{1}{2}\left(\omega_0\sqrt{1-g^{-4}} - \omega_0 - \frac{\Omega}{2}(g^2 + g^{-2})\right) & g > 1 \end{cases} \tag{3.26}$$

which is extensive in the ratio Ω/ω_0. Therefore, we normalize E_G by Ω/ω_0 and then take the limit $\Omega/\omega_0 \to \infty$,

---

[1] In this sense, by analogy with the Ehrenfest classification of phase transitions, we could refer to this QPT as second order. However, we will continue using the term continuous phase transitions as explained in the Introduction.

$$e_G(g) \equiv \lim_{\Omega/\omega_0 \to \infty} \frac{\omega_0}{\Omega} E_G(g) = \begin{cases} -\frac{\omega_0}{2} & 0 \leq g \leq 1 \\ -\frac{\omega_0}{4}(g^2 + g^{-2}) & g > 1 \end{cases} \tag{3.27}$$

and thus, its first and second derivative with respect to g become

$$\frac{d}{dg} e_G(g) = \begin{cases} 0 & 0 \leq g \leq 1 \\ -\frac{\omega_0}{2}(g - g^{-3}) & g > 1 \end{cases} \tag{3.28}$$

$$\frac{d^2}{dg^2} e_G(g) = \begin{cases} 0 & 0 \leq g < 1 \\ -\frac{\omega_0}{2}(1 + 3g^{-4}) & g > 1, \end{cases} \tag{3.29}$$

which clearly manifests the expected discontinuity at $g = 1$ since $\lim_{\epsilon \to 0^+} \frac{d^2}{dg^2} e_G$ $(1 + \epsilon)$ differs from $\lim_{\epsilon \to 0^-} \frac{d^2}{dg^2} e_G(1 + \epsilon)$. Ground-state energy $e_G(g)$ and its second derivative can be found in Fig. 3.2a.

3.2.2.2 Qubit and Bosonic Mode Observables

Beyond the properties of the ground-state energy, qubit and bosonic observables appear as essential signatures of the QPT, such as σ_z, σ_x or $\langle a^\dagger a \rangle$ as well as the variance of the quadratures of the bosonic mode. Having access to the actual ground state $|\varphi^0(g)\rangle$, either $|\varphi^0_{np}(g)\rangle$ or $|\varphi^0_{sp}(g)\rangle_\pm$ depending on g, allows us to obtain analytically these relevant quantities as a function of g. Hence, the ground-state expectation value of any observable $\mathcal{A}$ can be obtained from $\langle \varphi^0(g) | \mathcal{A} | \varphi^0(g) \rangle$. Due to the simple form of the qubit part of the ground state, $\langle \vec{\sigma} \rangle$ readily gives

$$\langle \sigma_x \rangle = \langle \varphi^0(g) | \sigma_x | \varphi^0(g) \rangle = \begin{cases} 0 & 0 \leq g \leq 1 \\ \mp\sqrt{1 - g^{-4}} & g > 1 \end{cases} \tag{3.30}$$

$$\langle \sigma_z \rangle = \langle \varphi^0(g) | \sigma_z | \varphi^0(g) \rangle = \begin{cases} -1 & 0 \leq g \leq 1 \\ -\frac{1}{g^2} & g > 1 \end{cases} \tag{3.31}$$

and $\langle \sigma_y \rangle = 0$. Remarkably, for $g > 1$, $\langle \sigma_x \rangle \neq 0$ which is a direct consequence of the broken symmetry. Recall that any state satisfying the Z_2 symmetry is compelled to $\langle \sigma_x \rangle = 0$, and thus, the particular sign of $\langle \sigma_x \rangle$ depends on the symmetry-breaking branch. In addition, the qubit gains population in its upper level due to QPT, being the state an eigenstate of σ_x in $g \to \infty$ as $\langle \sigma_z \rangle \to 0$ and $\langle \sigma_z \rangle \to \pm 1$. On the other hand, the computation of bosonic observables involve displaced and squeezed Fock states. Since $\langle a^\dagger a \rangle \propto \Omega/\omega_0$ in the superradiant phase, it is convenient to calculate the rescaled number of bosonic excitations, which reads

$$\begin{aligned} n_c &= \frac{\omega_0}{\Omega} \langle a^\dagger a \rangle = \frac{\omega_0}{\Omega} \langle \varphi^0(g) | a^\dagger a | \varphi^0(g) \rangle \\ &= \begin{cases} \frac{\omega_0}{\Omega} \frac{\cosh[2r_{np}(g)]-1}{2} & 0 \leq g \leq 1 \\ \frac{g^2 - g^{-2}}{4} + \frac{\omega}{\Omega} \frac{\cosh[2r_{sp}(g)]-1}{2} & g > 1 \end{cases} \end{aligned} \tag{3.32}$$

Taking the limit $\Omega/\omega_0 \rightarrow \infty$, n_c simply vanishes for $0 < g < 1$, and becomes $(g^2 - g^{-2})/4$ in the superradiant phase. While $\langle a^\dagger a \rangle$ diverges for $g > 1$ as a consequence of the infinitely large displacement, α_g, the pathological case at $g = 1$ stems from an infinitely squeezed state, which shows as well an infinitely large $\langle a^\dagger a \rangle$. Recall that r_{np} diverges at $g = 1$. An interesting feature of this QPT resides thus in the squeezing of the bosonic mode, as expected from the form of $|\varphi^0(g)\rangle$. In fact, this becomes more evident analyzing the variance of the quadratures of the bosonic field, $x = a + a^\dagger$ and $p = i(a^\dagger - a)$, i.e., $\Delta x = \sqrt{\langle x^2 \rangle - \langle x \rangle^2}$ and $\Delta p = \sqrt{\langle p^2 \rangle - \langle p \rangle^2}$, which result in

$$\Delta x = \begin{cases} (1 - g^2)^{-1/4} & 0 \leq g \leq 1 \\ (1 - g^{-4})^{-1/4} & g > 1 \end{cases} \tag{3.33}$$

$$\Delta p = \begin{cases} (1 - g^2)^{1/4} & 0 \leq g \leq 1 \\ (1 - g^{-4})^{1/4} & g > 1. \end{cases} \tag{3.34}$$

Hence, the QPT is accompanied by an infinite amount of squeezing, since $\Delta p \rightarrow 0$ and $\Delta x \rightarrow \infty$ at the critical point. Interestingly, $|\varphi^0(g)\rangle$ remains in a state of minimum uncertainty since $\Delta x \Delta p = 1 \, \forall g$. Moreover, the bosonic field also points out the spontaneous symmetry-breaking QPT, as seen in the previous section. For $g > 1$ the bosonic operators acquire a coherence since they are displaced $a \rightarrow a + \alpha$ with $\alpha \propto (\Omega/\omega_0)^{1/2}$, and thus, $\langle a \rangle \neq 0$. Recall that any state satisfying the Z_2 symmetry of the QRM must exhibit $\langle a \rangle = \langle a^\dagger \rangle = 0$. These signatures are plotted in Fig. 3.2.

3.2.2.3   Entanglement Entropy

So far, we have not investigated the role of quantum correlations among both subsystems, which can be quantified by the entanglement. Certainly, QPTs are recognized to leave also an imprint on the entanglement, responsible of long-range quantum correlations [59–64]. As in the multi-qubit version of the QRM, the Dicke model, we find that the entanglement entropy displays a singular behavior as a consequence of the QPT. We quantify the entanglement relying on the von Neumann entanglement entropy,

$$S_{vN} = -\text{Tr}\left[\rho_s \log_2 \rho_s\right] \tag{3.35}$$

where ρ_s denotes the reduced spin or harmonic oscillator density matrix. Note that $S_{vN} = 0$ for ρ_s resulting in a pure state, as it is the case for the normal phase. To the contrary, for $g > 1$ the symmetrized ground-state wave function reads

$$\left|\varphi^0_{sp}(g)\right\rangle_s = \frac{1}{\sqrt{2}}\left(\left|\varphi^0_{sp}(g)\right\rangle_+ \pm \left|\varphi^0_{sp}(g)\right\rangle_-\right) \tag{3.36}$$

with $\left|\varphi^0_{\mathrm{sp}}(g)\right\rangle_{\pm}$ given in Eq. (3.22). Then, tracing out the harmonic oscillator degree of freedom, we obtain the reduced spin density matrix (see Appendix C for more details of the calculation)

$$\rho_s = \frac{1}{2}\left((1 - g^{-2})\left|\uparrow\right\rangle\left\langle\uparrow\right| + (1 + g^{-2})\left|\downarrow\right\rangle\left\langle\downarrow\right|\right), \tag{3.37}$$

from which it follows that the von Neumann entanglement entropy becomes

$$S_{\mathrm{vN}}(g) = \begin{cases} 0 & 0 \leq g \leq 1 \\ -\frac{1-g^{-2}}{2}\log_2\left(\frac{1-g^{-2}}{2}\right) - \frac{1+g^{-2}}{2}\log_2\left(\frac{1+g^{-2}}{2}\right) & g > 1 \end{cases}, \tag{3.38}$$

which vanishes close to the critical point as $S_{\mathrm{vN}}(g) \approx -(g - 1)\log_2(g - 1)$. Note that in addition, there is a sudden growth of entanglement as the critical point is traversed, as it can be observed in Fig. 3.3a.

3.2.2.4 Ground-State Fidelity

Zanardi and Paunković pointed out another fingerprint of QPTs, which can be witnessed by the so-called ground-state fidelity [65]. This fidelity quantifies how sensitive is the ground state to small variations in the control parameter g. Formally, let $\left|\varphi^0(g)\right\rangle$ be the ground state of a certain critical Hamiltonian $H(g)$ which undergoes a QPT at the critical value g_c. Then, as a consequence of the QPT the ground state at g_c becomes orthogonal to $\left|\varphi^0(g)\right\rangle$ with $g \neq g_c$. This can be written as $F(g, \delta g) = |\langle\varphi^0(g)|\varphi^0(g - \delta g)\rangle|$, and thus, $F(g_c, \delta g)$ vanishes. As we have obtained the exact form of the ground state of H_{QRM} in the $\Omega/\omega_0 \to \infty$ limit (see Eqs. (3.21) and (3.22)), we can calculate this ground-state fidelity. In particular, for the QRM it reads

$$F(g, \delta g) = \left|\left\langle\varphi^0_{\mathrm{np}}(g)\middle|\varphi^0_{\mathrm{np}}(g - \delta g)\right\rangle\right| = \left|\langle\downarrow|\langle0|S^\dagger[r_{\mathrm{np}}(g)]S[r_{\mathrm{np}}(g - \delta g)]|0\rangle|\downarrow\rangle\right|. \tag{3.39}$$

Note that here we restrict ourselves to the normal phase, i.e., $0 \leq g \leq 1$ with $\delta g > 0$. Since the overlap between two squeezed vacuum states is $\langle0|S^\dagger[r_2]S[r_1]|0\rangle = (\cosh(r_1 - r_2))^{-1/2}$ with $r_1, r_2 \in \mathbb{R}$, and thus

$$F(g, \delta g) = \sqrt{\frac{2}{\left(\frac{1-g^2}{1-(g-\delta g)^2}\right)^{1/4} + \left(\frac{1-g^2}{1-(g-\delta g)^2}\right)^{-1/4}}}. \tag{3.40}$$

For a small variation close to the critical point $g_c = 1$, $|\delta g| \ll 1$, the previous expression can be approximated as

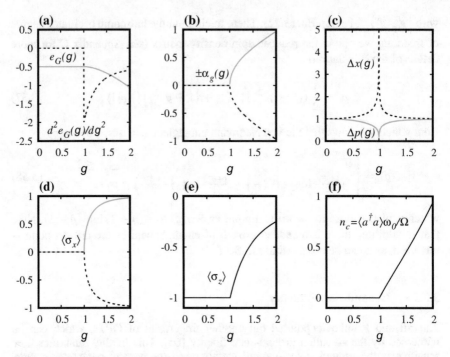

Fig. 3.2 Relevant ground-state signatures of the superradiant QPT underwent by the QRM in the $\Omega/\omega_0 \to \infty$ limit. In **a** the normalized ground-state energy $e_G(g)$ together with its second derivative, showing the discontinuity at the critical point $g = 1$. In **b** the acquired coherence of the bosonic field $\langle a \rangle = \pm \alpha_g(g)$ is represented, being non-zero for $g > 1$. The variance of the quadratures, shown in **c**, Δx and Δp reveal an infinitely squeezed ground-state at the critical point. The figures **e** and **f** show the qubit observables $\langle \sigma_x \rangle$ and $\langle \sigma_z \rangle$, respectively. The former discloses the spontaneous symmetry breaking, as $\langle a \rangle$ in **b**. In the superradiant phase the ground state shows an increasing population of the upper level of the qubit, as indicated by $\langle \sigma_z \rangle$. Finally, the rescaled number of bosonic excitations $n_c \equiv \omega_0/\Omega \langle a^\dagger a \rangle$ is plotted in **f**

$$F(g, \delta g) \approx 2^{1/2} \left(\frac{1 - g}{\delta g} \right)^{1/8}, \qquad (3.41)$$

that is, the fidelity vanishes at $g = g_c = 1$ as $F(g, \delta g) \propto (1 - g)^{1/8}$. A small variation of g close to the QPT produces a dramatic change in the ground state. Indeed, the ground state around g_c becomes orthogonal to the one at $g_c - \delta g$. Note that, while the previous result is only valid for the normal phase, a similar analysis in the superradiant phase will lead to the same power law. The ground-state fidelity is plotted in Fig. 3.3b.

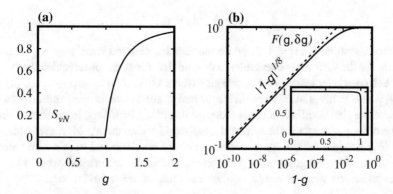

Fig. 3.3 Further signatures of the superradiant QPT of the QRM. In **a** the von Neumann entanglement entropy S_{vN} is plotted, according to Eq. (3.38), which shows an abrupt increase of entanglement as the critical point $g = 1$ is traversed. In **b** the ground-state fidelity, $F(g, \delta g)$ with $\delta g = 0.1$, is shown in a double logarithmic scale (see Eq. (3.40)), to illustrate its scaling as $g \to 1$, which precisely follows $|1 - g|^{1/8}$, plotted as a guide to the eyes (dotted line). Certainly, the ground-state fidelity vanishes at the critical point, while it remains close to 1 before reaching the critical point, as shown in the inset

3.2.2.5   Critical Exponents

As we have explained in the Introduction, the presence of a QPT entails a particular power-law scaling behavior close to the critical point, independently of microscopic details. In general, we refer to singular observables to the ones for which their expectation value can be written as $\langle \mathcal{A} \rangle \propto |g - g_c|^{\gamma_{\mathcal{A}}}$ for $|g - g_c| \ll 1$, where $\gamma_{\mathcal{A}}$ corresponds to the critical exponent. It is worth remarking that finding the critical exponents is a central task to identify its corresponding universality class. Moreover, beyond the elegance of this compact and universal behavior, the derived $\gamma_{\mathcal{A}}$ will serve us to corroborate the validity and consistence of the obtained results in the $\Omega/\omega_0 \to \infty$ limit.

From the previous results, we can extract the following power-law behavior of the observables close to the critical point, and thus, we can identify their corresponding critical exponent $\gamma_{\mathcal{A}}$,

$$\Delta \sim |g - g_c|^{1/2} \tag{3.42}$$

$$\Delta x \sim |g - g_c|^{-1/4} \tag{3.43}$$

$$\Delta p \sim |g - g_c|^{1/4} \tag{3.44}$$

$$\langle a \rangle \sim |g - g_c|^{1/2} \tag{3.45}$$

$$n_c \sim |g - g_c|^{1} \tag{3.46}$$

$$\langle \sigma_z \rangle + 1 \sim |g - g_c|^{1} \tag{3.47}$$

$$\langle \sigma_x \rangle \sim |g - g_c|^{1/2} \tag{3.48}$$

$$F(g, \delta g) \sim |g - g_c|^{1/8}. \tag{3.49}$$

As aforementioned, $z\nu = 1/2$, while the rest are denoted simply by $\gamma_{\mathcal{A}}$ where $\mathcal{A}$ stands for the different observables. We note that for some observables the power-law behavior only takes place at one side of the QPT, such as $\langle \sigma_z \rangle + 1$, n_c, $\langle a \rangle$ and $\langle \sigma_x \rangle$, whose value is strictly zero in the normal phase. However, the featured behavior at the other side is sufficient to characterize its critical exponent, in the same manner as the order parameter in the standard Ginzburg–Landau theory. Moreover, since $\langle a \rangle$ and $\langle \sigma_x \rangle$ account for the spontaneous symmetry breaking and have a zero value in one phase, they are good order parameters. Therefore, the critical exponent of the order parameter reads $\beta = 1/2$, equivalent to that of mean-field theory.

3.2.3 Finite-Frequency Scaling

Strict singular behavior, characteristic of phase transitions, takes only place in the thermodynamic limit. Indeed, finite-size systems do not display abrupt or sudden changes, i.e., the singularities are smoothed. However, how a smooth behavior evolves towards a sharp singularity when the system size increases is intimately related with its behavior in the thermodynamic limit. In Chap. 1 we have discussed this issue for a general system undergoing a phase transition, emphasizing the relevance of the so-called finite-size scaling theory [21, 22], which for the QRM we have dubbed finite-frequency scaling (FFS). Typically, critical exponents cannot be directly accessed unless the system can be exactly solved. Needless to say, exactly solvable many-body systems are an exception and finding a solution constitutes a major task [66]. Therefore, thanks to the FFS the critical exponents can be inferred by analyzing finite-size systems, and thus, a powerful tool when dealing with complex many-body systems [18], which we will comment in Chap. 6.

As we have seen, the QRM exhibits a QPT even in the absence of a thermodynamic limit but in a suitable parameter limit. Therefore, the size of the system remains fixed while increasing the ratio Ω/ω_0 leads to the emergence of the aforementioned critical signatures. As an example, we show in Fig. 3.4 how the spectral gap $\Delta(g)$ of the H_{QRM} closes at $g = 1$ as the ratio Ω/ω_0 increases. Accordingly, and because the size of the system remains constant, we denote this scaling theory as *finite-frequency* scaling when applied to the QPT of the QRM. Moreover, since we have provided an exact solution of the QRM that unveils the presence of a QPT we will not use this scaling theory to obtain the critical exponents but rather to confirm their value. In short, we recall that the finite-size scaling theory hypothesizes that if $\langle \mathcal{A} \rangle$ behaves as

$$\langle \mathcal{A} \rangle (L \to \infty, g) = \mathcal{A}_0 |g - g_c|^{\gamma_{\mathcal{A}}} \tag{3.50}$$

for $|g - g_c| \ll 1$ in the $L \to \infty$ limit, then, for a large but still finite L,

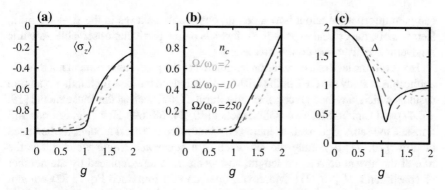

Fig. 3.4 Behavior of three relevant quantities of the QRM as the ratio Ω/ω_0 increases, as function of the dimensionless coupling g. The panels **a**, **b** and **c** correspond to $\langle\sigma_z\rangle$, $n_c = \omega_0/\Omega\langle a^\dagger a\rangle$ and energy gap Δ, respectively. In the $\Omega/\omega_0 \to \infty$ limit, the singular behavior plotted in Fig. 3.2 is recovered, that is, increasing the ratio Ω/ω_0 leads to the appearance of the superradiant QPT. The finite-size scaling theory can be applied to this QPT, where the ratio Ω/ω_0 plays the role of the conventional *size*. Therefore, we denote this scaling theory as finite-frequency scaling when applied to the QPT of the QRM

$$\langle \mathcal{A} \rangle (L, g) = \mathcal{A}_0 |g - g_c|^{\gamma_A} F_A \left(\frac{L}{\xi} \right) \tag{3.51}$$

$$= \mathcal{A}_0 |g - g_c|^{\gamma_A} F_A \left(|g - g_c|^\nu L \right) \tag{3.52}$$

where $F_A(x)$ is function which is only defined in terms of the asymptotic properties, i.e. it fulfills

$$\lim_{x \to \infty} F_A(x) = 1, \quad \text{and} \quad \lim_{x \to 0} F_A(x) \propto x^{-\gamma_A/\nu}. \tag{3.53}$$

Hence, in the $L \to \infty$ limit, the expected behavior of Eq. (3.50) is recovered, while for finite L, the singularity is washed out and the following scaling at g_c is predicted

$$\langle \mathcal{A} \rangle (L, g = g_c) \propto L^{-\gamma_A/\nu}. \tag{3.54}$$

Note that, independently of the observable $\mathcal{A}$, the function F_A depends solely on the ratio between the *length* of the system L and its correlation length ξ, which for a continuous QPT follows $\xi \propto |g - g_c|^{-\nu}$ [13, 67]. For this reason, F_A is known as a scaling function. The correctness of the finite-size scaling hypothesis can be proved by computing $\langle \mathcal{A} \rangle$ for different L and g, since $\langle \mathcal{A} \rangle (L, g)|g - g_c|^{-\gamma_A}$ should collapse into a single curve when plotted against $|g - g_c|^\nu L$, independently of the values of L and g alone. We will come back to these scaling functions in Sect. 3.2.3.2. Note however, that, since the finite-size scaling hypothesis is based on leading order in $1/L$, it is expected to capture correctly the behavior of the system for larger sizes. Moreover, as shown in Eq. (3.54), the scaling of $\langle \mathcal{A} \rangle (L, g = g_c)$ in terms of L enables the determination of $-\gamma_A/\nu$. In this way, from finite-size systems one can gain information about how a certain observable behaves in the $L \to \infty$ limit. Remarkably, the critical exponent γ_A depends on the particular observable $\mathcal{A}$, while ν governs the diverging correlation length ξ.

However, the lack of a correlation length makes the previous argument not directly applicable to study the QPT of the QRM. This is indeed the case for another quantum systems which have been recognized to exhibit a QPT, such as the Dicke model [19, 46, 47] and Lipkin–Meshkov–Glick model [20, 24, 68, 69]. However, one can still translate the same arguments to these zero-dimensional critical models, as discussed in Refs. [70, 71] assuming the existence of a *coherence length* which plays the role of the absent correlation length, and where L is now replaced by the number of constituents N [70, 71]. Moreover, one can still introduce Eq. (3.52) without resorting to any diverging length scale, where the scaling variable adopts the form $x = |g - g_c|^\nu N$. The validity of such an assumption will be then tested if finite-size scaling functions exist and the scaling variable applies in all the examined observables. We stress that, while one may argue whether such a coherence length does or not correspond to a proper length scale, it was proposed by analogy with spatially extended systems, solely intended to correctly grasp scaling features in zero-dimensional models. This is the very case for the Dicke and Lipkin–Meshkov–Glick models, which successfully follow the finite-size scaling theory as reported in [69, 72–74]. Nevertheless, the QPT in the QRM challenges even more the notion of finite-size scaling. The limit $\Omega/\omega_0 \to \infty$ provokes the exploration of the whole infinite Hilbert space by the low-energy states, which typically occurs in a conventional thermodynamic limit. In view of this, we hypothesize a finite-frequency scaling theory for the QRM replacing L by the ratio Ω/ω_0 in the previous equations. This theory provides de facto a good description of the QRM as it Ω/ω_0 increases, as we will support both theoretically and numerically. Therefore, the observables discussed in Sect. 3.2.2 are expected to obey the following scaling

$$\langle \mathcal{A} \rangle \, (\Omega/\omega_0, g = g_c) \propto \left(\frac{\Omega}{\omega_0} \right)^{\delta_A}, \tag{3.55}$$

where we have introduced the finite-frequency scaling exponent of $\mathcal{A}$ as $\delta_A \equiv -\gamma_A/\nu$, being γ_A the critical exponent that dictates the behavior of $\langle \mathcal{A} \rangle$ close to g_c in the $\Omega/\omega_0 \to \infty$ limit (see Eq. (3.50)). These critical exponents γ_A have been determined analytically for several observables from our effective model. However, one may have noticed that ν only appears in the QRM in combination with z, namely $z\nu = 1/2$, while we have used this critical exponent in the scaling variable $x = |g - g_c|^\nu \Omega/\omega_0$, by analogy with spatially extended systems. Nevertheless, as the system lacks correlation length, one could have proposed a different scaling variable, say $x' = |g - g_c|^{\nu'} \Omega/\omega_0$, such that $\nu' \neq \nu$. Certainly, if $F_A(x)$ exists, also $F_A(x^{\nu'/\nu}) = F_A(x')$, which would lead to $\delta'_A = -\gamma_A/\nu'$. The point is that, while the critical exponent ν may be considered ambiguous due to the absence of spatial dimension, it acquires a clear meaning and defined value upon the identification of the scaling given in Eq. (3.55). For example, if one might argue that the diverging

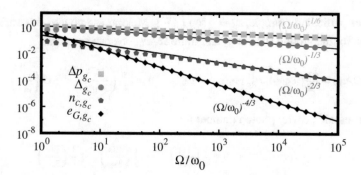

Fig. 3.5 Scaling of different quantities at $g = g_c$ as Ω/ω_0 increases, namely, from top to bottom Δp (squares), energy gap Δ (circles), the order parameter n_c (stars), and ground-state energy correction e_G (diamonds). The analytical results, depicted with solid lines, predict precisely the exact diagonalization results (points) for all observables. Note however, the deviation for n_{c,g_c} for $\Omega/\omega_0 \lesssim 10^2$, which can be better visualized in Fig. 3.6. The finite-frequency scaling exponents for each observable are indicated, as obtained from our effective theory, and the corresponding fitted exponents are collected in the Table 3.1

variance Δx plays the role of the absent correlation length in this system, it would lead to $\nu = 1/4$ and $z = 2$ (see Eq. (3.43)) and therefore, it would provide wrong δ_A scaling exponents, as we will see in the following. We will rely on finite-frequency scaling theory to determine the value of ν, as also done in the developments involving Dicke and Lipkin–Meshkov–Glick models [69, 72–74]. Finally, although one may resort directly to a numerical diagonalization of H_{QRM} to corroborate the validity of these predicted scaling laws, given in Eq. (3.55), and then find a common critical exponent ν, one can still go further with our effective theory to predict these scaling exponents in an analytical manner, as we present in the following (Fig. 3.5).

3.2.3.1 Variational Method for Finite-Frequency Scaling

Here we show how to obtain the finite-frequency scaling exponents δ_A for the considered observables. For that purpose, we shall start from the low-energy Hamiltonian with the first-order correction in ω_0/Ω. As we are interested in the scaling at the critical point, we may use the normal-phase Hamiltonian H_{np}^{Ω} given in Eq. (3.23). However, since the latter Hamiltonian cannot be diagonalized, we propose an Ansatz for the wave function $|\psi_0(s)\rangle = S[s]|0\rangle|\downarrow\rangle$ by comparison with the actual ground state in the $\Omega/\omega_0 \to \infty$ limit. Then, we resort to a variational method with a unique free parameter s, that will be determined minimizing the following energy functional

$$E_0(s) = \langle\psi_0(s)| H_{np}^{\Omega} |\psi_0(s)\rangle = \frac{\omega_0}{2}\cosh(2s) - \frac{\omega g^2}{4}e^{2s} + \frac{3\omega_0^2 g^4}{16\Omega}e^{4s} - \frac{\Omega}{2} + \frac{g^2\omega_0^2}{4\Omega}. \quad (3.56)$$

From $dE_0(s)/ds = 0$ and since $d^2E_0(s)/ds^2 > 0$ for any real s, the sought value of s is obtained solving $\frac{3g^4\omega_0}{2\Omega}e^{6s} + (1 - g^2)e^{4s} - 1 = 0$, which at the critical point $g_c = 1$

becomes particularly simple, $s_c = \frac{1}{6} \log \left(\frac{2\Omega}{3\omega_0} \right)$. Making use now of this variational solution for the ground state at g_c and finite Ω/ω_0, one can show that

$$e_{G,g_c}(\Omega/\omega_0) = \frac{\omega_0}{\Omega}(\langle\psi_0(s_c)| H_{np}^\Omega |\psi_0(s_c)\rangle - E_{G,np}(g_c)) = \frac{\omega_0}{4} \left(\frac{2\Omega}{3\omega_0} \right)^{-4/3} + \mathcal{O}\left(\frac{\omega_0^2}{\Omega^2} \right), \quad (3.57)$$

and for rescaled cavity photon number n_c,

$$n_{c,g_c}(\Omega/\omega_0) = \frac{\omega_0}{\Omega} \langle\psi(s_c)| a^\dagger a |\psi(s_c)\rangle \approx \frac{1}{6} \left(\frac{2\Omega}{3\omega_0} \right)^{-2/3} - \frac{1}{3} \left(\frac{2\Omega}{3\omega_0} \right)^{-1} + \frac{1}{6} \left(\frac{2\Omega}{3\omega_0} \right)^{-4/3} \quad (3.58)$$

It is worth emphasizing that, while the leading order scaling exponent is $-2/3$, higher-order terms may have an important impact in n_c as their scaling exponents are close to $-2/3$, namely, -1 and $-4/3$. Therefore, they may be non-negligible for moderate Ω/ω_0 values and may affect the leading-order scaling, as we will see later with the numerical verification of this scaling. Next, we find the leading order correction of the variance of cavity field quadratures,

$$\Delta x_{g_c}(\Omega/\omega_0) = e^{s_c} = \left(\frac{2\Omega}{3\omega_0} \right)^{1/6}, \quad \Delta p_{g_c}(\Omega/\omega_0) = e^{-s_c} = \left(\frac{2\Omega}{3\omega_0} \right)^{-1/6}. \quad (3.59)$$

We now assume that the mth excited states are $S[s_{\min}(\Omega/\omega_0, g)] |m\rangle$. Thus, the energy gap $\Delta_{n,m}(g_c)$ between the states $S(s_{\min}) |n\rangle$ and $S(s_{\min}) |m\rangle$ at the critical point is

$$\Delta_{n,m}(g, \Omega/\omega_0)\big|_{g=g_c} = \langle m| S^\dagger[s_c] H_{np}^\Omega S[s_c] |m\rangle - \langle n| S^\dagger[s_c] H_{np}^\Omega S[s_c] |n\rangle$$

$$= \omega_0 \frac{(m-n)(3+m+n)}{4} \left(\frac{2\Omega}{3\omega_0} \right)^{-1/3}. \quad (3.60)$$

The finite-frequency correction for the the excitation energy at $g = g_c$ is, then,

$$\Delta_{g_c}(\Omega/\omega_0) = \langle 1| S^\dagger[s_c] H_{np}^\Omega S[s_c] |1\rangle - \langle 0| S^\dagger[s_c] H_{np}^\Omega S[s_c] |0\rangle = \omega_0 \left(\frac{2\Omega}{3\omega_0} \right)^{-1/3}. \quad (3.61)$$

From the previous expression we can extract the value of critical exponent z. Note that $\delta_\Delta = -\gamma_\Delta/\nu$ with $\gamma_\Delta = z\nu = 1/2$ and thus, from Eq. (3.55), finite-frequency scaling $\Delta(\Omega/\omega_0, g_c) \propto (\Omega/\omega_0)^{-z}$, and therefore, $z = 1/3$ and $\nu = 3/2$.

Note however that the finite-frequency scaling for qubit observables, such as $\langle \sigma_z \rangle$, cannot be obtained simply considering $e^{-S} \approx \mathbb{1}$ as done for previous observables. To the contrary, one must calculate

$$\langle \sigma_z \rangle (\Omega/\omega_0, g = g_c) = \langle\downarrow| \langle 0| S^\dagger[s_c] U^\dagger \sigma_z U S[s_c] |0\rangle |\downarrow\rangle \quad (3.62)$$

where $U = e^{\lambda S_1 + \lambda^3 S_3}$, as discussed in Sect. 3.2.1, which brings us to

Table 3.1 Critical and finite-frequency scaling exponents at g_c for certain relevant quantities of the QPT in the QRM. The exponent ν must be equal for any observable, whose value is $\nu = 3/2$. The FFS exponents are obtained analytically through a variational method, and numerically by fitting the exact numerical results of the QRM. The fits were performed from $\Omega/\omega_0 = 10^3$ to $\Omega/\omega_0 = 10^5$

	Δ	n_c	$\langle\sigma_z\rangle + 1$	Δp
Critical exponent (γ_A)	1/2	1	1	1/4
FFS exponent ($\delta_A = -\gamma_A/\nu$)	−1/3	−2/3	−2/3	−1/6
ν	3/2	3/2	3/2	3/2
Fitted result for δ_A	−0.3317(4)	−0.629(5)	−0.665(1)	0.16661(1)

$$\langle\sigma_z\rangle\,(\Omega/\omega_0, g = g_c) \approx -1 + \frac{1}{3}\left(\frac{2\Omega}{3\omega_0}\right)^{-2/3}. \tag{3.63}$$

These scaling relations are consistent with the hypothesized finite-frequency scaling theory, supporting the value of the critical exponent $\nu = 3/2$, and the obtained exponents $\delta_{\sigma_z} = -2/3$, $\delta_{\Delta x} = 1/6$ and $\delta_{n_c} = -2/3$. In the next part we present a numerical verification of these exponents, as well as a corroboration of the validity of the low-energy effective Hamiltonians.

3.2.3.2 Numerical Confirmation

Exact numerical diagonalization of the H_{QRM} allows us to verify the derived critical exponents of the QPT, as well as the validity of the effective low-energy description of the QRM for $\Omega/\omega_0 \gg 1$. In this regard, the finite-frequency scaling theory gains considerable importance since it enables the comparison between numerical computed expectation values and the results from the effective theory. However, since the number of relevant Fock states blows up with a diverging Ω/ω_0, a numerical diagonalization of the H_{QRM} in this extreme regime becomes increasingly harder and costly from a computational point of view. The numerical simulations presented here were performed ensuring the convergence of ground state properties upon the inevitable Fock space truncation, achieving results up to $\Omega/\omega_0 \sim 10^5$. These large Ω/ω_0 values allow us to verify the predicted power-law scaling in a clean manner even for small exponents such as 1/6 as featured by Δx.

In the Table 3.1 we collect the obtained fitted scaling exponents from the exact numerical diagonalization, and compared with the predicted exponents. The agreement is excellent, supporting the correctness of our low-energy effective theory. The fits were performed in the range from $\Omega/\omega_0 = 10^3$ to $\Omega/\omega_0 = 10^5$ in order to avoid, as much as possible, sub-leading correction to the scaling exponents. Indeed, higher-order corrections become particularly important for n_{c,g_c}, as we anticipated after deriving its finite-frequency scaling in Eq. (3.58), which lead to a worse agreement between the fitted finite-frequency scaling exponent and its leading-order scaling predicted theoretically, $-2/3$, see Eq. (3.58). In Fig. 3.6 we illustrate the deviation

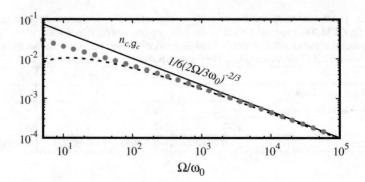

Fig. 3.6 Finite-frequency scaling of n_{c,g_c} at the critical point g_c obtained by exact numerical diagonalization of H_{QRM} (light-gray points), together with the predicted leading-order scaling $1/6(2\Omega/(3\omega_0))^{-2/3}$ (solid line) and with higher-order corrections (dashed line), $1/6(2\Omega/(3\omega_0))^{-2/3} - 1/3(2\Omega/(3\omega_0))^{-1} + 1/6(2\Omega/(3\omega_0))^{-4/3}$ (see Eq. (3.58)). For small Ω/ω_0 values the predicted scaling breaks down, while it is recovered approximately starting from $\Omega/\omega_0 = 10^3$

of the simple leading-order scaling for small Ω/ω_0 values, and the recovery of the power-law scaling starting at $\Omega/\omega_0 \approx 10^3$. The failure of a simple power-law scaling for small Ω/ω_0 values is expected, since the predicted relations have been attained considering just the first-order correction Hamiltonian H_{QRM}^Ω, and thus only valid for $\Omega/\omega_0 \gg 1$. Furthermore, it should be also mentioned that these scaling relations have been obtained using a variational method which turns out to provide an excellent description of the critical ground state in this regime. As a side remark we comment that the expected scaling of the ground-state fidelity at the critical point results in $F(g_c, \delta g) \propto \left(\frac{\Omega}{\omega_0}\right)^{-1/12}$, however, a numerical verification of such a small scaling exponent is difficult within the accessible Ω/ω_0 range.

Finally, in order to quantify the agreement between H_{QRM}^Ω and the actual H_{QRM} for $\Omega/\omega_0 \gg 1$, we propose a last numerical test. We compute numerically the ground-state energy, energy gap, Δp and n_c from H_{QRM}^Ω and compare them with the ones obtained from the H_{QRM}. This is plotted in the Fig. 3.7 for two values of the dimensionless coupling, $g = 1/2$ and $g = 1$. Certainly, these results show that the deviation from the exact solution is very small and decays fast as Ω/ω_0 increases. This further supports the validity of the presented results, and that H_{QRM}^Ω is indeed a very good approximation of H_{QRM} for $\Omega/\omega_0 \gg 1$.

3.2.3.3 Finite-Frequency Scaling Functions

As we have seen previously, the finite-frequency scaling hypothesis entails a power-law scaling of different quantities at the critical point as the ratio Ω/ω_0 increases. Yet, finite-frequency scaling relies on the existence of scaling functions, $F_A(x)$ with $x = \Omega/\omega_0|g - g_c|^\nu$, which are independent of the specific values of the ratio Ω/ω_0 and g.

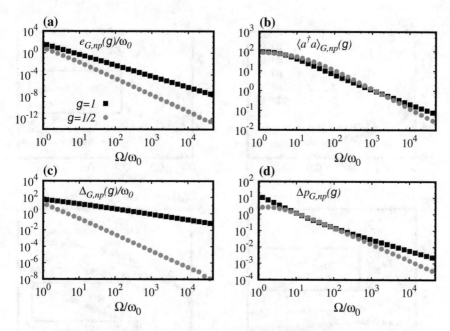

Fig. 3.7 Difference, in percentage, between the solutions from the effective Hamiltonian $H^{\Omega}_{\mathrm{QRM}}$ and the numerically exact H_{QRM} at $g = 1/2$ (light-gray circles) and $g = 1$ (black squares), for different quantities, **a** ground-state energy, **b** ground-state number of bosonic excitations, **c** energy gap and **d** variance of $p = i(a^{\dagger} - a)$ at the ground state. In particular, we plot $100 \times (O^{\mathrm{exact}} - O)/O^{\mathrm{exact}}$ where O^{exact} corresponds to the numerically exact value obtained from H_{QRM} and O its counterpart computed using $H^{\Omega}_{\mathrm{QRM}}$. The errors decay fast as a function of Ω/ω_0, indicating that $H^{\Omega}_{\mathrm{QRM}}$ is a very good approximation of the QRM for $\Omega/\omega_0 \gg 1$

Therefore, by scanning different Ω/ω_0 and g one obtains a set of data, $\langle A \rangle\,(\Omega/\omega_0, g)$, which collapses into a single curve upon a proper rescaling. These scaling functions, together with the critical exponents, are essential to disclose the universality class to which a phase transition belongs [17, 18]. Note that $F_A(x)$ together with the aforementioned scaling solely at the critical point, given in Eq. (3.55), appear as a handy tool to determine the critical exponents γ_A.

These scaling functions can be easily computed for different quantities by simply performing numerical diagonalization of the full H_{QRM} for various Ω/ω_0 and g values. Then, since $F_A(x) \equiv \langle A \rangle\,|g - g_c|^{-\gamma_A}$ with $x = \Omega/\omega_0|g - g_c|^{\nu}$, organizing the ground-state expectation values such that $F_A(x)$ is plotted as a function of x, it should result in data collapse into a single curve, and thus disclosing the actual form of the scaling function. This is precisely plotted in Fig. 3.8 for the energy gap Δ, rescaled boson number n_c, qubit population $\sigma_z + 1$ and variance of the bosonic position quadrature Δx. As a matter of fact, the collapse of different data points is evident, supporting the validity of finite-frequency scaling theory for the QRM. Moreover, the behavior of the scaling function as $x \to 0$ is shown in the insets, where the predicted power law is observed, i.e., $F_A(x) \to x^{-\gamma_A/\nu}$.

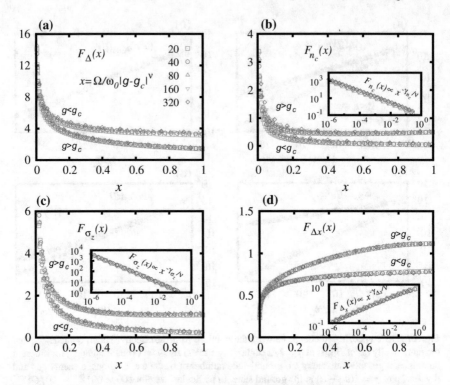

Fig. 3.8 Finite-frequency scaling functions of the QRM for different quantities $\mathcal{A}$, $F_{\mathcal{A}}(x)$ with $x = \Omega/\omega_0 |g - g_c|^\nu$ and $\nu = 3/2$. The panel **a** corresponds to the energy gap Δ, the rescaled number of bosonic excitations $n_c = \frac{\omega_0}{\Omega} \langle a^\dagger a \rangle$ is plotted in **b**, while $\sigma_z + 1$ and Δx in **c** and **d**, respectively. The different point styles correspond to various Ω/ω_0 values, as indicated in **a**. As expected from finite-frequency scaling theory, $\langle \mathcal{A} \rangle |g - g_c|^{-\gamma_{\mathcal{A}}} = F_{\mathcal{A}}(x)$, and thus the date collapse into a single curve, independently of the particular value of Ω/ω_0 or g. Note that the scaling functions are plotted for both sides, $g < g_c$ and $g > g_c$, and in addition panels **b**, **c** and **d** show the scaling of $F_{\mathcal{A}}(x)$ for $x \to 0$, which precisely follows the expected power-law $F_{\mathcal{A}}(x) \to x^{-\gamma_{\mathcal{A}}/\nu}$. Recall that $\gamma_{\sigma_z} = \gamma_{n_c} = 1$ and $\gamma_{\Delta x} = -1/4$

3.2.4 Universality Class

Throughout the present chapter we have mentioned repeatedly the similarities between the QRM and its multi-spin version, the so-called Dicke model [19], as well as with the Lipkin–Meshkov–Glick model [20]. Since it is well established that the Dicke and Lipkin–Meshkov–Glick models belong to the same universality class, we shall consider only the Dicke model in the following for the sake of brevity. The Dicke model undergoes a continuous superradiant QPT in the limit of $N \to \infty$ spins [46–49], however, we have so far not explicitly shown whether the QPT in the QRM belongs to the same universality class as the QPT of the Dicke or Lipkin–Meshkov–Glick models. Recall that within the theory of continuous phase transitions, two models that undergo critical behavior belong to the same universality class when they share critical exponents and finite-size scaling functions, as explained in Chap. 1. In the following we demonstrate that both models do belong to the same universality class. For that purpose, we shall consider he Dicke Hamiltonian, which can be written as

$$H_{\text{Dicke}} = \omega_a J_z + \omega_b a^\dagger a + g \frac{\sqrt{\omega_a \omega_b}}{2\sqrt{N}} J_x (a + a^\dagger) \qquad (3.64)$$

where the angular momentum operators $\vec{J} = (J_x, J_y, J_z)$ with $J_\alpha = \sum_i^N \sigma_\alpha/2$ describe the N spins coupled to one bosonic mode, as in the case of the QRM. Since J^2 commutes with H_{Dicke}, the analysis can be restricted to the highest angular momentum subspace, $J = N/2$. As shown in previous works, the Dicke undergoes a QPT at $g_c = 1$, that divides the normal ($g < g_c$) from the superradiant phase ($g > g_c$). The properties of these phases are similar to those of the QRM, as the ground state exhibits $\langle J_z \rangle + J = 0$ and $\langle a^\dagger a \rangle/N = 0$ for $g < g_c$, while for $g > g_c$, $\langle J_z \rangle + J \neq 0$ and $\langle a^\dagger a \rangle/N \neq 0$, as well as $\langle a \rangle \neq 0$ as a result of a spontaneous Z_2 symmetry breaking. Moreover, the QPT of the Dicke model is characterized by the same set of critical exponents as the ones obtained for the QRM, see Table 1 in Ref. [74]. Among others, we find that the Dicke model features $\nu = 3/2$, $z = 1/3$, $\gamma_{J_z} = 1$ and $\gamma_{n_c} = 1$, that is, in the limit $N \to \infty$ the energy gap closes at the critical point as $\Delta \propto |g - g_c|^{z\nu}$, and the observables J_z and $a^\dagger a$ behave as $\langle J_z \rangle/J + 1 \propto |g - g_c|^{\gamma_{J_z}}$ and $n_c \equiv \langle a^\dagger a \rangle/N \propto |g - g_c|^{\gamma_{n_c}}$. In addition, the universality class to which these models belong is identified upon equivalent finite-size scaling functions (finite-frequency for the QRM) and critical exponents. For that reason, we compute the finite-size scaling functions of the Dicke model, obtained from a numerical diagonalization of H_{Dicke}, and demonstrate that they are indeed equivalent to those of the QRM, as we show in Fig. 3.9 for the energy gap Δ and spin population J_z. Note that, since the equivalence among scaling functions is only required up to constant factors [17], to better illustrate this equivalence, we plot the finite-size scaling functions of the Dicke model as $K_1 \tilde{F}_A(K_2 \tilde{x})$ where $\tilde{x} = N|g - g_c|^\nu$ and $K_{1,2}$ are constant factors included here to reproduce its counterpart in the QRM, $F_A(x)$. In Fig. 3.9 we have considered $K_2 = 2$ and $K_1 = 1.275$, which provide a good agreement between $F_A(\tilde{x})$ (QRM) and $K_1 \tilde{F}_A(K_2 \tilde{x})$ (Dicke model).

Hence, the QRM undergoes a QPT which lies within the same universality class of the superradiant QPT of the Dicke or Lipkin–Meshkov–Glick models, that is, they are of a mean field type [75]. However, it is worth mentioning that the superradiant QPT in the Dicke model owns critical signatures which are absent in the QRM, and vice versa. This is for example the case for the amount of entanglement (see [54] for an extended analysis) or for the correlation functions $\langle J_\alpha^2 \rangle/N^2$ for $\alpha = x, y, z$ in the ground state [74] (see [68, 69, 72, 73] for a discussion regarding the Lipkin–Meshkov–Glick model). As shown in [61, 62], the N spins get entangled with the bosonic mode, which ultimately leads to a diverging entanglement entropy at the critical point, $S_{\text{vN}} \propto -\nu \log_2 |g - g_c|$, in the $N \to \infty$ limit. Nevertheless, since the QRM involves just a single spin these features are absent in its QPT, although it still shows an abrupt growth in S_{vN}.

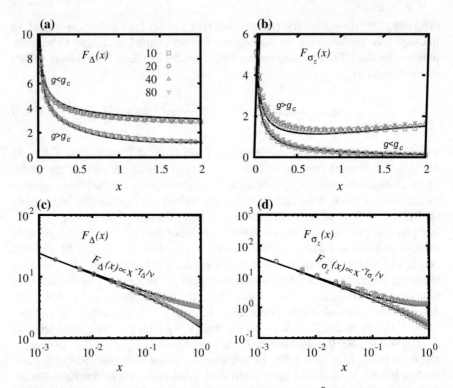

Fig. 3.9 Equivalence among the scaling functions of the Dicke model $\tilde{F}_A(\tilde{x})$ with $\tilde{x} = N|g - g_c|^\nu$ (finite-size) and the QRM, $F_A(x)$ (finite frequency). In **a** and **c**, the scaling function corresponds to the energy gap Δ, while in **b** and **d** to $Jz/J + 1$ for the Dicke model, and its equivalent for the QRM, $\sigma_z + 1$. Note that **c** and **d** are a zoom for small x values of **a** and **b**, respectively. The finite-frequency scaling functions of the QRM are depicted with a solid black line (obtained from the collapse of different data, as shown in Fig. 3.8), while the data computed for the Dicke model with different number of spins N is plotted with points. To better illustrate the equivalence between these scaling functions, we represent $K_1\tilde{F}_A(K_2\tilde{x})$ with $K_2 = 2$ and $K_1 = 1.275$

3.3   Excited-State Quantum Phase Transition

Besides the superradiant QPT exhibited by the QRM, this model features yet another kind of critical behavior, which is the subject of the present section. This feature accompanies the QPT but manifests itself in a collection of excited states of the Hamiltonian that governs the system. Thus, non-analytic signatures can also be found at non-zero excitation energies [25, 27, 28, 76, 77]. As we will see later on, the origin of this singularities is strongly related to the underlying properties of the corresponding semi-classical counterpart of the quantum system [23, 24, 27, 28]. Such critical phenomenon is known as excited-state quantum phase transition (ESQPT), whose main feature resides in a singularity in the density of states or in its derivatives at a certain critical excitation energy [27, 28]. The order of the derivative of the density

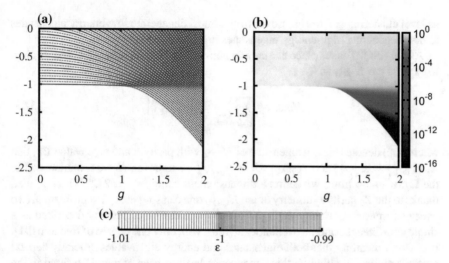

Fig. 3.10 Energy spectrum $\varepsilon_k^\pm$ of H_{QRM} as a function of the dimensionless coupling g for a ratio $\Omega/\omega_0 = 40$ (**a**). The solid black and dashed gray lines correspond to different parities. For $g \gtrsim 1$ and $\varepsilon \lesssim -1$ there is a pair of nearly degenerate ground state. In **b** we plot the energy difference $\Delta_k = |\varepsilon_k^+ - \varepsilon_k^-|$ as a function of ε and g for $\Omega/\omega_0 = 40$, which serves as a phase diagram where the crossover at $\varepsilon = -1$ is noticeable. There is an increase of the level density around $\varepsilon = -1$ which is a precursor of the diverging semiclassical density of states, as it can be seen in **c** where each point corresponds to a different energy eigenvalue of H_{QRM} for $\Omega/\omega_0 = 4 \times 10^3$ and $g = 1.2$

of states at which non-analytic behavior appears depends crucially on the number of degrees of freedom, being specially relevant in low-dimensional systems. More specifically, ESQPTs in a system with f degrees of freedom stem from the extremal points of its f-dimensional energy landscape, producing abrupt changes or divergences in the $(f-1)$-derivative of the density of states [28, 78]. As a paradigmatic example, we find again the already mentioned Dicke and Lipkin–Meshkov–Glick models [19, 20], with $f = 2$ and $f = 1$, respectively, which in the thermodynamic limit display a logarithmic divergence in the first derivative and in the density of states itself, respectively [68, 69, 79, 80]. Such singularities in the density of states have been recently observed in microwave photonic crystals [81, 82]. Besides a singular density of states, an ESQPT may have an impact into the system in diverse manners, both static [83, 84] and from a dynamical point of view [76, 77, 85–90].

Therefore, the QRM appears as a perfect candidate to exhibit an ESQPT as it undergoes a QPT while at the same time has only one degree of freedom, $f = 1$. Additionally, it is also interesting to investigate whether a finite-component system QPT is accompanied by an ESQPT, which is not evident or straightforward from the previous analysis. As an anticipation of the following developments and results, we will show that the QRM does display all the hallmarks of an ESQPT, including a logarithmic divergence of the semiclassical density of states at a certain critical energy, critical behavior of the mean-field observables and its corresponding precursors such as level clustering and the crossover from almost degenerate low-energy eigenstates to non-degenerate high-energy eigenstates in the superradiant phase.

As the ESQPT takes place at a certain non-zero excitation energy, it is convenient to express the H_{QRM} as

$$H_{QRM} = \sum_{k=0} \sum_{\alpha=\pm} E_k^\alpha \left|\varphi_k^\alpha\right\rangle\left\langle\varphi_k^\alpha\right| \tag{3.65}$$

where $\left|\varphi_k^\alpha\right\rangle$ denotes the kth eigenstate of H_{QRM} with parity α and eigenvalue E_k^α, that is, $\Pi\left|\varphi_k^\pm\right\rangle = \pm\left|\varphi_k^\pm\right\rangle$ and $H_{QRM}\left|\varphi_k^\pm\right\rangle = E_k^\pm\left|\varphi_k^\pm\right\rangle$. Moreover, as we are interested in the $\Omega/\omega_0 \to \infty$ limit, we define a dimensionless energy $\varepsilon_k^\pm \equiv 2E_k^\pm/\Omega$. Note that, thanks to the Z_2 parity symmetry of the H_{QRM}, one can split the QRM problem in two parity subspaces. Certainly, each parity subspace can be effectively described as a single non-linear harmonic oscillator and thus, with only one degree of freedom [91]. It is worth recalling that both eigenstates and energy eigenvalues crucially depend on the coupling g, although this dependence has not been explicitly written in the previous equation.

In the $\Omega/\omega_0 \to \infty$ limit the Z_2 symmetry is spontaneously broken in the superradiant phase ($g > g_c = 1$), and all the low-energy eigenstates become doubly degenerate, that is, $\varepsilon_k^+ = \varepsilon_k^-$, which would entail a regular density of states. Hence, the inspection of the an eventual ESQPT is not attainable within the derived low-energy effective Hamiltonians, presented previously. A first, yet naive approach, consists in computing the energy spectrum $\varepsilon_k^\pm$ to observe if there is a sign, reminiscence of an ESQPT. For that purpose, we choose $\Omega/\omega_0 = 40$ and compute the corresponding spectrum as a function of g, as well as the energy difference between the kth eigenstates of opposite parity, $\Delta_k = |\varepsilon_k^+ - \varepsilon_k^-|$, which is plotted in Fig. 3.10. The results suggest the existence of a transition between a doubly degenerate spectrum (for $\varepsilon < -1$) and non-degenerate eigenstates (for $\varepsilon > -1$) for any $g > 1$. Certainly, as a consequence of finite-frequency effects, the degeneracy between $\left|\varphi_k^+\right\rangle$ and $\left|\varphi_k^-\right\rangle$, with $\varepsilon_k^\pm < -1$, is lifted, although the difference Δ_k becomes vanishingly small for large Ω/ω_0. Moreover, in Fig. 3.10a one can advise a clustering of eigenstates around $\varepsilon = -1$, which becomes more clear in Fig. 3.10c that corresponds to $\Omega/\omega_0 = 4 \times 10^3$ and $g = 1.2$. The presence of a clustering of eigenstates implies an increase of the density of states, which takes place around the energy $\varepsilon = -1$. Hence, because the ESQPT is related to changes in the phase space under the semiclassical limit of the quantum system, a semiclassical analysis of the QRM is suitable, presented in the following.

3.3.1 Semiclassical Approach

The observations reported in the previous part require a proper analysis of the H_{QRM} at higher energies than considered in Sect. 3.2, which can be attained by means of a semiclassical approach. Note that this crucially differs from the previous quantum mechanical solution in the $\Omega/\omega_0 \to \infty$ limit, where we have seen that quantum fluctuations in the bosonic mode are of great relevance ($\omega_0 > 0$ while $\Omega \to \infty$). The semiclassical approach however simplifies the QRM to a classical harmonic oscillator ($\omega_0 \to 0$), which despite it discloses certain singular behaviors [53–55], quantum fluctuations cannot be captured, such as the diverging squeezing of the cavity field. The semiclassical version of the QRM is accomplished in a standard manner. For that, we first describe the harmonic oscillator by means of its position and momentum operators, $\hat{x}$ and $\hat{p}$, respectively,

$$\hat{x} = \frac{1}{\sqrt{2}}(a^\dagger + a), \tag{3.66}$$

$$\hat{p} = i\frac{1}{\sqrt{2}}(a^\dagger - a). \tag{3.67}$$

and then, we take the classical limit, i.e., $[\hat{x}, \hat{p}] \to 0$ [92], achieved replacing the operators $\hat{x}$ and $\hat{p}$ by continuous variables x' and p'. Therefore, the semiclassical version of the QRM reads

$$H_{\mathrm{scl}}(x', p') = \frac{\omega_0}{2}(x'^2 + p'^2) - \frac{\omega_0}{2} + \frac{\Omega}{2}\sigma_z - \lambda\sqrt{2}x\sigma_x. \tag{3.68}$$

The previous Hamiltonian acquires a more recognizable form upon diagonalization of the spin Hamiltonian and rescaling the oscillator variables, $x = \sqrt{\omega_0/\Omega}x'$ and $p = \sqrt{\omega_0/\Omega}p'$. Thus, the semiclassical Hamiltonian becomes

$$H_{\mathrm{scl}}^\pm(x, p)/\Omega = \frac{p^2}{2} + V_{\mathrm{eff}}^\pm(x), \quad \text{with} \quad V_{\mathrm{eff}}^\pm(x) = \frac{1}{2}x^2 \pm \frac{1}{2}\sqrt{1 + 2g^2x^2}. \tag{3.69}$$

where $V_{\mathrm{eff}}^\pm(x)$ represents an effective semiclassical potential [53–55], and $\pm$ stand for the two different spin branches. Recall that $g = 2\lambda/\sqrt{\omega_0\Omega}$ is the dimensionless coupling constant, and $g_c = 1$ corresponds to the QPT [30]. This effective potential is plotted in Fig. 3.11 for three different, and characteristic, g values, namely, $g = 1/2$, $g = 1$ and $g = 2$. For high energy $\sim\Omega$, the effective potential has always unique energy minimum located at $x = 0$, in contrast to the more interesting situation at lower energies, for which $V_{\mathrm{eff}}^-(x)$ presents the typical bifurcation from a unique minimum at $x = 0$ for $g < 1$ to two minima at $x = \pm\sqrt{(g^2 - g^{-2})/2}$ for $g > 1$. In other words, the QPT of the QRM appears in its semiclassical limit as a transition from a single- to double-well potential. As $V_{\mathrm{eff}}^+(x)$ describes the high energy subspace ($\varepsilon \geq 1$), and we are interested in the properties of the QRM around $\varepsilon_c = -1$, we restrict ourselves to the analysis of $V_{\mathrm{eff}}^-(x)$. Note that the effective potential already indicates that the classical orbits sharply change for $g > 1$ as a function of the energy E; while for $E < E_c \equiv -\Omega/2$ (or in terms of the dimensionless energy, $\varepsilon < \varepsilon_c$) the orbits remain in either of the two disconnected regions of the double wells, as soon as $E > E_c$ (or $\varepsilon > \varepsilon_c$) the orbits explore the whole phase space. Note that, since

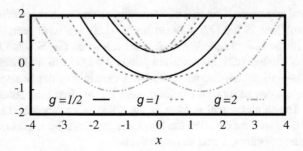

Fig. 3.11 Effective semiclassical potential $V_{\text{eff}}^{\pm}(x)$ for three different and characteristic coupling constants, $g = 1/2$ (normal phase), $g = 1$ (critical point) and $g = 2$ (superradiant phase), plotted with solid black, dotted gray and dotted-dashed light-gray lines, respectively. Note that the lower energy branch, $V_{\text{eff}}^{-}(x)$ exhibits the ubiquitous single- to double-well potential transition. The presence of a local maximum at $x = 0$ for $g > 1$ will be crucial for the ESQPT

$H_{\text{scl}}^{\pm}(x, p) = H_{\text{scl}}^{\pm}(-x, p)$, the localization of classical orbits for $\varepsilon > \varepsilon_c$ at one side, $x \lesssim 0$, hints at a spontaneous symmetry breaking. As a matter of fact, this strongly supports the observations of the QRM, as plotted in Fig. 3.10, where nearly doubly degenerate eigenstates appear for $g > 1$ and with eigenenergies $\varepsilon_k^{\pm} < \varepsilon_c = -1$.

As commented before, the phase-space structure in the semiclassical limit is intimately connected with the ESQPT [28], which in turn is related to the properties of the density of states $\nu(E, g)$ [93]. The semiclassical version of the quantum density of states, or level density, can be obtained making use of the Gutzwiller trace formula [93], or by means of an inverse Laplace transformation of the partition function [93], recently used in the realm of ESQPT for generalized Dicke models [79]. Here, however, we follow a standard yet suitable procedure to compute $\nu(E, g)$.

3.3.1.1 Density of States

The quantum density of states, or level density of a quantum system with spectrum $\{E_k^{\alpha}\}$ is simply given by

$$\nu_q(E, g) = \sum_{k=0}^{} \sum_{\alpha=\pm} \delta\left[E - E_k^{\alpha}\right]. \tag{3.70}$$

Recall that the dependence on g of $\nu_q(E, g)$ resides in the spectrum $\{E_k^{\alpha}\}$ and it is not explicitly written. In addition, it is convenient to introduce the accumulated number of states below an energy E, $N(E, g)$ or *number staircase function* [93], which reads

$$N(E, g) = \int_{-\infty}^{E} dE' \, \nu(E', g). \tag{3.71}$$

The semiclassical approximation of the quantum density of states of a system with f degrees of freedom is attained by integrating the phase-space shell at the corresponding energy E, that is,

$$\nu_{\text{scl}}(E, g) = \frac{1}{(2\pi)^f} \int d\vec{p} \, d\vec{x} \, \delta \left[E - H_{\text{scl}}(\vec{x}, \vec{p}, g) \right].$$ (3.72)

For the QRM, the semiclassical density of states can be simplified to

$$\nu_{\text{scl}}(\varepsilon, g) = \frac{1}{\omega_0 \pi} \frac{\partial}{\partial \varepsilon} \int dx \, dp \, \Theta \left[\varepsilon - p^2 - x^2 + \sqrt{1 + 2g^2 x^2} \right]$$ (3.73)

$$= \frac{\partial}{\partial \varepsilon} N(\varepsilon, g),$$ (3.74)

where we have introduced the dimensionless energy $\varepsilon \equiv 2E/\Omega$ and rescaled coordinates x and p. From Eq. (3.74) it follows that $N(\varepsilon, g)$ is obtained as the total phase-space volume enclosed in the shell of energy ε. Integrating over the momentum p, $\nu_{\text{scl}}(\varepsilon, g)$ becomes (see Appendix D for further details)

$$\nu_{\text{scl}}(\varepsilon, g) = \frac{2}{\omega_0 \pi} \int_{x_1}^{x_2} dx \, \frac{1}{\sqrt{\varepsilon - x^2 + \sqrt{1 + 2g^2 x^2}}},$$ (3.75)

where the integration limits are

$$x_1 = \sqrt{\varepsilon + g^2 - \sqrt{g^4 + 2\varepsilon g^2 + 1}} \; \Theta \left[\varepsilon_c - \varepsilon \right]$$ (3.76)

$$x_2 = \sqrt{\varepsilon + g^2 + \sqrt{g^4 + 2\varepsilon g^2 + 1}}.$$ (3.77)

Then, depending on the energy ε and coupling g, we can calculate the semiclassical density of states. In particular, we concentrate on obtaining analytical expressions for $\nu_{\text{scl}}(\varepsilon, g)$ for two important cases, namely, for $g = 1$ and ε_c and $g > 1$ for ε_c. The former corresponds to the QPT, while the latter will disclose the presence of an ESQPT. In the following, we present the derivation of the $\nu_{\text{scl}}(\varepsilon_c, g = 1)$, and then, in Sect. 3.3.2 we will discuss the signatures of the ESQPT. For the sake of brevity, we leave the detailed mathematical derivation of $\nu_{\text{scl}}(\varepsilon, g)$ for the Appendix D, and summarize the main findings here. In addition, we confirm the validity of this semiclassical approximation by comparing with the numerical results of the QRM. For that reason, after computing the energy spectrum of H_{QRM} for a given coupling g, it is mandatory to perform an average of delta-level density, $\nu_q(\varepsilon, g) = \sum_{k,\alpha} \delta \left[\varepsilon - \varepsilon_k^\alpha \right]$, which we denote as $\tilde{\nu}_q(\varepsilon, g)$ (quantum averaged density of states) and is attained as follows. We choose an energy window of M consecutive eigenstates, whose width is $\Delta \varepsilon = \varepsilon_{k+N} - \varepsilon_k$ with $k \geq 0$, then $\tilde{\nu}_q(\bar{\varepsilon}, g) = M/\Delta \varepsilon$ where $\bar{\varepsilon}$ denotes the middle energy within the chosen window. Note that, while M is left as a free parameter, its value must be large enough to provide a smooth average while at the same time small to retrieve the features of $\nu(\varepsilon, g)$ and prevent any numerical artifact.

In order to derive an analytic expression of the density of states for $g = 1$ close to the energy ε_c (recall that ε_c corresponds to the ground-state energy for $g = 1$).

As we are interested in the behavior close to ε_c, we shall consider $\varepsilon = \varepsilon + \delta\varepsilon$ being $0 < \delta\varepsilon \ll 1$. Since $\varepsilon > \varepsilon_c$, $x_1 = 0$ while the upper limit can be Taylor expanded as $x_2 = (2\delta\varepsilon)^{1/4} + O(\delta\varepsilon^{3/4})$. Then, we can perform the integration of Eq. (3.75) at leading order of $\delta\varepsilon$, which results in (see Appendix D for further details),

$$\nu_{\text{scl}}(\varepsilon_c + \delta\varepsilon, g = 1) \approx \frac{\Gamma(5/4)}{\Gamma(3/4)} \frac{2^{5/4}}{\omega_0\sqrt{\pi}} \delta\varepsilon^{-1/4}, \tag{3.78}$$

where $\Gamma(m)$ is the Euler gamma function, $\Gamma(z) = \int_0^\infty dx\, x^{z-1}e^{-x}$. Therefore, the semiclassical density of states diverges as $\nu_{\text{scl}}(\varepsilon, g = 1) \propto (\varepsilon - \varepsilon_c)^{-1/4}$ for $\varepsilon - \varepsilon_c \ll 1$, i.e., close to the ground-state energy. This divergence is yet another a signature of the QPT, which stems from the low-energy spectral collapse at the critical point (see Eq. (3.14)). On the other hand, one can show that for $g < 1$, the density of states does not show any singular behavior. In Fig. 3.12a we plot the semiclassical density of states as a function of the dimensionless energy ε for $g = 1/4$ and $g = 1$ obtained from Eq. (3.75), and compared with the quantum averaged density of states obtained numerically from the QRM, $\bar{\nu}_q(\varepsilon, g)$. The divergence of the density of states is corroborated by the numerical results, as shown in Fig. 3.12c, which support the power-law scaling close to ε_c. In the following, we discuss the signatures of the ESQPT, whose main feature resides in a logarithmic divergence on the density of states for any $g > 1$ at ε_c.

3.3.2   Signatures of the ESQPT

In this part we show certain signatures of the ESQPT in the QRM, starting with the logarithmic divergence of the semiclassical density of states at ε_c for any $g > 1$. As done for the $g = 1$ case, we shall calculate $\varepsilon = \varepsilon_c + \delta\varepsilon$ with $0 < \delta\varepsilon \ll 1$. However, since ε_c is no longer the ground-state energy, we may approach ε_c from the different side, i.e., $\delta\varepsilon$ may be negative. The procedure is equivalent as the one given here for $\delta\varepsilon > 0$, and left to the Appendix D, where we also provide a detailed derivation of the following results. Then, as $\varepsilon > \varepsilon_c$, $x_1 = 0$ and $x_2 = \sqrt{2(g^2 - 1)} + O(\delta\varepsilon)$, with a density of states

$$\nu_{\text{scl}}(\varepsilon_c + \delta\varepsilon, g) = \frac{2}{\omega_0\pi} \int_{x_1}^{x_2} dx\, \frac{1}{\sqrt{-1 + \delta\varepsilon - x^2 + \sqrt{1 + 2g^2x^2}}}. \tag{3.79}$$

The previous expression contains two possible singularities right at the integration limits, x_1 and x_2 when $\delta\varepsilon = 0$. For that reason, it is convenient to split the integration as $\int_{x_1}^{x_2} dx = \int_{x_1}^{x_m} dx + \int_{x_m}^{x_2} dx$ where the arbitrary x_m is chosen such that $0 < x_m \ll 1$ but still greater than $\delta\varepsilon$. In addition, one can show that the latter integral does not lead to any true singularity in ν_{scl} and just results in a constant shift, denoted by K. In this manner, we obtain

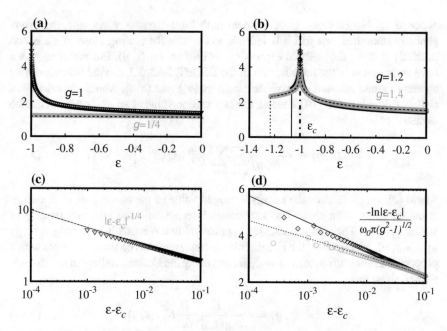

Fig. 3.12 Density of states of the QRM as a function of the dimensionless energy ε for different couplings, $g = 1/4$, 1 (**a**), 1.2 and 1.4 (**b**). The semiclassical density of states, $\nu_{\mathrm{scl}}(\varepsilon, g)$ is plotted with lines, while the points correspond to the quantum average density of states $\tilde{\nu}_q(\bar{\varepsilon}, g)$ for a H_{QRM} with $\Omega/\omega_0 = 10^3$ and $M = 10$ (see main text for further details). In **a** and **c** the divergence of the density of states is observed for $g = 1$ close to the ground-state energy, ε_c. For $g > 1$, ε_c no longer corresponds to the ground-state energy, and the density of states exhibits a logarithmic divergence, better illustrated in **d**. This feature locates the ESQPT

$$\nu_{\mathrm{scl}}(\varepsilon_c + \delta\varepsilon, g) = \frac{2}{\omega_0 \pi} \int_{x_1}^{x_m} dx \left(\frac{1}{\sqrt{\delta\varepsilon(g^2 - 1)x^2}} + O(x^4) \right) + K \qquad (3.80)$$

$$\approx \frac{1}{\omega_0 \pi \sqrt{g^2 - 1}} \ln \left(\frac{2x_m^2(g^2 - 1)}{\delta\varepsilon} \right) + K. \qquad (3.81)$$

Hence, $\nu_{\mathrm{scl}}(\varepsilon, g > 1)$ diverges at ε_c in logarithmic fashion, i.e., differently compared to the ground-state QPT case. In addition, we find the same behavior for $\varepsilon - \varepsilon_c \to 0^-$, and therefore we can write

$$\nu_{\mathrm{scl}}(\varepsilon, g > 1) \sim \frac{-\ln|\varepsilon - \varepsilon_c|}{\omega_0 \pi \sqrt{g^2 - 1}} \quad \text{for } |\varepsilon - \varepsilon_c| \ll 1, \qquad (3.82)$$

which is a clear indication of an ESQPT taking place in the QRM in the $\Omega/\omega_0 \to \infty$ limit. In Fig. 3.12b and d we plot $\nu_{\mathrm{scl}}(\varepsilon, g)$ computed from Eq. (3.75), together with the quantum averaged density of states, $\tilde{\nu}_q(\varepsilon, g)$, for $g = 1.2$ and $g = 1.4$. Indeed, both results nicely agree as well as confirm the presence of the divergence at ε_c. Nonetheless, since we deal with finite-frequency systems, the divergence is smoothed, but still it is feasible to observe the scaling close to ε_c, which precisely follows the analytic expression derived for $\nu_{\mathrm{scl}}(\varepsilon, g)$. Furthermore, as we have commented in the introduction of the ESQPT, Sect. 3.3, critical behaviors also appear in some relevant observables, such as $\langle \sigma_z \rangle$ and $\langle a^\dagger a \rangle$, which are calculated closely following Ref. [79]. Indeed, the expectation value of an observable $\mathcal{A}$ can be obtained as

$$\langle \mathcal{A} \rangle (\varepsilon, g) = \frac{1}{\nu_q(\varepsilon, g)} \sum_{k=0} \sum_{\alpha=\pm} \langle \varphi_k^\alpha | \mathcal{A} | \varphi_k^\alpha \rangle \delta \left[\varepsilon - \varepsilon_k^\alpha \right], \qquad (3.83)$$

where $\langle \mathcal{A} \rangle (\varepsilon, g)$ denotes the energy averaged value of the observable $\mathcal{A}$ at energy ε and coupling strength g. This expression is exact with the quantum density of states, $\nu_q(\varepsilon, g)$. Then, its semiclassical approximation consists in replacing $\nu_q(\varepsilon, g)$ by $\nu_{\mathrm{scl}}(\varepsilon, g)$. Moreover, if the Hamiltonian linearly depends on the observable with a proportional constant β, then $\mathcal{A} = \partial_\beta H$, and using the Hellmann–Feynman theorem, we can show that

$$\langle \mathcal{A} \rangle (\varepsilon, g) = -\frac{1}{\nu_{\mathrm{scl}}(\varepsilon, g)} \frac{\partial}{\partial \beta} N_{\mathrm{scl}}(\varepsilon, g) \qquad (3.84)$$

where $N_{\mathrm{scl}}(\varepsilon, g) = \int_{-\infty}^{\varepsilon} de \; \nu_{\mathrm{scl}}(e, g)$ is the accumulated number of states (see Appendix D for further details). Note that the dependence of $N_{\mathrm{scl}}(\varepsilon, g)$ on β is not explicitly written. In this manner, because $a^\dagger a = \partial_{\omega_0} H$ and $\sigma_z = 2\partial_\Omega H$, it follows

$$\langle a^\dagger a \rangle (\varepsilon, g) = -\frac{1}{\nu_{\mathrm{scl}}(\varepsilon, g)} \frac{\partial}{\partial \omega_0} N_{\mathrm{scl}}(\varepsilon, g) \qquad (3.85)$$

$$\langle \sigma_z \rangle (\varepsilon, g) = -\frac{2}{\nu_{\mathrm{scl}}(\varepsilon, g)} \frac{\partial}{\partial \Omega} N_{\mathrm{scl}}(\varepsilon, g), \qquad (3.86)$$

and therefore, a singular $\nu_{\mathrm{scl}}(\varepsilon, g)$ conveys singular behaviors in these mean-field expectation values. The theoretical prediction given by Eqs. (3.85) and (3.86) is plotted in Fig. 3.13, together with the numerically exact quantum expectation values for H_{QRM}, i.e., $\langle \varphi_k^\pm | a^\dagger a | \varphi_k^\pm \rangle$ and $\langle \varphi_k^\pm | \sigma_z | \varphi_k^\pm \rangle$. The good agreement between the semiclassical approximation and the quantum expectation observables for $\Omega/\omega_0 = 10^2$ and $\Omega/\omega_0 = 10^3$ can be seen in Fig. 3.13, which also unveil the presence of a sharp dip close to ε_c. It is worth noting that this singular behavior is present for any $g > 1$, in the same manner as in $\nu_{\mathrm{scl}}(\varepsilon_c, g)$.

Assuredly, an interesting feature of the ESQPT resides in that the eigenstates have a vanishing average qubit and boson population around ε_c for $g > 1$, that is, the eigenstates resemble a vacuum state, $|\phi_{0,\downarrow}\rangle = |0\rangle |\downarrow\rangle$. This fact can be understood as a consequence of wave-function localization around $x = 0$ [27, 87]. In particular, the overlap $P_{0,\downarrow}^{k,+} = |\langle \phi_{0,\downarrow} | \varphi_k^+ \rangle|^2$ reaches a maximum value for some k such that $\varepsilon_k^+ \approx \varepsilon_c$. Note that the spectrum of H_{QRM} can be labeled according to the parity symmetry, and

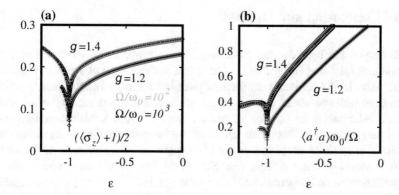

Fig. 3.13 Comparison between quantum and semiclassical results, obtained from Eqs. (3.85) and (3.86) for the boson number, $\langle a^\dagger a \rangle$ and $\langle \sigma_z \rangle$, respectively. In **a** we show $(\langle \sigma_z \rangle + 1)/2$ while in **b** the rescaled boson number $\langle a^\dagger a \rangle \omega_0 / \Omega$, for two coupling strengths, $g = 1.2$ and $g = 1.4$. The semiclassical result is depicted by a solid line, while the points correspond to the expectation value of the H_{QRM}, i.e., $|\varphi_k^{\pm}\rangle$ with energy $\varepsilon_k^{\pm}$; light gray for $\Omega/\omega_0 = 10^2$ and black for $\Omega/\omega_0 = 10^3$. Note that the points basically lie on top of the lines, i.e., on the semiclassical result

therefore $P_{0,\downarrow}^{k,-} \equiv 0$ because $|0\rangle\,|\downarrow\rangle$ belongs to the positive subspace. Nevertheless, considering $P_{1,\downarrow}^{k,-}$ one can draw the same conclusion for the negative parity subspace, i.e., $P_{1,\downarrow}^{k,-}$ becomes maximal for some k such that $\varepsilon_k^- \approx \varepsilon_c$. These overlaps are given in Fig. 3.14, for $g = 1.2$ and 1.4 with a frequency ratio $\Omega/\omega_0 = 10^3$. As one can observe, the eigenstates close to ε_c, where the ESQPT occurs, recall a vacuum state.

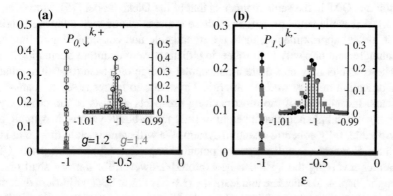

Fig. 3.14 Overlap probabilities $P_{0,\downarrow}^{k,+} = |\langle\downarrow|\langle 0|\,\varphi_k^+\rangle|^2$ (**a**) and $P_{1,\downarrow}^{k,-} = |\langle\downarrow|\langle 1|\,\varphi_k^-\rangle|^2$ (**b**) as a function of the eigenenergy ε for two different couplings, $g = 1.2$ and $g = 1.4$, plotted in black and gray, respectively. At the critical energy ε_c, the eigenstates show a maximal population of the vacuum state, $|0\rangle\,|\downarrow\rangle$ within the positive parity subspace, and $|1\rangle\,|\downarrow\rangle$ for the opposite. The insets show a zoom close to the critical energy ε_c

3.4   Conclusion and Outlook

This chapter has been devoted to demonstrate and support the existence of both a continuous QPT and ESQPT in the QRM in a suitable limit of system parameters, rather than in the conventional thermodynamic limit of infinitely many particles. Hence, in order to stress the difference, we call this novel manner of obtaining critical behaviors *finite-component* system phase transitions. Moreover, in spite of the different origin of this phase transition, we have shown that its signatures bear similarities to those of certain many-body quantum systems, such as Dicke and Lipkin–Meshkov–Glick models. The QRM can be then divided in two distinct phases, namely, normal and superradiant. Among the different signatures, we can highlight the diverging cavity field squeezing at the critical point, diverging number of bosonic excitation in the superradiant phase, as well as spontaneous symmetry breaking. Moreover, thanks to the derived analytical solution, we identified critical and finite-frequency scaling exponents, which coincide with those of the other models. Indeed, a closer inspection to the scaling functions reveals that the QPTs of the QRM and Dicke belong to the same universality class, i.e., they are of a mean field type. Therefore, the paradigmatic QRM, which consists of two constituent particles, qubit and harmonic oscillator, can display a superradiant QPT, as the Dicke model.

It is worth mentioning that the question of whether the critical behavior of Dicke and Lipkin–Meshkov–Glick models correspond actually to a QPT is under debate [75], as their $T = 0$ phase transitions can be captured by a mean-field analysis neglecting quantum fluctuations. However, these systems exhibit singularities in the ground state in agreement with continuous QPTs. In addition, we find the QRM that presents similar behavior as these totally connected models, as it shows the same type of critical phenomena. Hence, the phase transition in the QRM can be termed continuous QPT in the same manner as that of the Dicke model [75]. Moreover, we remark that while some properties of these systems can be well addressed under a semiclassical approximation, there are remarkable non-classical fingerprints, such as entanglement or cavity field squeezing, distinctive of quantum fluctuations.

These results may constitute a conceptual change on phase transitions, that is, a modification on how we look at critical systems, so far restricted to many-body systems. Indeed, one of the most exciting prospects consists in the capability of studying critical phenomena, inherent of truly many-body systems, in a small, thus controllable, fully coherent quantum system. We will come back to this point later, in Chap. 5, where we will propose a potential experimental platform where QRM dynamics involving the QPT can be explored. However, for that we shall first go through Chap. 4, to elucidate and learn the rich physics of nonequilibrium dynamics in this system. Therefore, we leave the outlook and future prospects on the dynamics and implementation of the QRM to the corresponding Chaps. 4 and 5.

Interestingly, the findings reported in this chapter may stimulate not only further inquiries into the QRM and the different footprints of criticality, but also at a deeper fundamental level, such as whether finite-component system QPTs represent a distinct manner of achieving phase transitions and how this approach reconciles with the conventional thermodynamic limit. This of a particular relevance in the Dicke model as it reduces simply to the QRM when only a single qubit is considered, so then, what does occur to the superradiant QPT of the Dicke model if the $\Omega/\omega_0 \to \infty$ limit is performed and vice versa, how does the superradiant QPT of the QRM break down when increasing the number of qubits? A conclusive answer to these issues may take advantage of the recent works on the existence of approximate integral of motion in the Dicke model [94].

In particular, one interesting direction of further research consists in finding whether quantum critical phenomena may appear in another models without scaling up the number of system components (i.e., without resorting to the conventional thermodynamic limit), and if so, determine under what conditions it occurs. In this regard, the work presented in [95] constitutes a big step forward in this direction. In Ref. [95] the concept of a *finite-component* system phase transition is extended to a Jaynes–Cummings lattice system, which presents distinct physics to that of the QRM, with different critical exponents. Therefore, it is certainly possible to observe rich phenomena of phase transitions without scaling up system size but rather tuning system parameters, as explained in this chapter for the QRM. In addition, it is worth mentioning that in [95] it is shown that a single Jaynes–Cummings model undergoes a QPT where, instead of a discrete parity as that of the QRM, a continuous symmetry is spontaneously broken, and consequently a Goldstone mode emerges. Hence, these two works demonstrate that *finite-component* system phase transitions are not restricted to a specific symmetry nor to a type of interaction.

Moreover, it has been recently shown that the QRM also features a phase transition when the bosonic mode undergoes dissipation, leading to different phases in terms of steady states [96]. As in the isolated case, this opens new avenues to investigate and explore dynamics of dissipative phase transitions in a small and fully controllable open quantum system, and at the same time, it demonstrates that finite-component system phase transitions do exist in the realm of open quantum systems. Among the multiple and interesting prospects triggered by these findings, one could mention the potential critical behavior of a QRM under different noise types [97], which in the Dicke model leads to the modification of critical exponents [98] and a rich steady-state phase diagram [99, 100]. That is, how the abrupt changes of the steady state depend on both amplitude and spectral noise properties, and whether the latter has an impact in the dissipative phase transition.

In short, we have unveiled a novel type of phase transitions that take place without scaling up the number of system constituents, and thus, it paves the way for the realization of critical models by means of small and fully controllable quantum systems. Therefore, phase transitions may be explored overcoming the complexity of coherently manipulating and storing an increasing number of quantum registers as required for conventional critical systems, and hence, opening the door for many intriguing possibilities.

References

1. M.O. Scully, M.S. Zubairy, *Quantum Optics* (Cambridge University Press, Cambridge, England, 1997)
2. M.A. Nielsen, I.L. Chuang, *Quantum Computation and Quantum Information* (Cambridge University Press, Cambridge, England, 2000)
3. I.I. Rabi, On the process of space quantization. Phys. Rev. **49**, 324 (1936). https://doi.org/10.1103/PhysRev.49.324
4. I.I. Rabi, Space quantization in a gyrating magnetic field. Phys. Rev. **51**, 652 (1937). https://doi.org/10.1103/PhysRev.51.652
5. I.I. Rabi, J.R. Zacharias, S. Millman, P. Kusch, A new method of measuring nuclear magnetic moment. Phys. Rev. **53**, 318 (1938). https://doi.org/10.1103/PhysRev.53.318
6. E.T. Jaynes, F.W. Cummings, Comparison of quantum and semiclassical radiation theories with application to the beam maser. Proc. IEEE **51**, 89 (1963). https://doi.org/10.1109/PROC.1963.1664
7. D. Braak, Q.-H. Chen, M.T. Batchelor, E. Solano, Semi-classical and quantum Rabi models: in celebration of 80 years. J. Phys. A Math. Theor. **49**, 300301 (2016), http://stacks.iop.org/1751-8121/49/i=30/a=300301
8. S. Haroche, Nobel lecture: controlling photons in a box and exploring the quantum to classical boundary. Rev. Mod. Phys. **85**, 1083 (2013). https://doi.org/10.1103/RevModPhys.85.1083
9. D.J. Wineland, Nobel lecture: superposition, entanglement, and raising Schrödinger's cat. Rev. Mod. Phys. **85**, 1103 (2013). https://doi.org/10.1103/RevModPhys.85.1103
10. S. Haroche, J.-M. Raimond, *Exploring the Quantum: Atoms, Cavities, and Photons* (Oxford University Press, Oxford, 2006)
11. G. Rempe, H. Walther, N. Klein, Observation of quantum collapse and revival in a one-atom maser. Phys. Rev. Lett. **58**, 353 (1987). https://doi.org/10.1103/PhysRevLett.58.353
12. C. Monroe, D.M. Meekhof, B.E. King, D.J. Wineland, A "Schrödinger cat" superposition state of an atom. Science **272**, 1131 (1996). https://doi.org/10.1126/science.272.5265.1131
13. S. Sachdev, *Quantum Phase Transitions*, 2nd edn. (Cambridge University Press, Cambridge, UK, 2011)
14. L.D. Carr (ed.), *Understanding Quantum Phase Transitions* (CRC Press, 2010)
15. M. Vojta, Quantum phase transitions. Rep. Prog. Phys. **66**, 2069 (2003), http://stacks.iop.org/0034-4885/66/i=12/a=R01
16. L.D. Landau, E.M. Lifshitz, *Statistical Physics*, 3rd edn. (Butterworth-Heinemann, Oxford, 1980)
17. H.E. Stanley, *Introduction to Phase Transitions and Critical Phenomena* (Clarendon Press, Oxford, 1971)
18. H.E. Stanley, Scaling, universality, and renormalization: three pillars of modern critical phenomena. Rev. Mod. Phys. **71**, S358 (1999). https://doi.org/10.1103/RevModPhys.71.S358
19. R.H. Dicke, Coherence in spontaneous radiation processes. Phys. Rev. **93**, 99 (1954). https://doi.org/10.1103/PhysRev.93.99
20. H. Lipkin, N. Meshkov, A. Glick, Validity of many-body approximation methods for a solvable model. Nucl. Phys. **62**, 188 (1965). https://doi.org/10.1016/0029-5582(65)90862-X
21. J.G. Brankov, *Introduction to Finite-Size Scaling* (Leuven University Press, Leuven, 1996)
22. M.E. Fisher, M.N. Barber, Scaling theory for finite-size effects in the critical region. Phys. Rev. Lett. **28**, 1516 (1972). https://doi.org/10.1103/PhysRevLett.28.1516
23. W.D. Heiss, M. Müller, Universal relationship between a quantum phase transition and instability points of classical systems. Phys. Rev. E **66**, 016217 (2002). https://doi.org/10.1103/PhysRevE.66.016217
24. F. Leyvraz, W.D. Heiss, Large-N scaling behavior of the Lipkin-Meshkov-Glick model. Phys. Rev. Lett. **95**, 050402 (2005). https://doi.org/10.1103/PhysRevLett.95.050402
25. P. Cejnar, M. Macek, S. Heinze, J. Jolie, J. Dobeš, Monodromy and excited-state quantum phase transitions in integrable systems: collective vibrations of nuclei. J. Phys. A: Math. Theor. **39**, L515 (2006), http://stacks.iop.org/0305-4470/39/i=31/a=L01

26. P. Cejnar, P. Stránský, Impact of quantum phase transitions on excited-level dynamics Phys. Rev. E **78**, 031130 (2008). https://doi.org/10.1103/PhysRevE.78.031130

27. M. Caprio, P. Cejnar, F. Iachello, Excited state quantum phase transitions in many-body systems. Ann. Phys. (N. Y.) **323**, 1106 (2008). https://doi.org/10.1016/j.aop.2007.06.011

28. P. Stránský, M. Macek, P. Cejnar, Excited-state quantum phase transitions in systems with two degrees of freedom: level density, level dynamics, thermal properties. Ann. Phys. (N. Y.) **345**, 73 (2014). https://doi.org/10.1016/j.aop.2014.03.006

29. P. Stránský, M. Macek, A. Leviatan, P. Cejnar, Excited-state quantum phase transitions in systems with two degrees of freedom: II. Finite-size effects. Ann. Phys. (N. Y.) **356**, 57 (2015). https://doi.org/10.1016/j.aop.2015.02.025

30. M.-J. Hwang, R. Puebla, M.B. Plenio, Quantum phase transition and universal dynamics in the Rabi model. Phys. Rev. Lett. **115**, 180404 (2015). https://doi.org/10.1103/PhysRevLett.115.180404

31. R. Puebla, M.-J. Hwang, M.B. Plenio, Excited-state quantum phase transition in the Rabi model. Phys. Rev. A **94**, 023835 (2016). https://doi.org/10.1103/PhysRevA.94.023835

32. C. Gerry, P. Knight, *Introductory Quantum Optics* (Cambridge University Press, Cambridge, England, 2004)

33. D.F. Walls, G.J. Milburn, *Quantum Optics*, 2nd edn. (Springer, Berlin, Heidelberg, 2008)

34. F. Beaudoin, J.M. Gambetta, A. Blais, Dissipation and ultrastrong coupling in circuit QED. Phys. Rev. A **84**, 043832 (2011). https://doi.org/10.1103/PhysRevA.84.043832

35. D.Z. Rossatto, C.J. Villas-Bôas, M. Sanz, E. Solano, Spectral classification of coupling regimes in the quantum Rabi model. Phys. Rev. A **96**, 013849 (2017). https://doi.org/10.1103/PhysRevA.96.013849

36. P. Forn-Díaz, J. Lisenfeld, D. Marcos, J.J. García-Ripoll, E. Solano, C.J.P.M. Harmans, J.E. Mooij, Observation of the Bloch-Siegert Shift in a qubit-oscillator system in the ultrastrong coupling regime. Phys. Rev. Lett. **105**, 237001 (2010). https://doi.org/10.1103/PhysRevLett.105.237001

37. T. Niemczyk, F. Deppe, H. Huebl, E.P. Menzel, F. Hocke, M.J. Schwarz, J.J. Garcia-Ripoll, D. Zueco, T. Hummer, E. Solano, A. Marx, R. Gross, Circuit quantum electrodynamics in the ultrastrong-coupling regime. Nat. Phys. **6**, 772 (2010). https://doi.org/10.1038/nphys1730

38. J. Casanova, G. Romero, I. Lizuain, J.J. García-Ripoll, E. Solano, Deep strong coupling regime of the Jaynes-Cummings model. Phys. Rev. Lett. **105**, 263603 (2010). https://doi.org/10.1103/PhysRevLett.105.263603

39. F. Yoshihara, T. Fuse, S. Ashhab, K. Kakuyanagi, S. Saito, K. Semba, Superconducting qubit-oscillator circuit beyond the ultrastrong-coupling regime. Nat. Phys. **13**, 44 (2017). https://doi.org/10.1038/nphys3906

40. P. Forn-Diaz, J.J. Garcia-Ripoll, B. Peropadre, J.-L. Orgiazzi, M.A. Yurtalan, R. Belyansky, C.M. Wilson, A. Lupascu, Ultrastrong coupling of a single artificial atom to an electromagnetic continuum in the nonperturbative regime. Nat. Phys. **13**, 39 (2017). https://doi.org/10.1038/nphys3905

41. D. Braak, Integrability of the Rabi model. Phys. Rev. Lett. **107**, 100401 (2011). https://doi.org/10.1103/PhysRevLett.107.100401

42. E. Solano, The dialogue between quantum light and matter. Physics **4**, 68 (2011). https://doi.org/10.1103/Physics.4.68

43. M.T. Batchelor, H.-Q. Zhou, Integrability versus exact solvability in the quantum Rabi and Dicke models. Phys. Rev. A **91**, 053808 (2015). https://doi.org/10.1103/PhysRevA.91.053808

44. A.J. Leggett, S. Chakravarty, A.T. Dorsey, M.P.A. Fisher, A. Garg, W. Zwerger, Dynamics of the dissipative two-state system. Rev. Mod. Phys. **59**, 1 (1987). https://doi.org/10.1103/RevModPhys.59.1

45. U. Weiss, *Quantum Dissipative Systems*, 3rd edn. (World Scientific, Singapore, 2008)

46. K. Hepp, E.H. Lieb, On the superradiant phase transition for molecules in a quantized radiation field: the dicke maser model. Ann. Phys. (N. Y.) **76**, 360 (1973). https://doi.org/10.1016/0003-4916(73)90039-0

47. Y.K. Wang, F.T. Hioe, Phase transition in the Dicke model of superradiance. Phys. Rev. A **7**, 831 (1973). https://doi.org/10.1103/PhysRevA.7.831

48. C. Emary, T. Brandes, Quantum chaos triggered by precursors of a quantum phase transition: the Dicke model. Phys. Rev. Lett. **90**, 044101 (2003a). https://doi.org/10.1103/PhysRevLett.90.044101

49. C. Emary, T. Brandes, Chaos and the quantum phase transition in the Dicke model. Phys. Rev. E **67**, 066203 (2003b). https://doi.org/10.1103/PhysRevE.67.066203

50. K.L. Hur, Entanglement entropy, decoherence, and quantum phase transitions of a dissipative two-level system. Ann. Phys. (N. Y.) **323**, 2208 (2008). https://doi.org/10.1016/j.aop.2007.12.003

51. G. Levine, V.N. Muthukumar, Entanglement of a qubit with a single oscillator mode. Phys. Rev. B **69**, 113203 (2004). https://doi.org/10.1103/PhysRevB.69.113203

52. A.P. Hines, C.M. Dawson, R.H. McKenzie, G.J. Milburn, Entanglement and bifurcations in Jahn-Teller models. Phys. Rev. A **70**, 022303 (2004). https://doi.org/10.1103/PhysRevA.70.022303

53. S. Ashhab, F. Nori, Qubit-oscillator systems in the ultrastrong-coupling regime and their potential for preparing nonclassical states. Phys. Rev. A **81**, 042311 (2010). https://doi.org/10.1103/PhysRevA.81.042311

54. L. Bakemeier, A. Alvermann, H. Fehske, Quantum phase transition in the Dicke model with critical and noncritical entanglement. Phys. Rev. A **85**, 043821 (2012). https://doi.org/10.1103/PhysRevA.85.043821

55. S. Ashhab, Superradiance transition in a system with a single qubit and a single oscillator. Phys. Rev. A **87**, 013826 (2013). https://doi.org/10.1103/PhysRevA.87.013826

56. J.R. Schrieffer, P.A. Wolff, Relation between the Anderson and Kondo Hamiltonians. Phys. Rev. **149**, 491 (1966). https://doi.org/10.1103/PhysRev.149.491

57. S. Bravyi, D.P. DiVincenzo, D. Loss, Schrieffer–Wolff transformation for quantum many-body systems. Ann. Phys. (N. Y.) **326**, 2793 (2011). https://doi.org/10.1016/j.aop.2011.06.004

58. T. Kato, *Perturbation Theory for Linear Operators* (Springer, Berlin, Heidelberg, 1980)

59. T.J. Osborne, M.A. Nielsen, Entanglement in a simple quantum phase transition. Phys. Rev. A **66**, 032110 (2002). https://doi.org/10.1103/PhysRevA.66.032110

60. G. Vidal, J.I. Latorre, E. Rico, A. Kitaev, Entanglement in quantum critical phenomena. Phys. Rev. Lett. **90**, 227902 (2003). https://doi.org/10.1103/PhysRevLett.90.227902

61. N. Lambert, C. Emary, T. Brandes, Entanglement and the phase transition in single-mode superradiance. Phys. Rev. Lett. **92**, 073602 (2004). https://doi.org/10.1103/PhysRevLett.92.073602

62. N. Lambert, C. Emary, T. Brandes, Entanglement and entropy in a spin-boson quantum phase transition. Phys. Rev. A **71**, 053804 (2005). https://doi.org/10.1103/PhysRevA.71.053804

63. J. Vidal, S. Dusuel, T. Barthel, Entanglement entropy in collective models. J. Stat. Mech. **2007**, P01015 (2007), http://stacks.iop.org/1742-5468/2007/i=01/a=P01015

64. L. Lepori, G. De Chiara, A. Sanpera, Scaling of the entanglement spectrum near quantum phase transitions. Phys. Rev. B **87**, 235107 (2013). https://doi.org/10.1103/PhysRevB.87.235107

65. P. Zanardi, N. Paunković, Ground state overlap and quantum phase transitions. Phys. Rev. E **74**, 031123 (2006). https://doi.org/10.1103/PhysRevE.74.031123

66. R.J. Baxter, *Exactly Solved Models in Statistical Mechanics* (Academic Press, London, 1989)

67. U.C. Taeuber, *Critical Dynamics: A Field Theory Approach to Equilibrium and Non-equilibrium Scaling Behavior* (Cambridge University Press, Cambridge, UK, 2014)

68. P. Ribeiro, J. Vidal, R. Mosseri, Thermodynamical limit of the Lipkin-Meshkov-Glick model. Phys. Rev. Lett. **99**, 050402 (2007). https://doi.org/10.1103/PhysRevLett.99.050402

69. P. Ribeiro, J. Vidal, R. Mosseri, Exact spectrum of the Lipkin-Meshkov-Glick model in the thermodynamic limit and finite-size corrections. Phys. Rev. E **78**, 021106 (2008). https://doi.org/10.1103/PhysRevE.78.021106

70. R. Botet, R. Jullien, P. Pfeuty, Size scaling for infinitely coordinated systems. Phys. Rev. Lett. **49**, 478 (1982). https://doi.org/10.1103/PhysRevLett.49.478
71. R. Botet, R. Jullien, Large-size critical behavior of infinitely coordinated systems. Phys. Rev. B **28**, 3955 (1983). https://doi.org/10.1103/PhysRevB.28.3955
72. S. Dusuel, J. Vidal, Finite-size scaling exponents of the Lipkin-Meshkov-Glick model. Phys. Rev. Lett. **93**, 237204 (2004). https://doi.org/10.1103/PhysRevLett.93.237204
73. S. Dusuel, J. Vidal, Continuous unitary transformations and finite-size scaling exponents in the Lipkin-Meshkov-Glick model. Phys. Rev. B **71**, 224420 (2005). https://doi.org/10.1103/PhysRevB.71.224420
74. J. Vidal, S. Dusuel, Finite-size scaling exponents in the Dicke model. Europhys. Lett. **74**, 817 (2006), http://stacks.iop.org/0295-5075/74/i=5/a=817
75. J. Larson, E.K. Irish, Some remarks on "superradiant" phase transitions in light-matter systems. J. Phys. A Math. Theo. **50**, 174002 (2017), http://stacks.iop.org/1751-8121/50/i=17/a=174002
76. A. Relaño, J.M. Arias, J. Dukelsky, J.E. García-Ramos, P. Pérez-Fernández, Decoherence as a signature of an excited-state quantum phase transition. Phys. Rev. A **78**, 060102 (2008). https://doi.org/10.1103/PhysRevA.78.060102
77. P. Pérez-Fernández, P. Cejnar, J.M. Arias, J. Dukelsky, J.E. García-Ramos, A. Relaño, Quantum quench influenced by an excited-state phase transition. Phys. Rev. A **83**, 033802 (2011). https://doi.org/10.1103/PhysRevA.83.033802
78. A. Relaño, J. Dukelsky, P. Pérez-Fernández, J.M. Arias, Quantum phase transitions of atom-molecule Bose mixtures in a double-well potential. Phys. Rev. E **90**, 042139 (2014). https://doi.org/10.1103/PhysRevE.90.042139
79. T. Brandes, Excited-state quantum phase transitions in Dicke superradiance models. Phys. Rev. E **88**, 032133 (2013). https://doi.org/10.1103/PhysRevE.88.032133
80. M.A. Bastarrachea-Magnani, S. Lerma-Hernández, J.G. Hirsch, Comparative quantum and semiclassical analysis of atom-field systems. I. Density of states and excited-state quantum phase transitions. Phys. Rev. A **89**, 032101 (2014). https://doi.org/10.1103/PhysRevA.89.032101
81. B. Dietz, F. Iachello, M. Miski-Oglu, N. Pietralla, A. Richter, L. von Smekal, J. Wambach, Lifshitz and excited-state quantum phase transitions in microwave Dirac billiards. Phys. Rev. B **88**, 104101 (2013). https://doi.org/10.1103/PhysRevB.88.104101
82. F. Iachello, B. Dietz, M. Miski-Oglu, A. Richter, Algebraic theory of crystal vibrations: singularities and zeros in vibrations of one- and two-dimensional lattices. Phys. Rev. B **91**, 214307 (2015). https://doi.org/10.1103/PhysRevB.91.214307
83. Z.-G. Yuan, P. Zhang, S.-S. Li, J. Jing, L.-B. Kong, Scaling of the Berry phase close to the excited-state quantum phase transition in the Lipkin model. Phys. Rev. A **85**, 044102 (2012). https://doi.org/10.1103/PhysRevA.85.044102
84. R. Puebla, A. Relaño, Non-thermal excited-state quantum phase transitions. Europhys. Lett. **104**, 50007 (2013), http://stacks.iop.org/0295-5075/104/i=5/a=50007
85. R. Puebla, A. Relaño, J. Retamosa, Excited-state phase transition leading to symmetry-breaking steady states in the Dicke model. Phys. Rev. A **87**, 023819 (2013). https://doi.org/10.1103/PhysRevA.87.023819
86. W. Kopylov, T. Brandes, Time delayed control of excited state quantum phase transitions in the Lipkin-Meshkov-Glick model. New J. Phys. **17**, 103031 (2015), http://stacks.iop.org/1367-2630/17/i=10/a=103031
87. L.F. Santos, F. Pérez-Bernal, Structure of eigenstates and quench dynamics at an excited-state quantum phase transition. Phys. Rev. A **92**, 050101 (2015). https://doi.org/10.1103/PhysRevA.92.050101
88. R. Puebla, A. Relaño, Irreversible processes without energy dissipation in an isolated Lipkin-Meshkov-Glick model. Phys. Rev. E **92**, 012101 (2015). https://doi.org/10.1103/PhysRevE.92.012101
89. G. Engelhardt, V.M. Bastidas, W. Kopylov, T. Brandes, Excited-state quantum phase transitions and periodic dynamics. Phys. Rev. A **91**, 013631 (2015). https://doi.org/10.1103/PhysRevA.91.013631

90. C.M. Lóbez, A. Relaño, Entropy, chaos, and excited-state quantum phase transitions in the Dicke model. Phys. Rev. E **94**, 012140 (2016). https://doi.org/10.1103/PhysRevE.94.012140

91. M.-J. Hwang, M.-S. Choi, Variational study of a two-level system coupled to a harmonic oscillator in an ultrastrong-coupling regime. Phys. Rev. A **82**, 025802 (2010). https://doi.org/10.1103/PhysRevA.82.025802

92. A. Messiah, *Quantum Mechanics* (Dover Publications, New York, 1961)

93. M. Brack, R.K. Bhaduri, *Semiclassical Physics* (Addison-Wesley, 1997)

94. A. Relaño, M.A. Bastarrachea-Magnani, S. Lerma-Hernández, Approximated integrability of the Dicke model. Europhys. Lett. **116**, 50005 (2016), http://stacks.iop.org/0295-5075/116/i=5/a=50005

95. M.-J. Hwang, M.B. Plenio, Quantum phase transition in the finite Jaynes-Cummings lattice systems. Phys. Rev. Lett. **117**, 123602 (2016). https://doi.org/10.1103/PhysRevLett.117.123602

96. M.-J. Hwang, P. Rabl, M.B. Plenio, Dissipative phase transition in the open quantum Rabi model. Phys. Rev. A **97**, 013825 (2018). https://doi.org/10.1103/PhysRevA.97.013825

97. E.G. Dalla Torre, E. Demler, T. Giamarchi, E. Altman, Quantum critical states and phase transitions in the presence of non-equilibrium noise. Nat. Phys. **6**, 806 (2010). https://doi.org/10.1038/nphys1754

98. D. Nagy, P. Domokos, Critical exponent of quantum phase transitions driven by colored noise. Phys. Rev. A **94**, 063862 (2016). https://doi.org/10.1103/PhysRevA.94.063862

99. S. Genway, W. Li, C. Ates, B.P. Lanyon, I. Lesanovsky, Generalized Dicke nonequilibrium dynamics in trapped ions. Phys. Rev. Lett. **112**, 023603 (2014). https://doi.org/10.1103/PhysRevLett.112.023603

100. P. Kirton, J. Keeling, Suppressing and restoring the Dicke superradiance transition by dephasing and decay. Phys. Rev. Lett. **118**, 123602 (2017). https://doi.org/10.1103/PhysRevLett.118.123602

Chapter 4
Quantum Rabi Model: Nonequilibrium

In the previous chapter we have demonstrated that the quantum Rabi model (QRM) undergoes a quantum phase transition (QPT) in a suitable parameter limit. Thanks to the derived effective description of the QRM in this limit, we have been able to identify critical exponents, scaling functions and the universality class to which its QPT belongs. In this respect, the scrutiny of the QRM carried out in Chap. 3 has been restricted to eigenstate properties, or in other words, the previous chapter has been devoted to investigate solely *equilibrium* features of this critical model. Unquestionably, equilibrium properties are essential to understand QPTs and ulti- mately, to interrogate about their universal properties. However, more intriguing aspects lie nowadays in the vast territory of nonequilibrium critical dynamics, where several questions remain to be disclosed [1–5]. Among them, the question to what extent universal equilibrium features of a system undergoing a QPT are inherited by nonequilibrium dynamics endures elusive, although significant efforts have been devoted to it [2, 6–13]. In this context we find the celebrated Kibble–Zurek (KZ) mechanism [14], which is the paradigmatic description of nonequilibrium physics involving phase transitions (see Introduction, Sect. 1.3 as well as Chap. 2 for a KZ analysis in the realm of classical continuous phase transitions). The KZ mechanism has become a milestone as it applies across a broad diversity of physical systems thanks to the universal features of continuous phase transitions [15–18], including as well QPTs [1, 7, 19–21]. Recall that its major achievement consists in the prediction of scaling laws for defect formation across a phase transition, which depend on the equilibrium critical exponents of the concerned phase transition.

Remarkably, although the KZ mechanism applies to QPTs [1, 19–21], prior to our work KZ arguments were considered to fail or not applicable when dealing with zero-dimensional models [11, 22]. We will make special emphasis on this particular subject throughout this chapter, even though different nonequilibrium scenarios are investigated. Indeed, while the strict $\Omega/\omega_0 \to \infty$ limit of the QRM is of interest in itself and allows us to gain valuable insights on the dynamics (Sect. 4.1), such as the appearance of KZ physics, how these particular nonequilibrium critical phenomena

© Springer Nature Switzerland AG 2018
R. Puebla, *Equilibrium and Nonequilibrium Aspects of Phase Transitions in Quantum Physics*, Springer Theses, https://doi.org/10.1007/978-3-030-00653-2_4

emerge from a QRM with finite, yet large, Ω/ω_0 value is a fascinating topic. This is precisely the subject of Sect. 4.2, where we extend the scope of finite-frequency scaling functions to nonequilibrium dynamics. Moreover, as in the equilibrium case, we show the equivalence between finite-frequency scaling functions of the QRM and those of the Dicke model in a nonequilibrium scenario, further supporting that both indeed belong to the same universality class, as stated in Chap. 3. Some of the results presented in this chapter have been published in [23, 24].

4.1 Nonequilibrium Dynamics and Kibble–Zurek Scaling Laws

In this part we study a simple time-dependent QRM, which among several sub-jects allows us to examine the KZ mechanism in this model. We recall that KZ arguments have been applied to QPTs with successful outcomes [1, 19–21], where quantum excitations play the role of long-lived topological defects in the traditional KZ mechanism [2, 8, 9]. Moreover, while in spatially extended quantum systems one can still define defects, see for example the paradigmatic Ising model (Sect. 1.2 and Chap. 6), in zero-dimensional models one must deal with quasiparticle excitations, as it is the case of the QRM. Furthermore, although KZ predictions can be applied to zero-dimensional quantum systems, numerical simulations have cast doubts on the validity of KZ arguments on these systems [11, 22]. In the following lines we present an exhaustive analysis of nonequilibrium dynamics following a quench towards the critical point which uncovers rich physics and reveals the soundness of KZ argu-ments. In particular, we shall consider a time-dependent Hamiltonian

$$H_{\text{QRM}}(t) = \omega_0 a^\dagger a + \frac{\Omega}{2}\sigma_z - \lambda(t)\sigma_x(a + a^\dagger), \qquad (4.1)$$

where $\lambda(t)$ depends linearly on time, unless stated otherwise, and the total time of the evolution or quench amounts to τ_Q. We start in Sect. 4.1.1 scrutinizing the $\Omega/\omega_0 \to \infty$ limit and considering as initial state the ground state of H_{QRM} at $\lambda(0)$. As we will show, KZ physics naturally arises in this situation. Then, in Sects. 4.1.2, 4.1.3 and 4.1.4, we extend the analysis to non-linear protocols, as well as to investigate the impact of different initial states and the persistence of universal scaling in sudden quenches, respectively. Finally, in Sect. 4.2 we address the impact of finite-frequency systems ($\Omega/\omega_0 < \infty$), that is, QRMs in which there is a precursor of the phase transition.

4.1.1 Infinite Limit: Kibble–Zurek Physics

In the strict $\Omega/\omega_0 \to \infty$ limit, the QRM Hamiltonian adopts a simple form, as we have shown in Chap. 3. Moreover, since we project the spin onto the low energy subspace, $\mathcal{H}_\downarrow$, the spin degree of freedom is not explicitly written. We first consider for simplicity a linear quench $g(t) = \lambda(t)/\lambda_c \propto t/\tau_Q$ as well as $0 \leq g(t) \leq g_c = 1$. Hence, although we do not traverse the critical point, this scenario is sufficient to analyze whether KZ scaling predictions apply to the critical dynamics of the QRM. In this manner, the considered nonequilibrium scenario reduces to the time-dependent Hamiltonian

$$H_{\mathrm{np}}(t) = \omega_0 a^\dagger a - \frac{\omega_0 g^2(t)}{4}\left(a + a^\dagger\right)^2 - \frac{\Omega}{2}. \tag{4.2}$$

We firstly take as initial state $|\psi(0)\rangle = |0\rangle$, which is the ground state of H_{np} at $g = 0$ and a quench $g(t) = g_f t/\tau_Q$ for $t \in [0, \tau_Q]$ with $0 \leq g_f \leq 1$. Recall that at the critical point $g_c = 1$, the QRM becomes gapless and the number of bosonic excitations becomes infinitely large, $\langle a^\dagger a \rangle \to \infty$, among other signatures. Therefore, one must examine carefully the dynamics of the system, as we will see later on. However, before discussing how to actually obtain $|\psi(t)\rangle$ or expectation values of observables under $H_{\mathrm{np}}(t)$, we review KZ arguments and how they apply to our problem.

The energy gap between different eigenstates of H_{QRM} vanishes in the $\Omega/\omega_0 \to \infty$ limit at $g_c = 1$ as $\epsilon_{\mathrm{np}}(g) \propto |g - g_c|^{z\nu}$ with $z\nu = 1/2$. In particular, the relevant energy gap for the dynamics turns out to be $\epsilon_{\mathrm{np}}(g) = 2\omega_0\sqrt{1 - g^2}$ since the first excited state cannot be accessed due to symmetry constraints. This energy gap conveys a time scale, $\tau = \epsilon_{\mathrm{np}}(g)^{-1}$, that sets the scale in which the system responses, and can be used for KZ arguments. On the other hand, the control parameter is linearly changed on time as $g(t) = g_f t/\tau_Q$, which in turn introduces another time scale. The main idea of KZ mechanism relies on introducing a dichotomy between adiabatic and impulsive dynamics. As sketched in Fig. 4.1, applying KZ arguments in terms of these time scales, namely, $\tau = (2\omega_0\sqrt{1 - g^2})^{-1}$ and the remaining time to reach

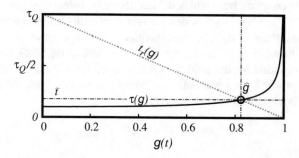

Fig. 4.1 Illustration of the boundary between adiabatic and impulsive dynamics as stated by KZ mechanism for the Rabi model. When $\tau(g) \ll t_r(g)$ the dynamics is essentially adiabatic; as the QPT is approached, $\tau(g) \gg t_r(g)$ and the dynamics becomes impulsive. In the figure $\tau(g)$ and $t_r(g)$ are depicted by black solid and gray dotted lines, respectively. The boundary $\hat{g}$ is obtained when both time scales match, which depends on the quench rate τ_Q^{-1}. If τ_Q is large enough, the boundary follows the scaling law $(1 - \hat{g}) \sim (\tau_Q\omega_0)^{-1/(z\nu+1)}$

the critical point $t_r = \tau_Q(1 - g)$, one can define the *freeze-out* instant or boundary between these two distinct dynamical regimes, adiabatic and impulsive, as $g(\hat{t}) \equiv \hat{g}$. That is, when both time scales match, KZ arguments predict that the state enters in the impulsive regime. In particular, when $1 - \hat{g} \ll 1 + \hat{g}$, i.e., $\hat{g}$ close to the critical point, then

$$1 - \hat{g} \approx \left(2^{3/2}\tau_q\omega_0\right)^{-1/(z\nu+1)}. \tag{4.3}$$

Additionally, the freeze-out time $\hat{t}$ associated with this coupling instant results in $\hat{t} = \tau_Q - \tau_Q^{z\nu/(z\nu+1)}/(2\omega_0^{2/3})$. Therefore, according to KZ arguments, the departure from equilibrium follows a universal power-law scaling characterized by the equilibrium critical exponents z and ν of the QPT. One should keep in mind however that such a freeze-out instant constitutes merely a working hypothesis, yet very helpful for deriving scaling laws. Evidently, a realistic system does not experience an abrupt transition between adiabatic and impulse dynamics at a particular instant, and therefore, these relations must be interpreted carefully. Nevertheless, one can test these first KZ scaling predictions by numerically solving the critical dynamics, and then estimating at what instant the system becomes less adiabatic. In this respect, although there is not a unique manner to quantify the loss of adiabaticity, we rely on the fidelity $F(t)$, which provides a good estimate [19]. This fidelity compares the evolved quantum state $|\psi(g(t))\rangle$ with the actual ground state at $g(t)$, $\left|\varphi_{\mathrm{np}}^0(g)\right\rangle = \mathcal{S}[r_{\mathrm{np}}(g)]|0\rangle$ (note that we have dropped out the spin degree of freedom as it remains in the $|\downarrow\rangle$ state), where $\mathcal{S}[x]$ represents the squeezing operator and $r_{\mathrm{np}}(g) = -1/4\ln(1 - g^2)$ corresponds to the squeezing parameter. That is, the fidelity reads

$$F(t) = \left|\left\langle\varphi_{\mathrm{np}}^0(g(t))\middle|\psi(t)\right\rangle\right|. \tag{4.4}$$

We estimate then the coupling at the freeze-out instant $\hat{g}$, for a particular driving time τ_Q, as the coupling from which the fidelity drops below a certain threshold f_k, $F(g(t) > \hat{g}) < f_k$. This also provides the estimator of the freeze-out time $\hat{t} = \tau_Q\hat{g}$. We remark that, since $H_{\mathrm{np}}(t)$ becomes singular at $g_c = 1$, the computation of $|\psi(t)\rangle$ is not straightforward. In particular, we obtain the fidelity, Eq. (4.4), solving the the Schrödinger equation for the unitary evolution operator. We refer to the interested readers to Appendix E for a detailed derivation, while here we quote the main results. The time-evolved state $|\psi(t)\rangle$ can be written as

$$|\psi(t)\rangle = e^{a(t)+i\Lambda}e^{b(t)a^{\dagger 2}}|0\rangle, \tag{4.5}$$

where Λ accounts for a constant factor, and thus $e^{i\Lambda}$ is just a global phase, and the coefficients $a(t)$ and $b(t)$ obey the following equation of motion

$$i\dot{a}(t) = -\frac{\omega_0 t^2}{2\tau_Q^2}b(t) \tag{4.6}$$

$$i\dot{b}(t) = -\frac{\omega_0 t^2}{4\tau_Q^2} + 2\omega_0\left(1 - \frac{t^2}{2\tau_Q^2}\right)b(t) - \frac{\omega_0 t^2}{\tau_Q^2}b^2(t), \tag{4.7}$$

whose initial conditions are $a(0) = b(0) = 0$ such that $|\psi(0)\rangle = |0\rangle$. Finally, since $|\psi(t)\rangle = \sum_{n=0}^{\infty}\alpha_n(t)\left|\varphi_{np}^n(g(t))\right\rangle$, the fidelity directly follows $F(g(t)) = |\alpha_0(t)|$, which turns out to be

$$F(t) = |\alpha_0(t)| = \frac{e^{\text{Re}[a(t)]}}{\left|(\cosh r_{np}(t)\left|\left(1 - 2e^{i\theta(t)}\right)\right|\,|b(t)|\tanh r_{np}(t))\right|^{1/2}}, \tag{4.8}$$

where $\theta(t) = \arctan(\text{Im}[b(t)]/\text{Re}[b(t)])$. Solving numerically the equations of motion for $a(t)$ and $b(t)$ with a particular driving time τ_Q, we determine $F(g(t))$ according to Eq. (4.8). In Fig. 4.2a we show the time-evolution of the fidelity $F(t)$ for three different quench rates τ_Q, from which we can obtain estimators of the freeze-out instant. For that, selecting a value f_k, an estimated $\hat{g}$ follows, as depicted in Fig. 4.2a. For the chosen f_k, we plot $1 - \hat{g}$ and $\tau_Q - \hat{t}$ as a function of τ_Q to corroborate the KZ scaling predictions. As we observe in Fig. 4.2b, independently of the f_k value, the scaling perfectly agrees with its KZ prediction for large enough τ_Q, namely, $1 - \hat{g} \sim \tau_Q^{-1/(z\nu+1)} = \tau_Q^{-2/3}$ and $\tau_Q - \hat{t} \sim \tau_Q^{z\nu/(z\nu+1)} = \tau_Q^{1/3}$. To the contrary, as τ_Q gets shorter, the scaling breaks down as the dynamics becomes sudden where KZ arguments are not longer valid, $\hat{g} \sim$ const and $\tau_Q - \hat{t} \approx \tau_Q$.

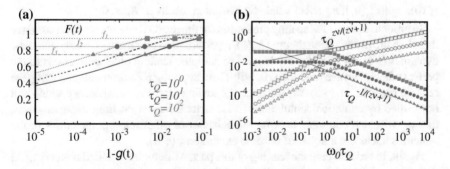

Fig. 4.2 First test of KZ scaling predictions. In **a** the time evolution of the fidelity $F(t)$ as a function of $1 - g(t)$ is plotted, considering three different rates, i.e., three τ_Q values. An estimation of the freeze-out instant $\hat{g}$ is obtained when $F(t)$ drops below a certain value f_k. Here, we chose $f_1 = 0.95$ (red), $f_2 = 0.85$ (blue) and $f_3 = 0.75$ (green), which allow us to obtain the scaling of $1 - \hat{g}$ (full points) and $\tau_Q - \hat{t}$ (open points) plotted in **b**. We confirm the KZ scaling prediction $1 - \hat{g} \sim \tau_Q^{-1/(z\nu+1)}$ and $\tau_Q - \hat{t} \sim \tau_Q^{z\nu/(z\nu+1)}$ for large τ_Q, with $z\nu = 1/2$. Note that for $\tau_Q \to 0$, KZ arguments are not valid and the dynamics results in a sudden quench, $\hat{g} \sim$ const while $\tau_Q - \hat{t} \sim \tau_Q$. The dotted lines in **b** are simply guides to the eyes following the power-laws $\tau_Q^{1/3}$ and $\tau_Q^{-2/3}$

This elementary inspection on the critical dynamics of the QRM has already served us to verify the first KZ scaling prediction. Indeed, while the distinction between adiabatic and impulsive dynamics is somehow arbitrary, the scaling law of such a frontier between these dynamical regimes (see Eq. (4.3)) plays a crucial role in subsequent predictions. Moreover, the previous analysis grants the freeze-out argument a physical meaning, which corroborates the intuitive and expected behavior of the dynamics, namely, the slower the quench, the more adiabatic the evolution. However, due to the zero energy gap at the critical point, the departure from equilibrium is unavoidable, although the coupling at which adiabatic dynamics breaks down approaches the critical point as the quench gets slower, i.e., $\hat{g} \to g_c$ as $\omega_0 \tau_Q \to \infty$.

The emblematic hallmark of the KZ mechanism resides, however, in the prediction that the formed excitations, created as a consequence of a phase transition, follow a universal scaling in terms of the quench rate. In the quantum realm one translates the arguments to quantum excitations. As in classical systems, one must quantify these excitations in a suitable manner, specifically for each system (see for example Introduction, Sect. 1.3, or Chap. 6, for a discussion on the Ising model). Since the QRM lacks spatial dimension, it is convenient to resort to the residual energy $E_r(\tau_Q)$ that quantifies the amount of quantum excitations formed during a ramp towards the critical point, providing a good measure of the adiabaticity of the evolution [7, 10, 11, 22, 25]. In our case, we rely on the residual energy at the end of the quench, which can be defined as

$$E_r(g(\tau_Q)) = \langle \psi(\tau_Q) | H_{np}(\tau_Q) | \psi(\tau_Q) \rangle - E_{GS}(g(\tau_Q)), \qquad (4.9)$$

that is, it gives account of the excess of energy with respect to the ground state. Since $|\psi(0)\rangle = \left| \varphi_{np}^0(0) \right\rangle$, a perfect adiabatic evolution results in $E_r \equiv 0$.

In order to derive a KZ scaling prediction for the residual energy, we first introduce the adiabatic perturbation theory (APT), a perturbation treatment of nearly adiabatic dynamics in terms of τ_Q^{-1}, very helpful to compute leading order corrections to the perfect adiabatic wave function [10, 26]. Then, applying KZ arguments and using the relation given in Eq. (4.3), we attain the KZ scaling for the residual energy which will be verified by numerical simulations in the strict $\Omega/\omega_0 \to \infty$ limit. Note however, that we will also comment later on the behavior of other relevant quantities of the system, such as the number of bosonic excitations, $\langle a^\dagger a \rangle$.

Again, in order to ease the reading of this part, we defer the detailed mathematical derivation of the APT to Appendix E, while quoting here the main outcomes. The time-evolved wavefunction obeying the Schrödinger equation can be written in terms of the instantaneous eigenstates of $H_{np}(t)$ as $|\psi(t)\rangle = \sum_n \alpha_n(t) e^{-i\Theta_n(t)} \left| \varphi_{np}^n(g(t)) \right\rangle$ where $\Theta_n(t) = \int_0^t \epsilon_n(t')dt'$ with $\epsilon_n(t) = n\omega_0\sqrt{1 - g^2(t)}$ the corresponding energy of the nth eigenstate. Then, the formal solution to the Schrödinger equation can be written as

$$\alpha_n(g) = -\sum_m \int_0^g dg' \alpha_m(g') \left\langle \varphi_{np}^n(g') \middle| \partial_{g'} \middle| \varphi_{np}^m(g') \right\rangle e^{i(\Theta_n(g')-\Theta_0(g'))}. \quad (4.10)$$

The previous expression simplifies significantly considering first $\alpha_0(0) = 1$ and second, slow quenches $\dot{g} \propto \tau_Q^{-1} \ll 1$. Indeed, substituting the specific form of the eigenstates, we find that, at leading order in τ_Q^{-1}, $\alpha_2(g)$ turns out to be the unique non-zero coefficient which reads

$$\alpha_2(g) \approx \frac{-i\dot{g}g}{2^{5/2}\omega_0 (1 - g^2)^{z\nu+1}} e^{i(\Theta_2(g)-\Theta_0(g))} + \mathcal{O}(\dot{g}^2). \quad (4.11)$$

Then, the residual energy E_r given in Eq. (4.19) can be recast as $E_r(g_f) = \sum_{n>0} \epsilon_n(g_f) |\alpha_n(g_f)|^2$ and thus, it can be approximated as $E_r(g_f) \approx \epsilon_2(g_f) |\alpha_2(g_f)|^2$ for nearly adiabatic dynamics. Therefore, by means of APT, assuming $\dot{g} \ll 1$ (or equivalently $\omega_0 \tau_Q \gg 1$), we obtain

$$E_r(g_f) \approx \tau_Q^{-2} \frac{g_f^4}{16\omega_0(1 - g_f^2)^{z\nu+2}}. \quad (4.12)$$

It is worth mentioning that $E_r \sim \tau_Q^{-2}$ corresponds to the standard scaling under APT [2]. However, the APT fails as soon as the quench involves the quantum critical point, that is, when $g_f \to 1$ since the adiabatic limit breaks down [7]. Hence, we need to resort to a different approach in this situation. Not surprisingly, KZ mechanism offers an answer. Specifically, making use of the simple dichotomy between adiabatic and impulsive dynamics, KZ mechanism states that, once the freeze-out instant is reached, the dynamics remains frozen as the system cannot react to external changes. Accordingly, KZ arguments predict that the residual energy at the critical point scales as

$$E_r(g_c) \approx E_r(\hat{g}) \sim \tau_Q^{-z\nu/(z\nu+1)} = \tau_Q^{-1/3}, \quad (4.13)$$

where we have used $1 - \hat{g} \sim \tau_Q^{-1/(z\nu+1)}$, derived in Eq. (4.3). The scaling of $\alpha_2(g)$ deserves special attention since $\alpha_2(\hat{g}) \sim \tau_Q^0$, in agreement with previous works for a zero-dimensional system [7–10]. Loosely speaking, the KZ scaling for a quantity that behaves as $\mathcal{A} \propto |g - g_c|^{\gamma_A}$ close to the critical point turns out to be $\mathcal{A} \sim \tau_Q^{-\gamma_A/(z\nu+1)}$, while in general for a d-dimensional model it becomes $\tau_Q^{-(\gamma_A+d\nu)/(z\nu+1)}$ [10]. At this stage it is important to remark that these scaling relations have been also predicted for other zero-dimensional systems, as Dicke or Lipkin–Meshkov–Glick models. Nevertheless, despite the considerable numerical efforts, these power-law scalings remained elusive, casting doubts on the validity of KZ arguments in these systems [11, 22]. Moreover, we recall that the KZ scaling prediction is relevant only if $\gamma_A/(z\nu + 1) < 2$, as otherwise becomes sub-leading and the standard τ_Q^{-2} scaling dominates. Indeed, in spatially extended systems with a dimension d above the so-called upper critical dimension, $d_u = z + 2/\nu$, the KZ scaling fades away [10, 25].[1] This, however, is not relevant for low-dimensional systems, as it is the case for the QRM, which features $z\nu = 1/2$ and therefore $z\nu/(z\nu + 1) = 1/3$.

In addition, as we have commented previously, the KZ universal scaling emerges in distinct relevant observables of the system, not only on the residual energy. As the spin degree of freedom is frozen in the lower subspace $|\downarrow\rangle$, we must handle bosonic observables that behave singularly, such as the number of bosonic excitations $\langle a^\dagger a \rangle$ and its squeezing, namely, $\Delta x = \sqrt{\langle x^2 \rangle - \langle x \rangle^2}$ with $x = a + a^\dagger$. Both observables show a singular behavior close to the critical point, see Chap. 3, Sect. 3.2, with their corresponding critical exponents $\gamma_{\Delta x} = -1/4$ and $\gamma_{n_{exc}} = 1$. Note however that the latter was defined as $n_{exc} \equiv \lim_{\Omega/\omega_0 \to \infty} \langle a^\dagger a \rangle \frac{\omega_0}{\Omega}$ in order to capture the superradiant nature of the QPT, since $\langle a^\dagger a \rangle \propto \Omega/\omega_0$ for $g > 1$. Here we focus on $\langle a^\dagger a \rangle$ in the normal phase, that does diverge as $g \to 1$ because the eigenstates consist of infinitely squeezed Fock states at the critical point, $\left| \varphi_{np}^n(g) \right\rangle = S[r_{np}(g)] |n\rangle$ with $r_{np}(g) = -1/4 \ln(1 - g^2)$. A simple calculation leads to $\left\langle \varphi_{np}^n(g) \right| a^\dagger a \left| \varphi_{np}^n(g) \right\rangle = \sinh^2(r_{np}(g)) + n \cosh(2r_{np}(g))$, which unveils that in the vicinity of the critical point $\left\langle \varphi_{np}^0(g) \right| a^\dagger a \left| \varphi_{np}^0(g) \right\rangle \propto |g - g_c|^{-1/2}$ and therefore $\gamma_{a^\dagger a} = -1/2$. Thus, for nearly adiabatic dynamics reaching the critical point we expect

$$\left\langle \psi(\tau_Q) \right| a^\dagger a \left| \psi(\tau_Q) \right\rangle \sim \tau_Q^{-\gamma_{a^\dagger a}/(z\nu+1)} = \tau_Q^{1/3}, \tag{4.14}$$

$$\Delta x(\tau_Q) \sim \tau_Q^{-\gamma_{\Delta x}/(z\nu+1)} = \tau_Q^{1/6}. \tag{4.15}$$

We now tackle the numerical confirmation of Eqs. (4.13), (4.14) and (4.15). It is then convenient to solve the unitary dynamics in the Heisenberg picture. Since the Hamiltonian in the $\Omega/\omega_0 \to \infty$ limit and for $0 \le g \le 1$ reads

$$H_{np}(g(t)) = \omega_0 a^\dagger a - \frac{\omega_0 g^2(t)}{4} \left(a + a^\dagger \right)^2 - \frac{\Omega}{2}, \tag{4.16}$$

the equation of motion is simply given by $i\dot{a}_H(t) = \left[a_H(t), H_{np,H}(t) \right]$ where the subscript H indicates operators in the Heisenberg picture, and $a_H(t) = u(t)a + v^*(t)a^\dagger$ with $u(0) = 1$ and $v(0) = 0$. The equality $|u(t)|^2 - |v(t)|^2 = 1$ guarantees that the commutation relation $\left[a_H(t), a_H^\dagger(t) \right] = 1$ is fulfilled. In this manner, the critical dynamics is attained after solving (see Appendix E)

---

[1] In this context, the upper critical dimension is defined as the dimension for which KZ scaling matches the standard APT scaling, τ_Q^{-2}, although it depends on the specific considered observable. Since the scaling of the residual energy for a non-zero dimensional system is $E_r \tau_Q^{-(d+z)\nu/(z\nu+1)}$, it follows $d_u = z + 2/\nu$. However, for the number of excitations, $n_{ex} \sim \tau_Q^{-d\nu/(z\nu+1)}$, one obtains a different upper critical dimension $\tilde{d}_u = 2z + 2/\nu$.

$$\frac{d}{dt}u(t) = -i\omega_0 \left[\left(1 - \frac{g^2(t)}{2}\right) u(t) - \frac{g^2(t)}{2} v(t) \right], \tag{4.17}$$

$$\frac{d}{dt}v(t) = i\omega_0 \left[\left(1 - \frac{g^2(t)}{2}\right) v(t) - \frac{g^2(t)}{2} u(t) \right], \tag{4.18}$$

while the residual energy can be computed as

$$E_r(g) = \langle 0| H_{\mathrm{np},H}(g) |0\rangle - E_{GS}(g)$$
$$= \omega_0 |v(t)|^2 - \frac{\omega_0 g^2(t)}{4} |u(t) + v(t)|^2 - \frac{\epsilon_{\mathrm{np}}(g) - \omega_0}{2}, \tag{4.19}$$

and the number of bosonic excitations results simply in $\langle \psi(\tau_Q)| a^\dagger a |\psi(\tau_Q)\rangle = \langle 0| a_H^\dagger(t) a_H(t) |0\rangle = |v(t)|^2$, while $\Delta x(t) = \sqrt{1 + 2|v(t)|^2 + u(t)v^*(t) + u^*(t)v(t)}$. In Fig. 4.3 the residual energy E_r is plotted as a function of the quench time, which precisely follows KZ scaling prediction when the critical point is reached, $\tau_Q^{-z\nu/(z\nu+1)} = \tau_Q^{-1/3}$. A fit to a power law function τ_Q^β in the range $\omega_0 \tau_Q \in [10^3, 10^6]$ provides $\beta = -0.333(1)$, in excellent agreement with the KZ prediction. On the contrary, when $g < g_c$ the energy gap opens, i.e., it becomes non zero and the APT can be successfully applied. As expected, for $\tau_Q \to 0$ the scaling breaks down as a consequence of the loss of adiabaticity, and E_r saturates. For quenches ending close to the critical point there is a crossover between both scaling laws, namely from $\tau_Q^{-z\nu/(z\nu+1)}$ to τ_Q^{-2}, as explicitly shown for $g = 0.999$ in Fig. 4.3. In these situations,

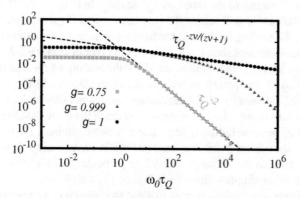

Fig. 4.3 Nonequilibrium scaling of the residual energy E_r for three remarkable cases, namely, at $g = 0.75 < g_c$ where the APT is applicable (light-gray squares), at $g = g_c = 1$ where APT fails and KZ mechanism arguments successfully explain the universal scaling law (black circles) and also for a quench ending close to the critical point, $g = 0.999$ (gray triangles), which show the crossover between universal KZ and standard APT scaling. The residual energy E_r is calculated from Eq. (4.19) solving the equations of motion, (4.17) and (4.18). For $g < g_c$, we observe the standard APT scaling, τ_Q^{-2} while when the critical point is reached, KZ scaling law emerges, $\tau_Q^{-z\nu/(z\nu+1)}$. The dashed black lines are guided to the eyes with the corresponding power-law scaling. Note that KZ scaling holds for slow quenches, $\omega_0 \tau_Q \gtrsim 10^0$, breaking down as τ_Q becomes shorter since the adiabatic condition is not satisfied, which corresponds to the limit of sudden quenches

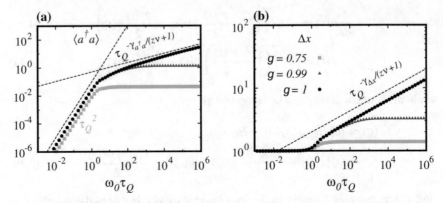

Fig. 4.4 Nonequilibrium scaling for the number of excitations $\langle a^\dagger a \rangle$ (**a**) and the variance Δx (**b**) at the end of the quench as a function of the total time τ_Q. As in Fig. 4.3, we consider three representative cases, namely $g = 0.75$ (light-gray squares), $g = 0.99$ (gray triangles) and $g = 1$ (black circles). For sufficiently slow quenches, the predicted KZ scaling holds for the quench ending at the critical point, $\langle \psi(\tau_Q) | a^\dagger a | \psi(\tau_Q) \rangle \sim \tau_Q^{-\gamma_{a^\dagger a}/(z\nu+1)} = \tau_Q^{1/3}$ and $\Delta x \sim \tau_Q^{-\gamma_{\Delta x}/(z\nu+1)} = \tau_Q^{1/6}$. For quenches ending at $g < g_c$, these quantities saturate once their corresponding value at g is attained, making it possible to observe traces of KZ scaling, if g is close enough to g_c. For fast quenches, $\langle a^\dagger a \rangle$ increases quadratically with time and the dynamics becomes nearly adiabatic. The dashed black lines are guides to the eyes with the corresponding power-law scaling

the eigenstates are close enough to be considered gapless for certain quench rates, while slower quenches will eventually resolve the small non-zero energy gap, and thus, shift the dynamics to the standard τ_Q^{-2} scaling. In terms of the KZ mechanism, the loss of adiabaticity will follow the universal law as if g_f were the strict critical point. Hence, KZ scaling holds until the freeze-out instant $\hat{g}$ surpasses the actual g_f, after which the dynamics abandon KZ physics in favor of nearly adiabatic dynamics well described by APT. In addition, we show the scaling of the number of excitations at the end of the quench as well as the squeezing Δx in Fig. 4.4, where we further support the validity of KZ arguments for distinct observables. For $g_f < g_c$ the scaling breaks down when these quantities reach the corresponding value at g_f, while at g_c they become infinitely large and it is never attained. Depending on how close g_f is to the critical point, the universal scaling may emerge before saturating to a specific value, as shown for $g_f = 0.99$. In particular for $\langle a^\dagger a \rangle$, the scaling τ_Q^2 for very rapid quenches simply follows from Eqs. (4.17) and (4.18) since $v(\tau_Q) \propto \tau_Q$ for $\omega_0 \tau_Q \ll 1$. These results demonstrate that KZ mechanism can be applied to critical dynamics even in a zero-dimensional system as it is the case of the QRM. In the following we provide an analysis of distinct scenarios, such as nonlinear ramps and the role of different initial states. The impact of finite-frequency systems is presented in Sect. 4.2.

4.1.2 Nonlinear Protocols

Nonlinear protocols are perhaps the most natural extension to the linear ramp towards the critical point [10, 25, 27]. Yet, the scaling relations are modified in a non-trivial fashion depending on the nonlinear coefficient $r > 0$ of the protocol $g(t)$, which reads[2]

$$g(t) = g_f \left(1 - \left(1 - \frac{t}{\tau_Q} \right)^r \right). \tag{4.20}$$

Naturally, for $r = 1$ the linear ramp studied previously is retrieved (see Fig. 4.5a). In addition to being an interesting theoretical extension, nonlinear protocols might be relevant in certain experimental platforms and also they might be of help to determine critical exponents of the system, although one must have prior knowledge of the location of the critical point. Note however that, although the functional form of $g(t)$ can be arbitrary, the procedure we present in the following lines can be applied unequivocally for any other nonlinear protocols, such as sinusoidal or hyperbolic functions [12]. Again, we take as initial coupling $g(0) = 0$ with $|\psi(0)\rangle = \left| \varphi_{np}^0(0) \right\rangle$ and $0 \leq g_f \leq g_c$. The main difference with respect to the linear case lies in that the rate $\dot{g}$ is not longer constant and depends on t,

$$\dot{g}(t) = \frac{g_f}{\tau_Q} r \left(1 - \frac{t}{\tau_Q} \right)^{r-1}. \tag{4.21}$$

As we are interested in nearly adiabatic quenches, as in the case of the linear ramp, we first obtain the leading order correction in τ_Q^{-1} by means of the APT, which becomes (see Appendix E for further details on the APT)

$$|\alpha_2(g)|^2 \approx \frac{g^2 g_f^2 r^2}{\tau_Q^2} \left(\frac{g_f}{g_f - g} \right)^{(2-2r)/r} \frac{1}{32\omega_0^2 (1 - g^2)^{2z\nu+2}}, \tag{4.22}$$

and allows us to estimate the residual energy as

$$E_r(g) \approx \frac{g^2 g_f^2 r^2}{\tau_Q^{2r}} \frac{(\tau_Q - t)^{2r-2}}{16\omega_0 (1 - g^2)^{z\nu+2}}. \tag{4.23}$$

Remarkably, as long as $g_f < g_c$, APT predicts $E_r \sim \tau_Q^{-2r}$. Recall that a KZ scaling prediction of the E_r crucially relies on the scaling between adiabatic and impulsive dynamics, i.e., on the scaling of $\hat{g}$. Although one can still work out the location of $\hat{g}$ when the two relevant time scales of the problem coincide, the fact that $\dot{g}(t)$ actually

---

[2]The form of the function $g(t)$ has been chosen by analogy with previous works, where $g(t) = (t/\tau_Q)^r$ but with $g(0) = 0$ the critical point [10, 25], whose equivalent nonlinear quench for $g(\tau_Q) = g_c = 1$ corresponds to that given in Eq. (4.20).

diverges at g_c for $r < 1$ and vanishes for $r > 1$ hinders this procedure.[3] Hence, we estimate the freeze-out instant as the coupling constant $\hat{g}$ at which the probability of excitation becomes of the order $\mathcal{O}(1)$, or in other words, when APT breaks down. Hence, considering $g_f = 1$ and from $\sum_{n>0} |\alpha_n(\hat{g})|^2 = 1$, we obtain

$$1 - \hat{g} \approx \left(\frac{16\tau_Q \omega_0}{r} \right)^{-r/(z\nu r+1)} , \tag{4.24}$$

where we can already observe that the nonlinear coefficient r modifies the exponent in a non-trivial fashion. We emphasize that, although for $r = 1$ the prefactor differs from the relation obtained in Eq. (4.3), we are merely interested in the scaling, that is, in how $\hat{g}$ scales with τ_Q. Then, applying KZ arguments once again, $E_r(g_c) \approx E_r(\hat{g})$, we predict

$$E_r(g_c) \approx 2\sqrt{2}\omega_0 \left(\frac{16\tau_Q \omega_0}{r} \right)^{-z\nu r/(z\nu r+1)} \sim \tau_Q^{-z\nu r/(z\nu r+1)} . \tag{4.25}$$

In the same manner, we obtain $\langle \psi(\tau_Q) | a^\dagger a | \psi(\tau_Q) \rangle \sim \tau_Q^{-\gamma_{a^\dagger a} r/(z\nu r+1)}$ and $\Delta x(\tau_Q) \sim \tau_Q^{-\gamma_{\Delta x} r/(z\nu r+1)}$ once the quench reaches the critical point. We confirm these scaling predictions by numerically solving the critical dynamics, although we only show explicitly the scaling of the residual energy E_r in Fig. 4.5b, c. Note that Eq. (4.25) overestimates the actual E_r by a factor $\mathcal{O}(1)$, as it also occurs in other systems applying KZ arguments [1, 14, 19]. However, the scaling exponent follows accurately the KZ prediction. Indeed, a fit of E_r to a power-law τ_Q^μ when $g_f = g_c$ for different r matches with $\mu = -z\nu r/(z\nu r + 1)$, see Fig. 4.5d.

4.1.3 Thermal States

The nonequilibrium scaling laws derived previously strongly rely on a ground state as initial state, $|\psi(0)\rangle = \left| \varphi_{\mathrm{np}}^0(g = 0) \right\rangle$. Therefore, it is pertinent to analyze whether these universal scaling laws survive when such a perfect ground state cannot be prepared, for example due to thermal effects. It is worth stressing that we only analyze the role of a thermal initial state, while it is assumed that thermal effects are irrelevant during the quench. Note that, since equilibrium properties even at finite temperature are dictated by the QPT in the surroundings of the critical point [28], nonequilibrium dynamics with thermal states may also follow the ground-state dynamics previously studied. In this respect, it is worth noting the works [8, 9] (see also the various

---

[3]One could nevertheless assume that the relevant time scale near g_c corresponds to the lowest non-vanishing (or non-diverging) time derivative of $g(t)$ at g_c, i.e., $d^m/dt^m g(t)|_{t=\tau_Q} \neq 0$ with $d^n/dt^n g(t)|_{t=\tau_Q} = 0$ for $n < m$. Although this argument provides the correct scaling for an integer r we resort here to a more transparent procedure.

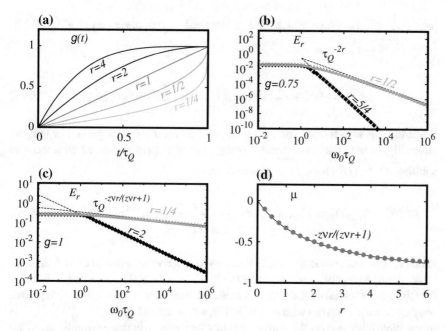

Fig. 4.5 Nonlinear quenches towards the critical point, as given in Eq. (4.20), for different r coefficients (**a**). The residual energy E_r scaling for nearly adiabatic quenches is modified: while for $g_f = 0.75 < g_c$ APT predicts τ_Q^{-2r} (**b**), at the critical point KZ arguments lead to $\tau_Q^{-z\nu r/(z\nu r+1)}$ (**c**), which perfectly agree with the numerical results. Moreover, a fit of E_r when $g_f = g_c$ to a power-law function τ_Q^{μ} in the region $\omega_0\tau_Q \in [10^1, 10^6]$ for different r agrees with the expected scaling $-z\nu r/(z\nu r+1)$, as shown in **d**

works within [10]), where the effect of finite temperature on the scaling laws was investigated, finding a modification of the scaling depending on the nature of the quasiparticles, namely, whether they obey fermion or boson particle statistics.

For that purpose it is convenient to consider first an nth eigenstate of $H_{\rm np}$ at $0 \leq g_0 < 1$ with $n \geq 0$, that is $|\psi(0)\rangle = \left|\varphi_{\rm np}^n(g_0)\right\rangle$. We denote by $E_r(n, g(t))$ the residual energy starting with the nth eigenstate, and due to the harmonic nature of the spectrum, it follows $E_r(n, g(t)) = (2n + 1)E_r(0, g(t))$ (see Appendix E). This relation is particularly useful for any diagonal ensemble. In particular, for a thermal state at g_0 with temperature β^{-1}, the state reads

$$\rho(g_0, \beta) = \frac{\sum_{n=0} e^{-\beta H_{\rm np}(g_0)} \left|\varphi_{\rm np}^n(g_0)\right\rangle\left\langle\varphi_{\rm np}^n(g_0)\right|}{\mathrm{Tr}\left[e^{-\beta H_{\rm np}(g_0)}\right]}, \tag{4.26}$$

where we have used the effective low-energy Hamiltonian of the QRM, and therefore, we assume that the spin splitting is still the largest energy scale in $H_{\rm QRM}$ and so, much larger than the thermal energy, i.e., $\Omega \gg 1/\beta$. In this manner, the residual energy adopts a simple form, $E_r^{\rm th}(\beta, g(t)) = \sum_n p_n E_r(n, g(t)) = \sum_n p_n(2n + 1)E_r(0, g(t))$, with p_n being the initial probability of populating the eigenstate

$\left| \phi^n_{np}(g_0) \right\rangle$ of $H_{np}(g_0)$, which for a thermal state reads $p_n = e^{-\beta n \omega_0 \sqrt{1-g_0^2}}$
$\left(e^{\beta \omega_0 \sqrt{1-g_0^2}} - 1 \right)$,

$$E_r^{th}(\beta, g(t)) = E_r(n = 0, g(t)) \coth\left(\frac{1}{2} \beta \omega_0 \sqrt{1-g_0^2} \right). \qquad (4.27)$$

Evidently, for zero temperature, $\beta \to \infty$, we recover $E_r(n = 0, g(t))$. For temperatures higher than the characteristic energy, but still small compared with the spin splitting, $\Omega \gg 1/\beta \gg \omega_0 \sqrt{1-g_0^2}$ we obtain

$$E_r^{th}(\beta \to 0, g(t)) \approx E_r(n = 0, g(t)) \frac{2}{\beta \omega_0 \sqrt{1-g_0^2}} \sim \beta^{-1} \tau_Q^{-z\nu r/(z\nu r+1)}, \qquad (4.28)$$

which reveals that, while the residual energy is largely enhanced, it does follow the same scaling as that of an initial ground state, $E_r(n = 0, g(t))$. Remarkably, because the QRM lacks spatial dimension, this scaling differs from the prediction for systems obeying bosonic statistics with $d > 0$ [8–10], where instead $E_r \sim \beta^{-1} \tau_Q^{-d\nu r/(z\nu r+1)}$ is found. Note however that the scaling in Eq. (4.28) appears in those spatially extended systems for each of the individual spatial modes, which upon summation leads to the previously mentioned scaling.

4.1.4   Sudden Quenches

So far we have considered nearly adiabatic ramps towards the critical point, however, sudden quenches also deserve to be mentioned as they also display a scaling law governed by equilibrium critical exponents of the system. Needless to say, sudden quenches appear in the opposite limit of nearly adiabatic ramps, and therefore, KZ scaling does not apply here. Indeed, since these quenches emerge as a limiting case when $\tau_Q \to 0$, we do not look into how the quantities scale with τ_Q but rather as a function of the quench amplitude $\Delta g = g_1 - g_0$ [8–10]. Recall that we are solely interested in the state of the system right after the quench, and as in the case of non-zero τ_Q, we do not concern ourselves with an eventual thermalization.

In this particular scenario, $\tau_Q \to 0$, it is straightforward to obtain the scaling of different quantities as the coefficients keep their initial value, $u = 1$ and $v = 0$. In particular, choosing $g_f = 1$ the residual energy follows (see Appendix E for details)

$$E_r^{sq} = (2n + 1)\omega_0 \frac{\sqrt{1+g_0}}{4} \Delta g^{1/2} \qquad (4.29)$$

for an initial state $|\psi(0)\rangle = \left| \varphi^n_{np}(g_0) \right\rangle$. Note that this scaling agrees with the general prediction, $E_r \propto |\Delta g|^{(d+z)\nu}$ [9, 10].

4.2 Finite-Frequency Case: Breakdown of KZ Scaling

As we have seen in Chap. 3, a finite-frequency ratio $\Omega/\omega_0 < \infty$ lifts the singular behavior in the same manner that it occurs for any finite-size system undergoing a conventional phase transition. On the other hand, the confirmed KZ scaling predictions are based on the strict $\Omega/\omega_0 \to \infty$ limit, and thus, it is imperative to investigate whether KZ scaling holds or if it can be observed at all in finite-frequency systems. Moreover, since any realistic implementation of the QRM is unavoidably limited to a finite Ω/ω_0, this question is of great relevance for an experimental observation of KZ physics. Indeed, we will make use of the theoretical framework developed here to probe the critical dynamics of the QRM in a realistic experimental platform (see Chap. 5). Loosely speaking, KZ scaling must be absent when Ω/ω_0 is too small, while as Ω/ω_0 increases, equilibrium scaling relations emerge, and therefore KZ scaling may be progressively recovered.

As in Sect. 4.1, we will address these questions solving the QRM unitary dynamics under a time-dependent protocol $g(t)$ with a total quench time τ_Q. However, we now include the first order correction in ω_0/Ω in the effective low-energy Hamiltonian, introduced in Chap. 3 (see Eq. (3.23)),

$$H_{\mathrm{np}}^{\Omega}(t) =$$

$$\omega a^\dagger a - \frac{\omega_0 g^2(t)}{4}\left(a + a^\dagger\right)^2 + \frac{\omega_0^2 g^4(t)}{16\Omega}\left(a + a^\dagger\right)^4 - \frac{\Omega}{2} + \frac{\omega_0^2 g^2(t)}{4\Omega}. \quad (4.30)$$

In order to calculate the first order correction to the dynamics in the $\Omega/\omega_0 \to \infty$ limit, we assume $a_H(t)$ to be linear in a and $a^\dagger$ and keep only linear terms in a and $a^\dagger$ in the normal-ordered Heisenberg equation of motion for $a_H(t)$. This procedure is suitable as long as the quartic term is small, that is, when it can be treated perturbatively. Then, the equations of motion for finite Ω/ω_0 adopt the following form (see Appendix E)

$$\frac{d}{dt}u(t) =$$

$$-i\omega_0\left[\left(1 - \frac{g^2(t)}{2}\right)u(t) - \frac{g^2(t)}{2}v(t)\right] - i\frac{3\omega_0 g^4(t)}{4\Omega}\left(u(t) + v(t)\right)|u(t) + v(t)|^2, \quad (4.31)$$

$$\frac{d}{dt}v(t) =$$

$$i\omega_0\left[\left(1 - \frac{g^2(t)}{2}\right)v(t) - \frac{g^2(t)}{2}u(t)\right] + i\frac{3\omega_0 g^4(t)}{4\Omega}\left(u(t) + v(t)\right)|u(t) + v(t)|^2, \quad (4.32)$$

where the initial condition is $u(0) = 1$ and $v(0) = 0$. The residual energy is obtained subtracting to $\langle 0| H_{np,H}^{\Omega}(t) |0\rangle$ the ground-state energy $E_{GS}^{\Omega}(g(t))$. As we have shown in Chap. 3, $E_{GS}^{\Omega}(g)$ is known at $g = 1$ (see Eq. 3.57), thus, the residual energy at $g = 1$ for a finite ratio Ω/ω_0 can be explicitly written as

$$E_r^{\Omega}(g = 1) = \omega_0 |v(\tau_Q)|^2 - \frac{\omega_0}{4}|u(\tau_Q) + v(\tau_Q)|^2 + \frac{3\omega_0^2}{16\Omega}|u(\tau_Q) + v(\tau_Q)|^4 - \frac{\omega_0}{4}\left(\frac{2\Omega}{3\omega_0}\right)^{-1/3}. \quad (4.33)$$

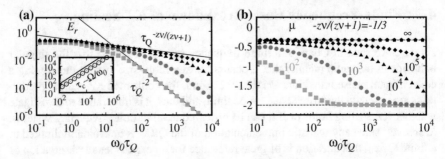

Fig. 4.6 Impact of finite-frequency Ω/ω_0 values in nonequilibrium scaling. In **a** we plot the residual energy E_r^{Ω} for different value Ω/ω_0 as a function of the quench time τ_Q ending at the critical point $g = 1$. The Ω/ω_0 values correspond to 10^2 (squares), 10^3 (circles), 10^4 (triangles) and 10^5 (diamonds), from light gray to black. The fitted scaling exponent τ_Q^{μ} as a function of the quench time τ_Q is plotted in **b** that illustrates the transition from τ_Q^{-2} (dynamics dictated by APT) to $\tau_Q^{-z\nu/(z\nu+1)}$ (KZ physics). The inset in **a** shows the expected scaling of the crossover time $\tau_c \sim \Omega/\omega_0$, where τ_c corresponds to the quench time at which the exponent becomes $\mu \approx -7/6$ (see main text for further details)

Taking the $\Omega/\omega_0 \to \infty$ limit we recover Eq. (4.19) in Sect. 4.1. In Fig. 4.6a we present the results for the residual energy $E_r^{\Omega}(g = 1)$ as a function of the quench time τ_Q. There is a crossover from the universal scaling $\tau_Q^{-1/3}$ to the conventional τ_Q^{-2} as the ratio Ω/ω_0 decreases, and for a given Ω/ω_0, as τ_Q becomes larger. The breakdown of the KZ scaling stems from the fact that, for any finite Ω/ω_0, the energy gap at the critical point is non zero, and thus, there is a minimum quench time $\tau_Q^* \propto (\epsilon_{np}^{\Omega})^{-1}$ after which the dynamics becomes adiabatic and well described by APT. Therefore, for quenches with $\tau_Q \geq \tau_Q^*$ APT scaling is recovered (see Eq. (4.12)). This is indeed a similar scenario as for a quench ending close but not exactly at the critical point, as we have discussed earlier. For larger Ω/ω_0, the energy gap decreases leading to a larger τ_Q^*, enabling the observation of KZ scaling in a broader region of τ_Q. Eventually, in the $\Omega/\omega_0 \to \infty$ limit, τ_Q^* becomes infinitely large and therefore KZ scaling holds for any quench time τ_Q in the nearly adiabatic regime, that is, $\omega_0\tau_Q \gtrsim 1$. The transition between these two dynamical regimes can be better observed in Fig. 4.6b, where we present the fitted scaling exponent μ, from $E_r^{\Omega} \propto \tau_Q^{\mu}$, as a function of τ_Q and for different Ω/ω_0 values. The fitted exponent μ for a quench time τ_Q is computed in a window $\tau_Q \in [\tau_Q/\Delta\tau_Q, \tau_Q\Delta\tau_Q]$ with $\log_{10}\Delta\tau_Q = 0.01$ such that we ensure an homogeneous sampling in a logarithmic scale.[4] The results show a clear crossover from the KZ scaling law, $\mu = -z\nu/(z\nu + 1) = -1/3$, to $\mu = -2$, as dictated by APT, and a convergence to KZ scaling as Ω/ω_0 increases. Remarkably, it is possible to witness KZ scaling for a large but still finite Ω/ω_0, such as 10^5 or 10^4 at short quench times, $\omega_0\tau_Q \approx 10$. This evidences the extremely large system sizes required

---

[4]Note that, in order to attain a reliable fit more quench times were computed, sampling more densely τ_Q than the data shown in Fig. 4.6a.

to attain KZ scaling. Note that in the $\Omega/\omega_0 \to \infty$ limit, μ adopts the discussed universal form, $\mu = -z\nu/(z\nu + 1) = -1/3$.

Moreover, since the opening of the energy gap is at the root of the crossover between these two dynamical regimes, it is not surprising that such a transition also manifests a power-law behavior related to the critical exponents of the system. Let τ_c denote the quench time at which the crossover between KZ and APT scaling occurs. Then, since the minimum energy gap sets a time scale $\hat{t} \sim \epsilon_{\mathrm{np}}^{-1}$ and because $\epsilon_{\mathrm{np}}(g = 1) \propto (\Omega/\omega_0)^{-z}$, it follows $\hat{t} \sim (\Omega/\omega_0)^z$. On the other hand, $\hat{t} \sim \tau_Q^{z\nu/(z\nu+1)}$ as derived using KZ arguments. The crossover time τ_c results when both match, $\tau_c^{z\nu/(z\nu+1)} \sim (\Omega/\omega_0)^z$, which leads to $\tau_c \sim (\Omega/\omega_0)^{(z\nu+1)/\nu} = \Omega/\omega_0$. Numerically, we estimate τ_c as the quench time at which the scaling exponent μ is half away from KZ and APT scaling, i.e., when $\mu = -7/6$ in Fig. 4.6b. Despite the rough estimate, it corroborates the scaling $\tau_c \sim \Omega/\omega_0$, shown in the inset of Fig. 4.6a. Finally, we emphasize that the same conclusions can be drawn from other relevant magnitudes such as $\langle a^\dagger a \rangle$ and Δx, which although we have not shown them explicitly here, their behavior is very similar to that presented in Fig. 4.4, where instead $\Omega/\omega_0 < \infty$ plays the role of $g_f < g_c$.

4.2.1 Nonequilibrium Finite-Frequency Scaling Functions

The emergence of interesting scaling relations in nonequilibrium dynamics is sparked by the existence of a critical point, or more precisely, due to a precursor of a critical point. As we have seen previously, the QRM displays a rich nonequilibrium behavior, in which KZ physics appear as a limiting case. Finite-frequency cases reveal that even the manner in which the dynamics abandons its universal scaling is dictated by equilibrium critical exponents. We can however go beyond the previous considerations and encompass, in a compact and elegant fashion, the nonequilibrium response of finite-frequency systems in a universal function in the same spirit as finite-size scaling functions for equilibrium [11, 29]. In this manner, the nonequilibrium scaling functions depend on two scaling variables, one exclusively dynamical and related to the quench time τ_Q, and the other as in the equilibrium case, $x = \Omega/\omega_0|g - g_c|^\nu$ (see Sect. 3.2). Assuming nearly adiabatic dynamics and a linear quench $g(t) = g_f t/\tau_Q$, the equation of motion can be cast in a universal form through a rescaling of parameters, exploiting the universal form of certain quantities such as the energy gap close to the critical point [11]. Indeed, upon this transformation, the dynamics does not longer depend on Ω/ω_0, τ_Q and the final coupling g_f alone, but on two scaling variables, $x \equiv \Omega/\omega_0|g - g_c|^\nu$ and $T \equiv \tau_Q(\Omega/\omega_0)^{-(1+z\nu)/\nu}$.

We present a brief derivation of this claim starting from the equation of motion in terms of the coefficients $\alpha_n(t)$ such that $|\psi(t)\rangle = \sum_n \alpha_n(t)e^{-i\int_0^t \epsilon_n(t')dt'} \left| \varphi_{\mathrm{np}}^n(g(t)) \right\rangle$, that is, from Eq. (4.10). Then, assuming that the dynamics is nearly adiabatic far away from the critical point, the coefficients $\alpha_n(t)$ remain constant until $g \sim g_c$, and thus, we can rely on the equilibrium scaling functions of the energy difference and transition amplitudes between nth and mth eigenstates, $\Delta_{n,m} = \epsilon_m(g) - \epsilon_n(g)$ and $\chi_{n,m} = \left\langle \varphi^n_{np}(g) \middle| \partial_g \middle| \varphi^m_{np}(g) \right\rangle$ respectively. While $\Delta_{n,m}$ has been discussed in Chap. 3 (see Eqs. (3.25) and (3.60)), the latter is derived here. In particular, $\chi_{n,m}$ can be obtained knowing that $\left| \varphi^n_{np}(g) \right\rangle = S[r_{np}(g)] |n\rangle |\downarrow\rangle$ with $r_{np} = -1/4 \ln(1 - g^2)$ for $0 \le g \le g_c$,

$$
\chi_{n,m} = \left\langle \varphi^n_{np}(g) \middle| \partial_g \middle| \varphi^m_{np}(g) \right\rangle = \left\langle \varphi^n_{np}(g) \middle| \frac{1}{2} \frac{\partial r_{np}(g)}{\partial g} \left(a^2 - (a^\dagger)^2 \right) \middle| \varphi^m_{np}(g) \right\rangle \tag{4.34}
$$

$$
= \frac{g}{4(1 - g^2)} \left(\sqrt{n(n-1)} \delta_{m+2,n} - \sqrt{m(m-1)} \delta_{m-2,n} \right).
$$

Therefore, $\chi_{n,m} \sim |g - g_c|^{\gamma_\chi}$ with $\gamma_\chi = -1$. As we are interested in the behavior of these quantities close to the critical point and for finite Ω/ω_0 values, we introduce the finite-frequency scaling functions $\Delta_{n,m} = (\Omega/\omega_0)^{-z} \Phi_\Delta(x)$ and $\chi_{n,m} = (\Omega/\omega_0)^{-\gamma_\chi/\nu} \Phi_\chi(x) = (\Omega/\omega_0)^{1/\nu} \Phi_\chi(x)$ where x is the scaling variable $x \equiv \Omega/\omega_0 |g - g_c|^\nu$ (see Chap. 3). Note that we have written the dependence on Ω/ω_0 explicitly. Making use of the previous expressions, Eq. (4.10) transforms into

$$
\alpha_n(x) = -\sum_m \int dx \frac{1}{\nu} x^{1/\nu - 1} \alpha_m(x) \Phi_\chi(x) \, e^{-i \frac{\tau_Q}{\nu g_f} (\Omega/\omega_0)^{-(1+z\nu)/\nu} \int dx \, x^{1/\nu - 1} \Phi_\Delta(x)}.
$$

$$
\tag{4.35}
$$

Therefore, defining $T \equiv \tau_Q (\Omega/\omega_0)^{-(1+z\nu)/\nu}$, the above expression depends only on $x_f = \Omega/\omega_0 |g_f - g_c|^\nu$ and T, and so does the evolved state $|\psi(t)\rangle$. Recall that $\nu = 3/2$ and $z = 1/3$, and thus $T = \tau_Q (\Omega/\omega_0)^{-1}$. Using this powerful result we can construct nonequilibrium scaling functions for several relevant quantities, which must collect both KZ physics and APT scaling as well as the crossover between them.

We first consider the residual energy E_r^Ω as it has played a central role in the previous parts of this chapter, which can be recast as

$$
E_r^\Omega = \sum_n |\alpha_n(\tau_Q)|^2 \Delta_{0,n}(g_f) \approx \left(\frac{\Omega}{\omega_0} \right)^{-z} S_{E_r}(x, T), \tag{4.36}
$$

valid only if $|g_f - g_c| \ll 1$ and where we have used that $\alpha_n(\tau_Q)$ is a function of x and T, from Eq. (4.35). The first consequence of this expression directly leads us to observe that $S_{E_r}(x_f, T) \equiv (\Omega/\omega_0)^z E_r^\Omega$ turns out to be a universal scaling function, which fulfills $S_{E_r} \sim T^{-z\nu/(z\nu+1)}$ for $T \ll 1$ and $S_{E_r} \sim T^{-2}$ for $T \gg 1$. The simplest case, $g_f = g_c$ and thus $x_f = 0$, suggests that plotting $(\Omega/\omega_0)^z E_r^\Omega$ as a function of $T = \tau_Q (\Omega/\omega_0)^{-(1+z\nu)/\nu}$ should collapse in a single curve, independently of the individual values of τ_Q or Ω/ω_0. However, we remark that τ_Q must be large enough to ensure nearly adiabatic dynamics, as the procedure is only valid when excitations are caused in the vicinity of the critical point. This is plotted in Fig. 4.7 for several

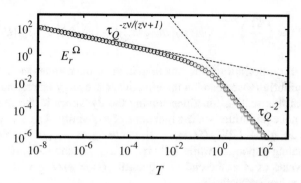

Fig. 4.7 Nonequilibrium finite-frequency scaling function of the residual energy E_r^{Ω} obtained as $S_{E_r} \equiv E_r^{\Omega} (\Omega/\omega_0)^z$ under a linear protocol $g(t) = t/\tau_Q$, for different quench times and Ω/ω_0 values. Then S_{E_r} results in a universal function when plotted against $T \equiv \tau_Q (\Omega/\omega_0)^{-(1+z\nu)/\nu}$. The residual energy was calculated through the low-energy effective Hamiltonian H_{np}^{Ω} (see Eq. (4.33)). Different values from $\Omega/\omega_0 = 10^1$ until 10^8 were considered, each of them plotted in a gray scale, ranging from light-gray (10^1) to black (10^8). The dashed lines are guides to the eyes fixing the characteristic KZ and APT power-law scaling

τ_Q and Ω/ω_0 values, relying on Eq. (4.33) and for $\omega_0\tau_Q > 1$. As we clearly observe, this nonequilibrium scaling function $S_{E_r}(x, T)$ encompasses both dynamical regimes (KZ and APT), enabling the observation of their corresponding scaling as well as the crossover between them. As a consequence, we can estimate that the dynamics of a QRM for a particular Ω/ω_0 value will lie in the KZ when $T \lesssim 10^{-4}$, that is, when $\tau_Q \lesssim 10^{-4}\Omega/\omega_0$ but still $\omega_0\tau_Q \gtrsim 1$ due to the nearly adiabatic condition. This sets a constrain for the observation of KZ physics with a finite Ω/ω_0 system, which is consistent with Fig. 4.6, where $\Omega/\omega_0 \gtrsim 10^5$ is needed to witness the universal scaling exponent $\mu = -z\nu/(z\nu + 1)$. Furthermore, the function $S_{E_r}(x, T)$ reveals that the resulting dynamics from different Ω/ω_0 are equivalent when they share the same value $x_f = \Omega/\omega_0|g_f - g_c|^{\nu}$. That is, when the protocol ends at $g_f = g_c - (x\omega_0/\Omega)^{1/\nu}$, plotting $E_r^{\Omega}(\Omega/\omega_0)^z$ the data collapses into a single and universal curve. Note, however, that $|g_f - g_c| \ll 1$ to ensure the validity of the universal features of the phase transition and thus, we expect worse data collapse for larger x and smaller Ω/ω_0 values.

In a straightforward manner, same arguments as for E_r can be applied to argue the existence of nonequilibrium scaling functions for other quantities, such as $\langle a^{\dagger}a \rangle \omega_0/\Omega$ or Δp, which we denote as S_{n_c} and $S_{\Delta p}$. Interestingly, we can obtain a nonequilibrium scaling function for $\langle \sigma_z \rangle$, which may be extremely advantageous from an experimental point of view, as spin degrees of freedom are commonly more accessible than bosonic observables. Since $\langle \varphi_{np}^n(g)| \sigma_z |\varphi_{np}^n(g) \rangle \approx \langle \varphi_{np}^0(g)| \sigma_z |\varphi_{np}^0(g) \rangle$ to a very good approximation for $\Omega/\omega_0 \gg 1$, and $\langle \varphi_{np}^0(g)| \sigma_z |\varphi_{np}^0(g) \rangle \approx -1 + 1/3(2\Omega/(3\omega_0))^{-\gamma_{\sigma_z}/\nu}$ with $\gamma_{\sigma_z} = 1$ (see Chap.3, Sect. 3.2),

$$S_{\sigma_z}(x, T) \equiv \left(\frac{\Omega}{\omega_0}\right)^{\gamma_{\sigma_z}/\nu} \left|\langle\psi(\tau_Q)| \sigma_z |\psi(\tau_Q)\rangle - \langle\varphi_{np}^0(g)| \sigma_z |\varphi_{np}^0(g)\rangle\right| \qquad (4.37)$$

becomes a nonequilibrium scaling function, in the same manner as $S_{E_r}(x, T)$.

As for equilibrium, we abandon the effective low-energy Hamiltonian and show the validity of this scaling functions solving the dynamics in the full H_{QRM}. We compute the nonequilibrium scaling functions of a quantity $\mathcal{A}$ as $S_{\mathcal{A}} = (\Omega/\omega_0)^{-\delta_{\mathcal{A}}}$ $|\langle\mathcal{A}\rangle (\tau_Q, \Omega/\omega_0, g_f) - \langle\mathcal{A}\rangle_{GS} (\Omega/\omega_0, g_f)|$ where $\delta_{\mathcal{A}} = -\gamma_{\mathcal{A}}/\nu$ denotes the finite-frequency scaling exponent, while $\langle\mathcal{A}\rangle (\tau_Q, \Omega/\omega_0, g_f)$ and $\langle\mathcal{A}\rangle_{GS} (\Omega/\omega_0, g_f)$ the expectation value of $\mathcal{A}$ at the end of a quench $g(t) = g_f t/\tau_Q$ with Ω/ω_0 and its ground-state value, respectively.

The resulting nonequilibrium scaling functions are shown in Fig. 4.8 for $\langle\sigma_z\rangle$, the residual energy E_r^Ω, the squeezing Δp and n_c for three different values of x, namely, 0, 1/2 and 1. The collapse of the data into single curves (one for each value of x) strongly supports the existence of these nonequilibrium scaling functions merging different dynamical regimes and bearing the equilibrium critical exponents.

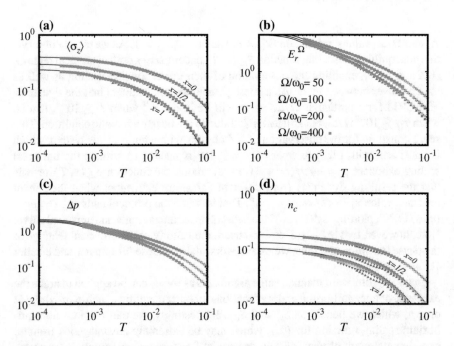

Fig. 4.8 Nonequilibrium finite-frequency scaling functions of the QRM, for three different x values, namely, $x = 0$, 1/2 and 1 from top to bottom, and as function of the rescaled quench time $T \equiv \tau_Q(\Omega/\omega_0)^{-(1+z\nu)/\nu}$ for four different values of Ω/ω_0, 50, 100, 200 and 400, plotted with different colors and point shapes. In **a** we plot $S_{\sigma_z}(x, T)$, in **b** $S_{E_r}(x, T)$, while **c** and **d** correspond to $S_{\Delta p}(x, T)$ and $S_{n_c}(x, T)$. The data falls into different curves depending on x and T, i.e. the data collapse, which strongly supports the suitability of these scaling functions (see main text for further details)

In this respect, these scaling functions provide great information regarding the critical features of the system, in and out of equilibrium. This procedure may be extremely helpful to verify the presence of the QPT and test the validity of equilibrium critical exponents in experimental setups, as evolution times are much shorter than required for adiabatic preparation. We finally comment that we have presented a similar idea in Chap. 2 in the realm of Ginzburg–Landau theory and a linear to zigzag phase transition in a Coulomb crystal.

4.2.2 Universality Class

As we have discussed for the equilibrium case in Chap. 3, the finite-frequency scaling functions of the QRM are equivalent to those of the Dicke [30] or Lipkin–Meshkov–Glick models [31], a clear indication that their QPTs belong to the same universality class. Hence, one may question whether the nonequilibrium features of the QRM, discussed previously, are representative of its universality class. In this regard, it is worth noting the work done in Ref. [11], where the nonequilibrium scaling functions were used to disclose that the dynamics of different models that lie in the same equilibrium universality class, as Dicke and Lipkin–Meshkov–Glick models, are not equivalent during the complete nonequilibrium protocol $g(t)$. However, in our case, we are concerned solely with dynamical features at the end of the protocol and in the vicinity of the critical point, which turn out to be indeed universal, that is, equivalent and also include the QRM. Again, as for equilibrium scaling functions, we resort to the Dicke model since the equivalence with the Lipkin–Meshkov–Glick model is well established when $|g_f - g_c| \ll 1$ [11]. For convenience, we rewrite the Dicke Hamiltonian

$$H_{\text{Dicke}} = \omega_a J_z + \omega_b a^\dagger a + g \frac{\sqrt{\omega_a \omega_b}}{2\sqrt{N}} J_x (a + a^\dagger), \qquad (4.38)$$

where the N two-level systems are described in terms of the collective angular momenta $J_\alpha = \sum_i^N \sigma_\alpha^i / 2$. Note that H_{Dicke} exhibits a QPT at $g_c = 1$ in the $N \to \infty$ limit [32–35]. As it is customary in the Dicke model, we consider the resonant condition $\omega_a = \omega_b$ and then solve numerically its critical unitary dynamics under the protocol $g(t) = g_f t / \tau_Q$ for various τ_Q and $N = 2J$ values. The initial state is, as in the QRM, the ground state at $g(0) = 0$, namely, $|\psi(0)\rangle = |J, -J\rangle |0\rangle$ such that $J^2 |J, -J\rangle = J(J+1) |J, -J\rangle$ and $J_z |J, -J\rangle = -J |J, -J\rangle$. Note that, since both critical exponents and equilibrium finite-size scaling functions are equivalent to those of the QRM, the procedure presented in 4.2.1 is also valid for the Dicke model, although the nonequilibrium scaling function for a quantity $\mathcal{A}$, denoted by $\tilde{S}_\mathcal{A}$, depends now on $\tilde{x} = N|g - g_c|^\nu$ and $\tilde{T} = \tau_Q N^{-(1+z\nu)/\nu}$. The equivalence between the critical dynamics of these two models is supported by the fact that both share the same nonequilibrium scaling functions up to constant factors, as shown in Fig. 4.9. There, we compare the $S_{\sigma_z}(x, T)$ and $S_{E_r}(x, T)$ of the QRM with those of the

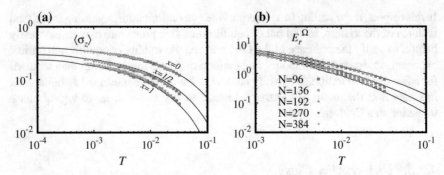

Fig. 4.9 Nonequilibrium finite-size scaling functions of the Dicke model (points) compared with those of the QRM (solid lines) for σ_z (**a**) and residual energy (**b**). Different size systems are plotted with different colors and point shapes. In order to better illustrate the equivalence between these scaling functions, we plot $K_1 \tilde{S}_{J_z}(\tilde{x}, \tilde{T} K_2)$ and $K_1 \tilde{S}_{E_r}(\tilde{x}, \tilde{T} K_2)$ with $K_{1,2}$ constant factors that account for microscopic differences between both models. Note that although they share critical exponents, the scaling variables for the Dicke model read $\tilde{x} = N|g - g_c|^\nu$ and $\tilde{T} = \tau_Q N^{-(1+z\nu)/\nu}$. In **a** we have chosen $K_1 = 5$ and $K_2 = 1/2$ that correctly reproduce the three different values of x considered here ($x = 0$, $1/2$ and 1), while in **b** a good agreement is found for $K_1 = 10$. The equivalence between these models, that fall into the same universality class, extends also into a nonequilibrium scenario

Dicke model, $K_1 \tilde{S}_{J_z}(\tilde{x}, \tilde{T} K_2)$ and $K_1 \tilde{S}_{E_r}(\tilde{x}, \tilde{T} K_2)$ with $K_{1,2}$ constant factors explicitly included here account for different microscopic details of the Dicke model [36]. Choosing $K_1 = 5$ and $K_2 = 1/2$ we obtain a very good agreement for σ_z and J_z for each of the x sheets ($x = 0$, $1/2$ and 1), while for the residual energy we found that $K_1 = 10$ reproduces well that of the QRM.

These results further support that the superradiant QPT of the QRM belongs to the same universality class as those of Dicke and Lipkin–Meshkov–Glick models. Moreover, this equivalence holds even in a nonequilibrium scenario and therefore, the critical dynamics in the QRM is representative of its universality class.

4.3 Conclusion and Outlook

This chapter has been devoted entirely to an in-depth analysis of nonequilibrium dynamics of the QRM involving the QPT, with special emphasis on nearly adiabatic dynamics in the spirit of the KZ mechanism. The question of to what extent equilibrium universal features are captured in a nonequilibrium scenario has attracted the attention of researchers during the last decades and significant efforts have been invested in its inspection [1, 2, 7, 10, 25]. In this chapter we have not only shown that KZ scaling holds even in a zero-dimensional system, in contrast to the common belief prior to this work [11, 22], but hopefully, we have also convinced the reader that the QRM constitutes a formidable system where distinct nonequilibrium scenarios can be tested. Among them, nonequilibrium finite-frequency scaling functions deserve special mention, which include the KZ scaling as a limiting case, and show a functional form that is shared within models belonging to the same universality class, such as the Dicke or Lipkin–Meshkov–Glick models. Note that these functions have been also discussed in Chap. 2 in the context of classical continuous phase transitions. In addition, as any experimental realization of the QRM is limited to a finite value of Ω/ω_0, these nonequilibrium finite-frequency scaling functions appear as a promising tool to experimentally probe the critical dynamics of a superradiant QPT, which is the subject of the following Chapter.

Despite the detailed analysis presented here, it is worth recalling that we have just provided a first glance at nonequilibrium critical features of the QRM, as we have addressed the simplest case, namely, the unitary dynamics of an isolated QRM driven by an external time-dependent coupling. Even in this case, there are more topics which might deserve a short mention, such as the role of more general initial states, optimal nonlinear protocols [37, 38] or the design of the so-called shortcuts to adiabaticity [39]. These techniques, besides being of theoretical interest, can lead to outstanding advantages in experimental platforms as, for example, in the search for fast preparation of valuable states. This becomes crucial if ones pursues the ground state at the critical point, a necessary condition to verify equilibrium scaling functions, since the required time to adiabatically prepare it under a naive linear ramp grows exponentially due to the closing of the energy gap. In the context of a driven isolated QRM, we mention the work done in [40], which closely follows the developments presented here considering instead a generalized QRM, that is, a QRM where rotating and counter-rotating terms have different weights, and finding non-trivial modifications to scaling exponents.

The inclusion of a non-negligible interaction with an environment definitely affects the critical dynamics, not only due to an eventual modification of equilibrium critical exponents [41, 42] but also as a consequence of environment-induced transitions. Note for example that the latter may lead to universal scaling in terms of the temperature of the bath [43–46] and potentially to an anti-KZ regime, in which slower quenches generate larger number of excitations [47], although more aspects may be explored [48–51]. In short, the inspection of system-environment interactions in a critical system joins the endless list of attractive subjects to investigate in the realm of quantum critical phenomena, being of particular significance in any experimental platform. We remark that neither these effects nor thermalization have been considered in this chapter, which constitute exciting directions of future research.

The QRM might also be a suitable model to explore and study the bonds between nonequilibrium KZ physics and concepts of thermodynamics, such as irreversible work done and entropy production, as well as ultimately the connection with the fluctuation theorems, namely, Jarzynksi equality and Tasaki-Crooks theorem [52, 53] (see Ref. [54] for a review).

References

1. J. Dziarmaga, Dynamics of a quantum phase transition and relaxation to a steady state. Adv. Phys. **59**, 1063 (2010). https://doi.org/10.1080/00018732.2010.514702
2. A. Polkovnikov, K. Sengupta, A. Silva, M. Vengalattore, Colloquium: Nonequilibrium dynamics of closed interacting quantum systems. Rev. Mod. Phys. **83**, 863 (2011). https://doi.org/10.1103/RevModPhys.83.863
3. J. Klinder, H. Keßler, M. Wolke, L. Mathey, A. Hemmerich, Dynamical phase transition in the open Dicke model. Proc. Natl. Acad. Sci. **112**, 3290 (2015). https://doi.org/10.1073/pnas.1417132112
4. S. Braun, M. Friesdorf, S.S. Hodgman, M. Schreiber, J.P. Ronzheimer, A. Riera, M. del Rey, I. Bloch, J. Eisert, U. Schneider, Emergence of coherence and the dynamics of quantum phase transitions. Proc. Natl. Acad. Sci. **112**, 3641 (2015). https://doi.org/10.1073/pnas.1408861112
5. J. Eisert, M. Friesdorf, C. Gogolin, Quantum many-body systems out of equilibrium. Nat. Phys. **11**, 124 (2015). https://doi.org/10.1038/nphys3215
6. A. Polkovnikov, Universal adiabatic dynamics in the vicinity of a quantum critical point. Phys. Rev. B **72**, 161201 (2005). https://doi.org/10.1103/PhysRevB.72.161201
7. A. Polkovnikov, V. Gritsev, Breakdown of the adiabatic limit in low-dimensional gapless systems **4**, 477 (2008). https://doi.org/10.1038/nphys963
8. C. De Grandi, V. Gritsev, A. Polkovnikov, Quench dynamics near a quantum critical point: application to the sine-Gordon model. Phys. Rev. B **81**, 224301 (2010a). https://doi.org/10.1103/PhysRevB.81.224301
9. C. De Grandi, V. Gritsev, A. Polkovnikov, Quench dynamics near a quantum critical point. Phys. Rev. B **81**, 012303 (2010b). https://doi.org/10.1103/PhysRevB.81.012303
10. A.K. Chandra, A. Das, B.K.C. (Eds.), *Quantum Quenching, Annealing and Computation* (Springer, Berlin Heidelberg, 2010)
11. O.L. Acevedo, L. Quiroga, F.J. Rodríguez, N.F. Johnson, New dynamical scaling universality for quantum networks across adiabatic quantum phase transitions. Phys. Rev. Lett. **112**, 030403 (2014). https://doi.org/10.1103/PhysRevLett.112.030403
12. A. Chandran, A. Erez, S.S. Gubser, S.L. Sondhi, Kibble-Zurek problem: universality and the scaling limit. Phys. Rev. B **86**, 064304 (2012). https://doi.org/10.1103/PhysRevB.86.064304
13. G. Nikoghosyan, R. Nigmatullin, M.B. Plenio, Universality in the dynamics of second-order phase transitions. Phys. Rev. Lett. **116**, 080601 (2016). https://doi.org/10.1103/PhysRevLett.116.080601
14. A. del Campo, W.H. Zurek, Universality of phase transition dynamics: topological defects from symmetry breaking. Int. J. Mod. Phys. A **29**, 1430018 (2014). https://doi.org/10.1142/S0217751X1430018X
15. I. Chuang, R. Durrer, N. Turok, B. Yurke, Cosmology in the laboratory: defect dynamics in liquid crystals. Science **251**, 1336 (1991). https://doi.org/10.1126/science.251.4999.1336
16. K. Pyka, J. Keller, H.L. Partner, R. Nigmatullin, T. Burgermeister, D.M. Meier, K. Kuhlmann, A. Retzker, M.B. Plenio, W.H. Zurek, A. del Campo, T.E. Mehlstäubler, Topological defect formation and spontaneous symmetry breaking in ion Coulomb crystals. Nat. Commun. **4**, 2291 (2013). https://doi.org/10.1038/ncomms3291
17. S. Ulm, J. Roßnagel, G. Jacob, C. Degünther, S.T. Dawkins, U.G. Poschinger, R. Nigmatullin, A. Retzker, M.B. Plenio, F. Schmidt-Kaler, K. Singer, Observation of the Kibble-Zurek scaling law for defect formation in ion crystals. Nat. Commun. **4**, 2290 (2013). https://doi.org/10.1038/ncomms3290
18. N. Navon, A.L. Gaunt, R.P. Smith, Z. Hadzibabic, Critical dynamics of spontaneous symmetry breaking in a homogeneous Bose gas. Science **347**, 167 (2015). https://doi.org/10.1126/science.1258676
19. W.H. Zurek, U. Dorner, P. Zoller, Dynamics of a quantum phase transition. Phys. Rev. Lett. **95**, 105701 (2005). https://doi.org/10.1103/PhysRevLett.95.105701

20. B. Damski, The simplest quantum model supporting the Kibble-Zurek mechanism of topological defect production: Landau-Zener transitions from a new perspective. Phys. Rev. Lett. **95**, 035701 (2005). https://doi.org/10.1103/PhysRevLett.95.035701

21. B. Damski, W.H. Zurek, How to fix a broken symmetry: quantum dynamics of symmetry restoration in a ferromagnetic Bose-Einstein condensate. New J. Phys. **10**, 045023 (2008) http://stacks.iop.org/1367-2630/10/i=4/a=045023

22. T. Caneva, R. Fazio, G.E. Santoro, Adiabatic quantum dynamics of the Lipkin-Meshkov-Glick model. Phys. Rev. B **78**, 104426 (2008). https://doi.org/10.1103/PhysRevB.78.104426

23. M.-J. Hwang, R. Puebla, M.B. Plenio, Quantum phase transition and universal dynamics in the Rabi model. Phys. Rev. Lett. **115**, 180404 (2015). https://doi.org/10.1103/PhysRevLett.115.180404

24. R. Puebla, M.-J. Hwang, J. Casanova, M.B. Plenio, Probing the dynamics of a superradiant quantum phase transition with a single trapped ion. Phys. Rev. Lett. **118**, 073001 (2017). https://doi.org/10.1103/PhysRevLett.118.073001

25. A. Dutta, G. Aeppli, B.K. Chakrabarti, U. Divakaran, T.F. Rosenbaum, D. Sen, *Quantum Phase Transitions in Transverse Field Spin Models: From Statistical Physics to Quantum Information* (Cambridge University Press, Cambridge, 2015)

26. G. Rigolin, G. Ortiz, V.H. Ponce, Beyond the quantum adiabatic approximation: adiabatic perturbation theory. Phys. Rev. A **78**, 052508 (2008). https://doi.org/10.1103/PhysRevA.78.052508

27. D. Sen, K. Sengupta, S. Mondal, Defect production in nonlinear quench across a quantum critical point. Phys. Rev. Lett. **101**, 016806 (2008). https://doi.org/10.1103/PhysRevLett.101.016806

28. S. Sachdev, *Quantum Phase Transitions*, 2nd edn. (Cambridge University Press, Cambridge, 2011)

29. M. Kolodrubetz, B.K. Clark, D.A. Huse, Nonequilibrium dynamic critical scaling of the quantum Ising chain. Phys. Rev. Lett. **109**, 015701 (2012). https://doi.org/10.1103/PhysRevLett.109.015701

30. R.H. Dicke, Coherence in spontaneous radiation processes. Phys. Rev. **93**, 99 (1954). https://doi.org/10.1103/PhysRev.93.99

31. H. Lipkin, N. Meshkov, A. Glick, Validity of many-body approximation methods for a solvable model. Nucl. Phys. **62**, 188 (1965). https://doi.org/10.1016/0029-5582(65)90862-X

32. K. Hepp, E.H. Lieb, On the superradiant phase transition for molecules in a quantized radiation field: the dicke maser model. Ann. Phys. (N. Y.) **76**, 360 (1973). https://doi.org/10.1016/0003-4916(73)90039-0

33. Y.K. Wang, F.T. Hioe, Phase transition in the Dicke model of superradiance. Phys. Rev. A **7**, 831 (1973). https://doi.org/10.1103/PhysRevA.7.831

34. C. Emary, T. Brandes, Quantum chaos triggered by precursors of a quantum phase transition: the Dicke model. Phys. Rev. Lett. **90**, 044101 (2003a). https://doi.org/10.1103/PhysRevLett.90.044101

35. C. Emary, T. Brandes, Chaos and the quantum phase transition in the Dicke model. Phys. Rev. E **67**, 066203 (2003b). https://doi.org/10.1103/PhysRevE.67.066203

36. H.E. Stanley, *Introduction to Phase Transitions and Critical Phenomena* (Clarendon Press, Oxford, 1971)

37. S. van Frank, M. Bonneau, J. Schmiedmayer, S. Hild, C. Gross, M. Cheneau, I. Bloch, T. Pichler, A. Negretti, T. Calarco, S. Montangero, Optimal control of complex atomic quantum systems. Sci. Rep. **6**, 34187 (2016). https://doi.org/10.1038/srep34187

38. R. Barankov, A. Polkovnikov, Optimal nonlinear passage through a quantum critical point. Phys. Rev. Lett. **101**, 076801 (2008). https://doi.org/10.1103/PhysRevLett.101.076801

39. E. Torrontegui, S. Ibáñez, S. Martínez-Garaot, M. Modugno, A. del Campo, D. Guéry-Odelin, A. Ruschhaupt, X. Chen, J.G. Muga, in *Advances in Atomic, Molecular, and Optical Physics*, vol. 62, ed. by E. Arimondo, P.R. Berman, C.C. Lin (Academic Press, 2013) pp. 117–169. https://doi.org/10.1016/B978-0-12-408090-4.00002-5

40. L.-T. Shen, Z.-B. Yang, H.-Z. Wu, S.-B. Zheng, Quantum phase transition and quench dynamics in the anisotropic Rabi model. Phys. Rev. A **95**, 013819 (2017). https://doi.org/10.1103/PhysRevA.95.013819
41. D. Nagy, P. Domokos, Nonequilibrium quantum criticality and non-Markovian environment: critical exponent of a quantum phase transition. Phys. Rev. Lett. **115**, 043601 (2015). https://doi.org/10.1103/PhysRevLett.115.043601
42. D. Nagy, P. Domokos, Critical exponent of quantum phase transitions driven by colored noise. Phys. Rev. A **94**, 063862 (2016). https://doi.org/10.1103/PhysRevA.94.063862
43. D. Patanè, A. Silva, L. Amico, R. Fazio, G.E. Santoro, Adiabatic dynamics in open quantum critical many-body systems. Phys. Rev. Lett. **101**, 175701 (2008). https://doi.org/10.1103/PhysRevLett.101.175701
44. D. Patanè, L. Amico, A. Silva, R. Fazio, G.E. Santoro, Adiabatic dynamics of a quantum critical system coupled to an environment: scaling and kinetic equation approaches. Phys. Rev. B **80**, 024302 (2009). https://doi.org/10.1103/PhysRevB.80.024302
45. P. Nalbach, Adiabatic-Markovian bath dynamics at avoided crossings. Phys. Rev. A **90**, 042112 (2014). https://doi.org/10.1103/PhysRevA.90.042112
46. P. Nalbach, S. Vishveshwara, A.A. Clerk, Quantum Kibble-Zurek physics in the presence of spatially correlated dissipation. Phys. Rev. B **92**, 014306 (2015). https://doi.org/10.1103/PhysRevB.92.014306
47. A. Dutta, A. Rahmani, A. del Campo, Anti-Kibble-Zurek behavior in crossing the quantum critical point of a thermally isolated system driven by a noisy control field. Phys. Rev. Lett. **117**, 080402 (2016). https://doi.org/10.1103/PhysRevLett.117.080402
48. S. Suzuki, T. Nag, A. Dutta, Dynamics of decoherence: universal scaling of the decoherence factor. Phys. Rev. A **93**, 012112 (2016). https://doi.org/10.1103/PhysRevA.93.012112
49. A. Bayat, T.J.G. Apollaro, S. Paganelli, G. De Chiara, H. Johannesson, S. Bose, P. Sodano, Nonequilibrium critical scaling in quantum thermodynamics. Phys. Rev. B **93**, 201106 (2016). https://doi.org/10.1103/PhysRevB.93.201106
50. S. Yin, P. Mai, F. Zhong, Nonequilibrium quantum criticality in open systems: the dissipation rate as an additional indispensable scaling variable. Phys. Rev. B **89**, 094108 (2014). https://doi.org/10.1103/PhysRevB.89.094108
51. S. Yin, C.-Y. Lo, P. Chen, Scaling behavior of quantum critical relaxation dynamics of a system in a heat bath. Phys. Rev. B **93**, 184301 (2016). https://doi.org/10.1103/PhysRevB.93.184301
52. C. Jarzynski, Nonequilibrium equality for free energy differences. Phys. Rev. Lett. **78**, 2690 (1997). https://doi.org/10.1103/PhysRevLett.78.2690
53. G.E. Crooks, Entropy production fluctuation theorem and the nonequilibrium work relation for free energy differences. Phys. Rev. E **60**, 2721 (1999). https://doi.org/10.1103/PhysRevE.60.2721
54. M. Campisi, P. Hänggi, P. Talkner, Colloquium: Quantum fluctuation relations: foundations and applications. Rev. Mod. Phys. **83**, 771 (2011). https://doi.org/10.1103/RevModPhys.83.771

Chapter 5
Superradiant QPT with a Single Trapped Ion

In Chaps. 3 and 4 we have examined equilibrium and nonequilibrium features of the quantum phase transition (QPT) exhibited by the quantum Rabi model (QRM). As we have emphasized, this finite-component system QPT differs from conventional phase transitions in many-body systems as it appears without scaling up system constituents. The present chapter exploits this fact, proposing a single trapped-ion experiment in which universal features of a superradiant QPT can be examined. This chapter completes, and also complements, the previous Chaps. 3 and 4 involving the QRM.

An experimental realization of a finite-component system QPT would open the door to the inspection of the dynamics involving a QPT in a small, fully controlled quantum system with a high degree of coherence without the necessity of scaling up the number of system components. It is therefore worth pursuing a potential implementation of the QRM in the parameter regime in which QPT becomes noticeable. The QRM naturally arises in a variety of platforms, as commented in Chap. 3. Among them, trapped ions [1, 2], circuit and cavity QED [3, 4], optomechanical systems [5], cold atoms [6, 7] and color centers in membranes [8] deserve special mention. Despite each of the platforms may exhibit certain advantages with respect to the rest, trapped ions are perhaps the most versatile setup up to date featuring high coherent quantum control, tunability of system parameters and the possibility of performing high-fidelity measurements [1, 2, 9–11]. In addition, as reported in [12, 13] and studied here in Chap. 2, trapped-ions have been demonstrated to be a suitable platform to study classical phase transitions.

We will then show the suitability of a single trapped-ion experiment to explore the features of a superradiant QPT. The trapped-ion QRM realization allows to access the extreme parameter regime required for the observation of the QPT, which may not be trivially achieved in another platforms. It is worth noting that the observation of critical phenomena typically requires a system comprising many constituents. Despite the formidable advances in trapped-ion technologies, quantum control, preservation of quantum coherence, and state measurements involving many ions still constitutes

© Springer Nature Switzerland AG 2018   123
R. Puebla, *Equilibrium and Nonequilibrium Aspects of Phase Transitions
in Quantum Physics*, Springer Theses, https://doi.org/10.1007/978-3-030-00653-2_5

a major challenge [10, 11] (see however the very recent works [14, 15]). This drawback can be overcome considering the QRM, which merely deals with a coherent interaction between a qubit and a single bosonic mode.

Trapped ions, as any other platform, is inevitably prone to experimental imperfections and uncontrolled interactions with the environment. Although the QRM can be well realized in a trapped-ion setup, these realistic imperfections may spoil the pursued universal traits of a superradiant QPT. It is therefore pertinent to examine the effect of different noise sources, and whether such universal behavior survives. Indeed, we find that nonequilibrium scaling functions can be still observed since they involve much shorter quench times than those required for adiabatic preparation, and thus, they are more resilient against noise than their equilibrium counterparts. Moreover, while the first part of this chapter, Sect. 5.1 is devoted to the inspection of the validity of the trapped-ion model to realize the QRM in the needed parameter regime, the second part, Sect. 5.2, is entirely dedicated to provide a strategy to cope with magnetic-field fluctuations. These fluctuations are typically the main source of decoherence in various trapped-ion setups. For that purpose, we will apply continuous dynamical decoupling techniques [16, 17], theoretically analyzed [18–21] and whose experimental feasibility and suitability has been already reported in [22, 23]. Here we extend the applicability of continuous dynamical decoupling methods for the implementation of a robust, noise resilient, yet tunable QRM.

Part of the results and materials collated in this chapter have been published in [24, 25], while similar ideas developed by the author [26, 27] have been omitted in this thesis.

5.1 Proposed Trapped-Ion Experiment

A single trapped-ion, subject to classical radiation, is sufficient to realize the QRM [28, 29] in several parameter regimes. As explained in Chap. 2, Sect. 2.2, individual ions can be trapped by means of either inhomogeneous static electric and magnetic fields (Penning trap) [30] or due to time-dependent electromagnetic fields (radio-frequency or Paul trap) [31]. In the following we will introduce the basic ingredients we need to construct a QRM from a trapped-ion Hamiltonian, to later discuss its validity to retrieve the aimed dynamics involving its superradiant QPT.

5.1.1 Trapped-Ion Platform and QRM

Consider a single atomic ion, where two internal electronic states, denoted by $|g\rangle$ and $|e\rangle$ are separated by a frequency ω_I. The ion, in turn is placed in the trap, which to a very good approximation may be considered to produce a harmonic confinement potential, with frequency ν. Recall that ν has been used in previous chapters to denote a critical exponent, however, for consistency with customary trapped-ion notation,

ν corresponds in this chapter to the frequency of the trap. Note that, while the ion is trapped along any direction, we are just interested in the motion along one axis, say x-axis. In addition, we neglect the so-called micromotion when describing the ion position $\mathbf{r}$ (see Chap. 2, Sect. 2.2 for a discussion, as well as the review [2]). The interaction of the trapped ion with distinct classical light fields is well described by the following Hamiltonian ($\hbar = 1$),

$$H_{\text{TI}} = \frac{\omega_I}{2}\sigma_z + \nu a^{\dagger}a + \sum_j \frac{\Omega_j}{2}\sigma_x \left[e^{i(\mathbf{k}_j\cdot\mathbf{r}-\omega_j t-\phi_j)} + \text{H.c.}\right], \qquad (5.1)$$

where the wave-vector $\mathbf{k} = (k_x, k_y, k_z)$, amplitude or Rabi frequency Ω_j, frequency ω_j and initial phase ϕ_j characterize the jth classical radiation source. It is important to remark that the previous Hamiltonian is only valid as long as $\omega_j \approx \omega_I$ which allows to reduce the full, generally complex, internal state structure of the ion to a simple two-level system, $|g\rangle$ and $|e\rangle$, such that the qubit is described by the usual spin-$\frac{1}{2}$ algebra, namely $\sigma_x = |e\rangle\langle g| + |g\rangle\langle e|$, $\sigma_y = -i\,|e\rangle\langle g| + i\,|g\rangle\langle e|$ and $\sigma_z = |e\rangle\langle e| - |g\rangle\langle g|$. Furthermore, the light fields are considered to have a detuning such that only the motion along the x-axis is excited or allowed (as we have seen in Chap. 2, by tuning trap frequencies one can effectively achieve a one-dimensional system). Therefore, we can simply consider $\mathbf{k}_j \cdot \mathbf{r} = k_j x$ where the position of the ion in the harmonic trap can be written as $x = (2m\nu)^{-1/2}\left(a + a^{\dagger}\right)$, where m denotes the ion mass. Then, it is insightful to introduce the so-called Lamb-Dicke parameter $\eta_j = k_j(2m\nu)^{-1/2}$ which accounts for the ratio between the extension of the ground-state motional wave function and the wavelength of the applied radiation. The amplitude or Rabi frequency Ω_j of the jth light field depends on the specific coupling between the electronic ion states and the electromagnetic field $\mathbf{E}_0$, typically involving dipolar or quadrupolar transitions. In this manner, $\Omega_j/2 = e\,\langle g|\,\mathbf{E}_0 \cdot \mathbf{r}\,|e\rangle$ or $\Omega_j/2 = e(k/2)\,\langle g|\,x(\mathbf{E}_0 \cdot \mathbf{r})\,|e\rangle$ with e the electron charge, respectively. Note however that, states coupled through quadrupolar transitions are typically used to encode qubits since they exhibit smaller spontaneous emission rates, i.e., longer coherence time, than dipolar transitions, which are commonly used for measurements. Yet, there are other schemes relying on adiabatic elimination of intermediates states, ultimately leading to a Hamiltonian of the form of Eq. (5.1), as it is the case of the so-called Raman scheme. Briefly, in this scheme the qubit is encoded by two ground states, connected near resonance via dipolar coupling by electromagnetic fields with a third excited state. Since this auxiliary state is short lived and sufficiently detuned, its population is negligible and can be adiabatically eliminated [32]. Then, the parameters in H_{TI} correspond to differences between the parameters of the two light fields, (see [2] for more details). In either of these schemes, the effective coupling between radiation and electronic states is captured by the last term in Eq. (5.1).

It is worth recalling that, since the ions possess many different internal levels, there is not a unique manner to encode a qubit. However, one demands long coherence time, and thus, the sought internal states must feature a very slow spontaneous emission rate (long lifetime), that is, they are either part of the electronic ground state or metastable states, for which transitions to lower states are forbidden. Typically, a qubit is encoded following two cases, namely, either using different electronic states, as typically done in a $^{40}Ca^+$ [33, 34], or by means of hyperfine levels of the electronic ground state, as in $^{171}Yb^+$ [22, 35]. In the following we will resort to the former, $^{40}Ca^+$, where the metastable states $|S_{1/2}, m_j = 1/2\rangle$ and $|D_{5/2}, m_j = 3/2\rangle$ are normally utilized to encode the qubit, which are separated by $\omega_I = 2\pi \times 4 \cdot 10^{14}$ Hz that corresponds to an optical wavelength of 729 nm. In addition, to better understand the forthcoming approximations, it is helpful to keep in mind typical parameter values for trapped-ion experiments. For example, trap frequencies ν amount to few MHz, the amplitudes or Rabi frequencies of the light fields lie in the range of kHz and thus $\omega_j \approx \omega_I \gg \nu \gg \Omega_j$ (although $|\omega_j - \omega_I|$ may be of the order of ν), while the Lamb-Dicke parameter results typically in $\eta \sim 10^{-1}$ or 10^{-2} [33, 34].

The main task now consists in attaining a QRM from H_{TI}. For that purpose, we closely follow the procedure reported in [28], which we present briefly in the following lines. Note that, while a single trapped-ion experiment may explore the QRM in distinct parameter regimes, as the deep-strong coupling [36], it is certainly not straightforward to accomplish a faithful realization of the QRM in the sought extreme parameter regime. The strategy consists in (i) moving to an interaction picture with respect to the free energy terms, $H_{TI,0} = \omega_I \sigma_z/2 + \nu a^\dagger a$ such that $H_{TI} = H_{TI,0} + H_{TI,1}$, (ii) approximating the resulting Hamiltonian using a rotating-wave approximation (RWA) (see Appendix B for an explanation) and considering the Lamb-Dicke regime, $\eta\sqrt{\langle(a + a^\dagger)^2\rangle} \ll 1$, and finally (iii) identifying a suitable interaction picture in which the trapped-ion Hamiltonian adopts the form of the QRM.

In the interaction picture, $H_{TI,1}^I \equiv \mathcal{U}_{TI,0}^\dagger H_{TI,1} \mathcal{U}_{TI,0}$ with $\mathcal{U}_{TI,0} = e^{-i \int dt' H_{TI,0}(t')}$ the time-evolution propagator, the Hamiltonian adopts the following form

$$H_{TI,1}^I = \sum_j \frac{\Omega_j}{2} \left(\sigma^+ e^{i\omega_I t} + \sigma^- e^{i\omega_I t}\right) \left[e^{i\eta_j(ae^{-i\nu t} + a^\dagger e^{i\nu t})} e^{-i\omega_j t} e^{-\phi_j} + \text{H.c.}\right]. \quad (5.2)$$

Then, since we are dealing with near resonant irradiation, $|\omega_I - \omega_j| \ll |\omega_I + \omega_j|$ and because $\Omega_j \ll \omega_I$, we can safely apply a RWA to eliminate the counter-rotating terms, i.e., those rotating with frequency $\omega_I + \omega_j$. This step is known as optical RWA. Moreover, requiring $\eta\sqrt{\langle(a + a^\dagger)^2\rangle} \ll 1$, that is, within the Lamb-Dicke regime, the exponential terms can be expanded as

$$e^{i\eta(a+a^\dagger)} \approx I + i\eta(a + a^\dagger) - \frac{\eta^2}{2}(a + a^\dagger)^2 + \mathcal{O}(\eta^3(a + a^\dagger)^3). \quad (5.3)$$

Hence, to a very good approximation the trapped-ion Hamiltonian becomes

$$H_{TI,1}^I \approx \sum_j \frac{\Omega_j}{2} \left[\sigma^+ \left\{I + i\eta(ae^{-i\nu t} + a^\dagger e^{i\nu t})\right\} e^{i(\omega_I - \omega_j)t} e^{-i\phi_j} + \text{H.c.}\right]. \quad (5.4)$$

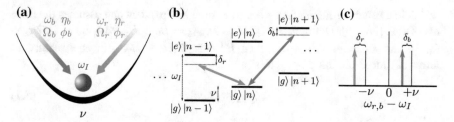

Fig. 5.1 **a** Schematic representation of a trapped-ion irradiated by two bichromatic lasers, driving detuned blue- and red-sidebands. Two internal electronic states, separated by a frequency ω_I, encode the qubit, while the motion of ion represents the bosonic degree of freedom, with frequency ν. In **b** we sketch the transitions driven by red- and blue-sidebands, as they create interaction terms of the type $(\sigma^+ a + \text{H.c.})$ (Jaynes-Cummings) and $(\sigma^+ a^\dagger + \text{H.c.})$ (anti-Jaynes-Cummings), respectively. The laser frequencies can be slightly detuned from these sidebands, i.e., $\omega_{r,b} = \omega_I \mp \nu - \delta_{r,b}$, as represented in **c**. See main text for further details

The final step consists in tuning the laser frequency ω_j such that it coincides with a multiple of the motional frequency ν. For example, for a single resonant driving (denoted by the subscript c), $\omega_c = \omega_I$, the Hamiltonian adopts the form of a *carrier excitation*, $H_{\text{TI},1}^I \approx \Omega_c/2(\sigma^+ e^{-i\phi_c} + \sigma^- e^{-i\phi_c})$. This is accomplished after neglecting terms rotating with frequency ν or higher, known as vibrational RWA, valid as long as the amplitude of the neglected terms is much smaller than the frequency at which they rotate. In particular, for the the carrier resonance we find the condition $\Omega\eta \ll \nu$, that holds in this setup. However, a more interesting situation appears when tuning the frequencies to $\omega_{r,b} = \omega_I \mp \nu$, since it results in a coherent interaction between the qubit and the motion, namely

$$H_{\text{TI},1}^I \approx \frac{\Omega_r}{2}\left[\sigma^+ a e^{-i\phi_r} + \text{H.c.}\right] + \frac{\Omega_b}{2}\left[\sigma^+ a^\dagger e^{-i\phi_b} + \text{H.c.}\right], \qquad (5.5)$$

where the first (last) term corresponds to a *red-sideband (blue-sideband) excitation*, producing transitions of the type $|g\rangle |n\rangle \leftrightarrow |e\rangle |n-1\rangle$ ($|g\rangle |n\rangle \leftrightarrow |e\rangle |n+1\rangle$). Note that the previous Hamiltonian is valid for $\Omega_{r,b} \ll \nu$. Recognizably, the interaction created by a red-sideband excitation is formally equivalent to a Jaynes-Cummings, while the blue-sideband leads to an anti-Jaynes-Cummings term, which essentially entangles the motion of the ion with its internal degrees of freedom, or in other words, the ion motion depends on the state $|e\rangle$ or $|g\rangle$. Recall that these interactions are similar to those investigated in the realm of cavity QED [3, 37].

For the realization of the QRM with a single trapped-ion we consider two detuned red- and blue-sidebands, namely, $\omega_{r,b} = \omega_I \mp \nu - \delta_{r,b}$, such that $|\delta_{r,b}| \ll \nu$. As we have explained previously, H_{TI} transforms into

$$H_{\text{TI},1}^I \approx -\frac{\eta\Omega}{2}\left[\sigma^+(a e^{i\delta_r t} + a^\dagger e^{i\delta_b t}) + \text{H.c.}\right], \qquad (5.6)$$

which in a suitable rotating frame becomes time independent and assumes the form of a QRM [28, 29]. Defining the QRM as $\tilde{H}_{\text{QRM}} = \tilde{H}_0 + \tilde{H}_1$, with $\tilde{H}_0 = \tilde{\Omega}/2\sigma_z + \tilde{\omega}_0 a^\dagger a$ and $\tilde{H}_1 = -\tilde{\lambda}\sigma_x(a + a^\dagger)$, then $\tilde{H}_1^I \equiv \tilde{\mathcal{U}}_0^\dagger \tilde{H}_1 \tilde{\mathcal{U}}_0 \approx H_{\text{TI},1}^I$, or equivalently, in terms of time-evolution propagators

$$\tilde{\mathcal{U}}_{\text{QRM}} \approx \mathcal{U}_E \mathcal{U}_{\text{TI}} \tag{5.7}$$

with $\mathcal{U}_E \equiv \tilde{\mathcal{U}}_0 \mathcal{U}_{\text{TI},0}^\dagger = e^{-it((\tilde{\Omega}-\omega_I)/2\sigma_z + (\tilde{\omega}_0-\nu)a^\dagger a)}$. The parameters of the simulated QRM are related to those of the trapped-ion setup as

$$\tilde{\Omega} = \frac{1}{2}\left(\delta_b + \delta_r\right), \quad \tilde{\omega}_0 = \frac{1}{2}\left(\delta_b - \delta_r\right), \quad \text{and} \quad \tilde{\lambda} = \frac{\eta\Omega}{2}. \tag{5.8}$$

We emphasize that the QRM parameters are denoted in this chapter by $\tilde{\Omega}$, $\tilde{\omega}_0$ and $\tilde{\lambda}$, not to be confused with the utilized trapped-ion frequencies. Note that we have considered equal Lamb-Dicke parameters and Rabi frequencies for both drivings, $\eta_{r,b} \equiv \eta$ and $\Omega_{r,b} \equiv \Omega$, as well as $\phi_{r,b} = 3\pi/2$. See Fig. 5.1 for a schematic representation of the setup, the excited transitions and the detuned frequencies of the bichromatic light fields.

It is worth mentioning that, while higher-order sidebands may be excited by tuning $\omega_{nr,nb} = \omega_I \mp n\nu$, their amplitudes become decreasingly smaller, $\Omega_{nr,nb}\eta^n/2(n!)$ and at the same time, the vibrational RWA deteriorates. Nevertheless, one can still find suitable parameters that allow the inspection of models with non-linear spin-boson interaction terms [26, 38].

The superradiant QPT of the QRM emerges in the limit $\tilde{\Omega}/\tilde{\omega}_0 \gg 1$ at the critical point $\tilde{g} = 2\tilde{\lambda}/\sqrt{\tilde{\omega}_0\tilde{\Omega}} = 1$. Hence, because the realization of the QRM in the trapped-ion setup is crucially constrained by the Lamb-Dicke regime and the validity of the considered RWAs, it is not evident that H_{TI} can reproduce the QRM in such a demanding parameter regime. In particular, as we have discussed in Chaps. 3 and 4, the number of bosonic excitations increases proportionally to $\tilde{\Omega}/\tilde{\omega}_0$, leading to a potential departure from the Lamb-Dicke regime, while at the same time reaching the critical point requires large Rabi frequencies, compromising the validity of the vibrational RWA. We will comment later on that a detuned carrier excitation, coming from each of the detuned sidebands drivings, appears as the leading-order contribution deteriorating the correct functioning of the QRM realization.

Finally, besides the actual validity of the previous approximations, preparation of an initial motional vacuum is relevant for our purposes. We briefly comment that applying well-established cooling techniques (resolved sideband cooling) the state can be brought close to the motional ground state $|0\rangle$, while the internal qubit states can be prepared easily by generating pulse sequences. Indeed, considering simply a red-sideband driving, and a decay rate $\Gamma_c \ll \nu$ such that produces transitions $|e\rangle \to |g\rangle$, one can attain a extremely small phonon population $\langle a^\dagger a \rangle \approx (\Gamma_c/\nu)^2 \ll 1$ [2].

5.1.2 Protocol and Trapped-Ion Simulations

In order to probe the QPT, we rely on an adiabatic ramp towards the critical point, starting from $\tilde{g}(0) = 0$ to $\tilde{g}(\tau_Q) \approx \tilde{g}_c$, as we have investigated in Chap. 4. Indeed, as $\tilde{\Omega}/\tilde{\omega}_0$ is finite, for sufficiently slow passages, the dynamics becomes to all effects adiabatic and the equilibrium scaling functions can be retrieved (see Chap. 3, Sect. 3.2.3), although the required time increases with increasing $\tilde{\Omega}/\tilde{\omega}_0$. This involves preparing an initial state $|\psi(0)\rangle = |0\rangle \, |g\rangle$ and then linearly increasing in time the Rabi frequencies, $\Omega(t) = \Omega_f t/\tau_Q$ with $\Omega_f = \tilde{g}_f \sqrt{\delta_b^2 - \delta_r^2}/(2\eta)$ with $0 \le \tilde{g}_f \le \tilde{g}_c = 1$. Recall that Rabi frequencies can be modulated in a time-dependent fashion. In this manner, for different quench rates τ_Q, final coupling $\tilde{g}_f$ and frequency ratio $\tilde{\Omega}/\tilde{\omega}_0$, we can reconstruct the finite-frequency scaling functions, both in equilibrium F_A (for $\tau_Q \gg 1$) and nonequilibrium $S_A(x, T)$ in the same manner as discussed in Chaps. 3 and 4. In the following we only consider σ_z as it can be measured with high fidelity in trapped ions, while measurements on the bosonic degree of freedom are typically more difficult to perform [2]. Nevertheless, if different observables can be experimentally accessed, the procedure is essentially the same as for σ_z, although one must remember that $A \propto |\tilde{g} - \tilde{g}_c|^{\gamma_A}$ is required. Recall also from Chap. 4 that the scaling variables for the QRM are $x \equiv \tilde{\Omega}/\tilde{\omega}_0 |\tilde{g} - \tilde{g}_c|^{3/2}$ and $T \equiv \tau_Q (\tilde{\Omega}/\tilde{\omega}_0)^{-1}$.

We performed the numerical simulations with realistic and feasible parameters, namely, $\tilde{\omega}_0 = 2\pi \times 200$ Hz, and $2\pi \times 10 \le \tilde{\Omega} \le 2\pi \times 80$ kHz, which enables the exploration of reasonably high frequency ratios $50 \le \tilde{\Omega}/\tilde{\omega}_0 \le 400$. Note that the previous parameters correspond to the frequencies of the realized QRM. The maximum Rabi frequency, depending on the specific value of $\tilde{\Omega}/\tilde{\omega}_0$ results in $2\pi \times 23.6 \le \Omega_f \le 2\pi \times 66.6$ kHz considering $\eta = 0.06$, and a trap frequency $\nu = 2\pi \times 1.36$ MHz [33]. In order to retrieve equilibrium scaling functions, we consider a quench rate $\tau_Q = 50 \times 2\pi/(\tilde{\omega}_0) = 250$ ms, while for the nonequilibrium function much faster protocols can be performed, $0.1 \le \tau_Q \tilde{\omega}_0/(2\pi) \le 2$, that is, $0.5 \le \tau_Q \le 10$ ms. Under the adiabatic protocol, $\tau_Q = 50 \times 2\pi/(\tilde{\omega}_0)$, the overlap between the time-evolved trapped-ion state $|\psi(g(t))\rangle$ and the actual QRM ground state $\left| \varphi_{np}^0(g) \right\rangle$ becomes $| \langle \psi(g_c)| \varphi_{np}^0(g_c) \rangle | \approx 0.99$ for $\tilde{\Omega}/\tilde{\omega}_0 = 400$, while for smaller $\tilde{\Omega}/\tilde{\omega}_0$ and $g < g_c$ it is highly enhanced. Recognizably, the quench time to approximately meet adiabatic dynamics is too long compared to typical decoherence times and noise rates affecting the setup. We leave however the discussion of the potential impact of noises to 5.1.4, while in 5.2 we will discuss how to cope with a typical and relevant source of decoherence, namely, magnetic-field fluctuations.

The numerical simulations of the trapped ion system have been performed assuming solely the optical RWA, that is, simulating the following Hamiltonian

$$H_{\mathrm{TI}}^I = \sum_{j=r,b} \frac{\Omega_j}{2} \left[\sigma^+ e^{i\eta_j (ae^{-i\nu t} + a^\dagger e^{i\nu t})} e^{i(\omega_I - \omega_j)t} e^{-i\phi_j} + \mathrm{H.c.} \right], \qquad (5.9)$$

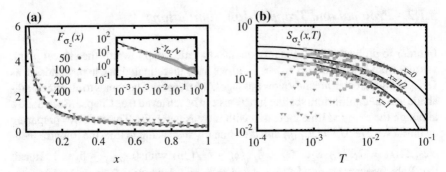

Fig. 5.2 Trapped ion reconstruction of the finite-frequency scaling functions for σ_z of the QRM to probe the superradiant QPT. The ideal scaling functions of the QRM are depicted with solid black lines, while the trapped-ion results are displayed with color points for the different $\tilde{\Omega}/\tilde{\omega}_0$ values, namely, 50, 100, 200 and 400. In **a** we represent the equilibrium scaling function $F_{\sigma_z}(x)$, and in the inset we show it in logarithmic scale to indicate its scaling behavior. The nonequilibrium finite-frequency scaling function $S_{\sigma_z}(x, T)$ is plotted in **b**, for $x = 0, 1/2$ and 1. Trapped-ion results were obtained from Eq. (5.9) considering a quench time $\tau_Q = 50 \times 2\pi/(\tilde{\omega}_0) = 250$ ms in **a** and $0.1 \leq \tau_Q \tilde{\omega}_0/(2\pi) \leq 2$ in **b**. The deviation for large values of $\tilde{\Omega}/\tilde{\omega}_0$ obscures the observation of the universality in the dynamics of the QRM. See main text for further details

where the parameters and their relation with those of the QRM have been given above (also collected in Table 5.1, presented later under the name of zeroth layer, notation which will become more evident in Sect. 5.2). The numerical results have been plotted in Fig. 5.2, together with the ideal QRM scaling functions. In (a) we represent the equilibrium finite-frequency scaling function $F_{\sigma_z}(x)$ of the QRM and its reconstruction by means of trapped-ion simulations. As one can observe, the agreement between the QRM and trapped-ion simulations deteriorates as $\tilde{\Omega}/\tilde{\omega}_0$ increases, as the data no longer collapse into the single curve $F_{\sigma_z}(x)$. Speeding up the passage does not help neither, as one can deduce from Fig. 5.2b for the nonequilibrium finite-frequency scaling function $S_{\sigma_z}(x, T)$. Indeed, $\tilde{\Omega}/\tilde{\omega}_0 = 400$ provides the worse results, leading to scattered points, and thus hindering the observation of the scaling function. Recall that different values of x convey a distinct final coupling according to $x = \tilde{\Omega}/\tilde{\omega}_0 |\tilde{g}_f - \tilde{g}_c|^{3/2}$.

As aforementioned, a detuned carrier excitation appears as the main spurious contribution for the correct functioning of the required approximations to attain a QRM, becoming dominant for increasing $\tilde{\Omega}/\tilde{\omega}_0$. As we proposed in [25] and explain in the following, the effect of such a carrier excitation can be overcome utilizing a standing-wave disposition of the light fields.

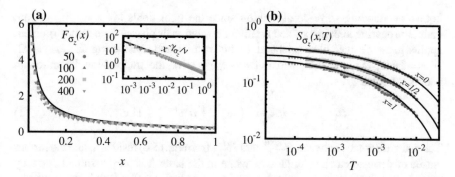

Fig. 5.3 Trapped ion reconstruction of the finite-frequency scaling functions for σ_z of the QRM to probe the superradiant QPT. The ideal scaling functions of the QRM are depicted with solid black lines, while the trapped-ion results based on a standing-wave configuration, are displayed with color points for the different $\tilde{\Omega}/\tilde{\omega}_0$ values, namely, 50, 100, 200 and 400. Compare the results with those plotted in Fig. 5.2 for a running-wave scheme. In **a** we represent the equilibrium scaling function $F_{\sigma_z}(x)$, and in the inset we show it in logarithmic scale to indicate its scaling behavior. The nonequilibrium finite-frequency scaling function $S_{\sigma_z}(x, T)$ is plotted in **b**, for $x = 0$, $1/2$ and 1. Trapped-ion results were obtained from Eq. (5.9) using standing-wave light fields driving detuned red- and blue-sidebands with 8% error between $\Omega_{r,b}$ and $\Omega_{rs,bs}$. In **a** the quench time amounts to $\tau_Q = 50 \times 2\pi/(\tilde{\omega}_0) = 250$ ms, and $0.1 \leq \tau_Q \tilde{\omega}_0/(2\pi) \leq 2$ in **b**. See main text for further details

5.1.3 Standing-Wave Configuration

The validity of Eqs. (5.6) and (5.7) relies on a number of approximations, namely, optical and vibrational RWAs as well as on the Lamb-Dicke regime. The main contribution to the breakdown of the realization of the QRM stems from a detuned carrier excitation produced by detuned red- and blue-sidebands, of the form $H_{d.c.} = \sum_{j=r,b} \frac{\Omega_j}{2} [i\sigma^+ e^{i(\pm\nu+\delta_j)t} + \text{H.c.}]$. Note that, for $\tilde{\Omega}/\tilde{\omega}_0 = 400$, the Rabi frequencies results in $\Omega_{r,b} = 2\pi \times 66.6$ kHz while $\delta_b = 2\pi \times 80.2$ kHz and $\delta_b = 2\pi \times 79.8$ kHz. Therefore, $|\pm\nu - \delta_{r,b}|/\Omega_{r,b} \approx 20$ which may cause already relevant and undesired excitations (see Appendix B). In order to correctly realize a QRM with large $\tilde{\Omega}/\tilde{\omega}_0$ values, we resort to a standing-wave disposition of the light fields to eliminate these detuned carrier excitations [39]. For that, we introduce two additional radiation sources, denoted by rs and bs subscripts, such that $\omega_{rs} = \omega_r = \omega_I - \nu - \delta_r$ and $\omega_{bs} = \omega_b = \omega_I + \nu - \delta_b$. In general, the combination of two detuned red-sideband drivings leads to (up to η^2)

$$H_{\text{TI},1}^I \approx \frac{1}{2} \left[\sigma^+ e^{i(\nu+\delta_r)t} \left(\Omega_r e^{-i\phi_r} + \Omega_{rs} e^{-i\phi_{rs}} \right) \right.$$
$$\left. + i\sigma^+ a e^{i\delta_r t} \left(\eta_r \Omega_r e^{-i\phi_r} + \eta_{rs} \Omega_{rs} e^{-i\phi_{rs}} \right) + \text{H.c.} \right]. \quad (5.10)$$

Hence, choosing $\Omega_r = \Omega_{rs}$ with $\eta_r = -\eta_{rs}$ and $\phi_r = \phi_{rs} + \pi$, the detuned carrier excitation is canceled out (actually, all the excitation terms accompanied by odd powers η^{2n+1}). This relation between the Lamb-Dicke parameters and phase is known as standing-wave configuration since the light fields $\mathbf{k}_r \cdot \mathbf{r} = -\mathbf{k}_{rs} \cdot \mathbf{r}$, are counterprogating keeping a fixed relation between their phases. The same argument applies to the detuned blue-sideband. In this manner, and considering $\Omega_{r,b,rs,bs} \equiv \Omega$, $\phi_{r,b} = 3\pi/2$, $\phi_{rs,bs} = \pi/2$, and $\eta_{r,b} = -\eta_{rs,bs} = \eta$, the trapped-ion Hamiltonian adopts the form

$$H_{\mathrm{TI},1}^{I,SW} \approx -\eta\Omega \left[\sigma^+ \left(ae^{i\delta_r t} + a^\dagger e^{i\delta_b t}\right) + \mathrm{H.c.}\right]. \tag{5.11}$$

The main difference between $H_{\mathrm{TI},1}^{I,SW}$ and $H_{\mathrm{TI},1}^{I}$ from Eq. (5.6) resides in that the former is free of detuned carrier excitations, while in the latter it is only assumed that they will produce a vanishing impact. In addition, the resulting coupling in the standing-wave scheme is doubled with respect to the running wave scheme, $\tilde{\lambda} = \eta\Omega$, while the qubit and mode frequency of the simulated QRM are equal, $\tilde{\omega}_0 = (\delta_b - \delta_r)/2$ and $\tilde{\Omega} = (\delta_b + \delta_r)/2$.

A standing-wave configuration of the light fields is however experimentally challenging to attain, and furthermore, it may also lead to significant imperfections in its implementation [40]. For that reason, and in order to illustrate that the exploration of the superradiant QPT in the QRM is possible even in the presence of certain imperfections, we perform numerical simulations of the trapped-ion Hamiltonian including errors in the Rabi frequencies, that is, $\Omega_{r,b} \neq \Omega_{rs,bs}$. Therefore, the spurious carrier excitation is not completely canceled out, and at the same time, the interaction terms are not perfectly realized. The numerical simulations up to 8% of error among $\Omega_{r,b}$ and $\Omega_{rs,bs}$, provide a good agreement with the targeted QRM. Note that these conditions are experimentally feasible, as reported in [40]. The results using a standing-wave configuration are shown in Fig. 5.3, with the same format as in Fig. 5.2 for the running-wave scheme. As we observe, and in contrast to the running-wave scheme, the QRM finite-frequency scaling functions are now well reproduced. The parameters used for the simulations are the same as in Sect. 5.1.2 with the exception of the Rabi frequency, which halves, i.e., $2\pi \times 11.8 \leq \Omega_f \leq 2\pi \times 33.3$ kHz. However, although the trapped-ion simulations disclose that the critical dynamics of a superradiant QPT can be explored in a single trapped-ion setup, the potential impact of distinct noise sources has been so far not examined, such as spin dephasing and motional heating. The rest of this chapter is devoted to the inspection and analysis of particular noise sources in a realistic trapped-ion implementation.

5.1.4 Effects of Noise

The quench time considered for the adiabatic preparation of the ground state, required to retrieve the equilibrium scaling function $F_{\sigma_z}(x)$, lies definitely in a region in which the effect of noise sources cannot be disregarded ($\tau_Q = 250$ ms). Moreover, as $\langle\sigma_z\rangle$ adopts values close to -1, any eventual spurious $|g\rangle \rightarrow |e\rangle$ transition will cause a substantial impact in $S_{\sigma_z}(x, T)$. It is therefore pertinent to analyze whether the

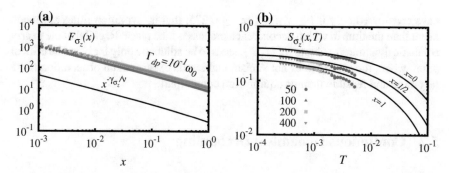

Fig. 5.4 Equilibrium and nonequilibrium finite-frequency scaling functions for σ_z of the QRM under the presence of realistic noise sources. The ideal scaling functions of the QRM are depicted with solid black lines, while the results including noise (see Eq. (5.12)) are depicted with color points for the different $\tilde{\Omega}/\tilde{\omega}_0$ values, namely, 50, 100, 200 and 400 In we represent the equilibrium scaling function $F_{\sigma_z}(x)$, largely affected by noise hindering the observation of universal equilibrium properties of the superradiant QPT. The nonequilibrium finite-frequency scaling function $S_{\sigma_z}(x, T)$ however turns out to be much more robust against noise, which is plotted in **b**, for $x = 0, 1/2$ and 1. the quench times are shortened to $0.1 \cdot 2\pi/(\tilde{\omega}_0) \leq \tau_Q \leq 0.275 \cdot 2\pi/(\tilde{\omega}_0)$, enabling the observation of the universal function $S_{\sigma_z}(x, T)$. See main text for further details

universal scaling functions can be observed under realistic experimental conditions. For that, we rely on a phenomenological master equation that governs the adiabatic evolution of a H_{QRM}, that is,

$$\dot{\rho} = -i[H_{\text{QRM}}(t), \rho] + \Gamma_{\text{dp}}\mathcal{L}[\sigma_z] + \Gamma_c\mathcal{L}[\sigma_-] + \Gamma_a\mathcal{L}[a] + \Gamma_h\mathcal{L}[a^\dagger],$$

where $\mathcal{L}[x] = x\rho x^\dagger - x^\dagger x\rho/2 - \rho x^\dagger x/2$ represents the customary Lindbladian superoperators [41]. Typical, yet optimistic, noise rates can be estimated as $\Gamma_{\text{dp}} = 2\pi \times 20$ Hz and $\Gamma_{c,a,h} = 2\pi \times 10$ Hz. Therefore, according to the considered trapped-ion parameters, it corresponds to $\Gamma_{\text{dp}}/\tilde{\omega}_0 = 0.1$ and $\Gamma_{c,a,h}/\tilde{\omega}_0 = 0.05$. We then solve the dynamics with these parameters to reconstruct the universal scaling functions $F_{\sigma_z}(x)$ and $S_{\sigma_z}(x, T)$. In Fig. 5.4 we plot the results, which indicate that, while the equilibrium scaling function $F_{\sigma_z}(x)$ is strongly modified by the noise, its nonequilibrium counterpart $S_{\sigma_z}(x, T)$ is much more robust to the effect of noise. The data for different $\tilde{\Omega}/\tilde{\omega}_0$ does not collapse onto the predicted universal function $F_{\sigma_z}(x)$ since the quench time is much longer than the coherence time of 50 ms. Hence, the universality of equilibrium properties is lost. Recall that although the scaling remains ($\sim x^{-\gamma_{\sigma_z}/\nu}$), data collapse is an indispensable requirement for the existence of $F_{\sigma_z}(x)$ and so for the QPT (see Introduction, Sect. 1.1.2 for a discussion). A closer inspection reveals that a noise rate $\Gamma_{\text{dp}}/\tilde{\omega}_0 \lesssim 10^{-3}$, together with $\Gamma_{c,a,h} = \Gamma_{\text{dp}}/2$, would be required for its experimental observation. On the other hand, the nonequilibrium function $S_{\sigma_z}(x, T)$ is a good candidate to probe the universal dynamics of a superradiant QPT. The resilience of this function stems from the relatively short quench times compared to noise rates. Indeed, in Fig. 5.4b we have considered only $0.1 \leq \tau_Q \tilde{\omega}_0/(2\pi) \leq 0.275$, that is, evolution times are much faster than the time in which decoherence processes take place. Recall that the nearly adiabatic dynamics to obtain $S_{\sigma_z}(x, T)$ needs to be adiabatic only far from the critical point, where the energy does not vanish, making its experimental observation more favorable and feasible than its equilibrium counterpart.

5.2   Continuous Dynamical Decoupling

The development of theoretical and experimental schemes to overcome the impact of different noise sources, and thus preserving quantum coherence of a particular system, is essential for the correct functioning of technologies exploiting quantumness. As aforementioned, among the different theoretical methods to prolong quantum coherence, we find decoherence-free subspaces [42] and dynamical decoupling [16]. In this regard, it is worth noting that these methods can handle distinct noise scenarios. In particular, dynamical decoupling emerges as a promising tool to handle noise exhibiting finite-width spectral density. Succinctly, in its continuous configuration, the method consists in creating a dressed basis whose energy gap is sufficiently large such that the effect of noise is eliminated. In this section we focus on examining the applicability of dynamical decoupling techniques in the reign of ion traps, allowing us to cope with a main source of decoherence in these setups, namely, spin dephasing. Here, we show that dynamical decoupling methods can be applied to a single trapped-ion setup realizing the QRM, prolonging its quantum coherence, and ultimately leading to a faithful inspection of the universal critical dynamics of the QRM. Moreover, although continuous dynamical decoupling has been already proposed [18–21] and experimentally realized [22, 23] in trapped ions, this technique allows us to apply decoupling in a *concatenated* manner, that is, introducing radiation sources consecutively to eliminate further noise sources. This is known as *concatenated continuous decoupling* (CCD), which has been proposed and experimentally demonstrated in a nitrogen-vacancy center in diamond [43].

Therefore, while the development of a robust and noise resilient implementation of a QRM with a single trapped ion can certainly be advantageous for the inspection of universal critical dynamics, it might also serve to explore various appealing parameter regimes of the QRM, such as the deep-strong coupling regime [27, 36] or the regime in which QRM reduces to the Dirac equation [33, 34, 44, 45]. We start in Sect. 5.2.1 explaining the basic operating principle of continuous dynamical decoupling, and then in Sect. 5.2.2, we discuss how this can be carried out to realize a robust QRM with a single trapped ion. In addition, in Sect. 5.2.3 we discuss the feasibility of a CCD scheme in the realm of trapped ions to attain a QRM. Finally, we comment that the developed techniques can be beneficial for the implementation of different spin-boson models, as for example a robust two-photon QRM in a trapped-ion setup [26, 38].

5.2.1 Basic Operating Principle

In the following lines we explain the basic operating principle of continuous dynamical decoupling. We are interested on the decoherence of a qubit caused by the unavoidable presence of experimental imperfections or uncontrolled interactions with the environment. Let simply consider a pure dephasing noise, $H = \omega_0 \sigma_z/2 + \delta_m(t)\sigma_z/2$. We do not consider transverse noise components because their effect is much less relevant than the dephasing noise in a trapped ion, $\delta_m(t)\sigma_z/2$. Recall that neglecting transverse noise effectively leads to an infinitely large T_1, while T_2 depends on the fluctuating term $\delta_m(t)$. However, as we will restrict ourselves to evolution times $\sim T_2$, this is justified. We will come back to this issue later.

Introducing a continuous driving field with amplitude Ω_D and frequency ω_D, the Hamiltonian becomes

$$H = \frac{\omega_0}{2}\sigma_z + \frac{\delta_m}{2}\sigma_z + \Omega_D \cos(\omega_D t)\sigma_x. \tag{5.12}$$

In a rotating frame with respect to $\omega_0 \sigma_z/2$, choosing $\omega_D = \omega_0$ and invoking a RWA, valid if $\Omega_D \ll \omega_0$, the Hamiltonian adopts a simple form

$$H^I \approx \frac{\delta_m(t)}{2}\sigma_z + \frac{\Omega_D}{2}\sigma_x. \tag{5.13}$$

Since $\delta_m(t)$ can be decomposed in Fourier space as $\delta_m(t) = \int d\omega e^{i\omega t}\tilde{\delta}_m(\omega)$, the previous Hamiltonian can be well approximated by $\Omega_D/2\sigma_x$ when $|\tilde{\delta}_m(\omega)| \ll |\omega \pm \Omega_D|$. Hence, if Ω_D lies in a region in which the noise spectrum is negligible, the qubit will be protected against the noise $\delta_m(t)/2\sigma_z$, being encoded instead in a dressed basis $\Omega_D/2\sigma_x$. In particular, the amplitude of the light field, Ω_D takes a double role: it determines the energy gap of the dressed basis, while at the same time eliminates the noise. In order to achieve a successful elimination of the noise, we have relied on a RWA, and therefore, because a RWA presents slightly different performance depending on the considered initial states, the proposed method inherits its dependence. This indeed is of particular relevance when dealing with states for which the noise does only introduce a global phase (in this case, $|e\rangle$ or $|g\rangle$), since the inclusion of a continuous dynamical decoupling protection rotates the basis, enabling the noise to produce transitions.

In Fig. 5.5 we show a specific noise model for the stochastic fluctuation $\delta_m(t)$, namely, an Orstein-Uhlenbeck (OU) noise [46, 47], which is a Markov and Gaussian stochastic process with an exact update formula and non-zero correlation time [48, 49]. This model correctly reproduces the observed exponential decay of coherences [32], and it is widely used to model these fluctuations [18, 19, 21, 43]. We explain in the following its application to a trapped-ion setup. Briefly, the expecta-

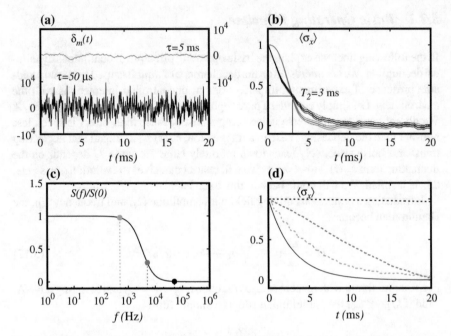

Fig. 5.5 Illustration of the basic operating principle of the continuous dynamical decoupling to cope with a noise $\delta_m(t)\sigma_z/2$, which is modeled as an Orstein-Uhlenbeck (OU) process. In **a** two evolution runs for $\delta_m(t)$ with relaxation time $\tau = 50$ µs (black) and $\tau = 5$ ms (gray) and intensity such that $T_2 = 3$ ms. After averaging 1000 stochastic trajectories, the expectation value $\langle\sigma_x(t)\rangle$ for an initial state $|\uparrow\rangle_x = (|e\rangle + |g\rangle)/\sqrt{2}$, as shown in **b** for both characteristic types of noise, $\tau \ll T_2$ and $\tau \gtrsim T_2$, whose main trait consists in an exponential or Gaussian decay, respectively. By definition, $\langle\sigma_x(T_2)\rangle = e^{-1}$. In **c** we show the noise spectrum $S(f)$, and three corresponding driving amplitudes Ω_D which result in the curves shown in **d** (from light-gray to black dashed lines). The larger Ω_D, the more is the noise suppressed, attaining larger coherence times. Recall that $\Omega \ll \omega_0$ must be also fulfilled. See main text for details

tion value $\langle\sigma_x(t)\rangle$ for an initial state $|\uparrow\rangle_x = (|e\rangle + |g\rangle)/\sqrt{2}$ decays as a consequence of this fluctuation. Selecting driving amplitudes Ω_D for which the noise spectrum becomes small and the RWA can be applied, $\langle\sigma_x(t)\rangle$ remains closer to the ideal and noiseless value, $\langle\sigma_x(t)\rangle = 1$ (see Fig. 5.5c and d) and the further developments.

5.2.2 Trapped-Ion Setup and Continuous Dynamical Decoupling

In this part we comment how to apply the previously explained method to obtain a robust QRM with a single trapped ion. As we have seen, the previous method copes with pure dephasing noise affecting a qubit. Hence, continuous dynamical decoupling appears as a promising tool to protect the ion internal states from magnetic fluctuations, which is relevant when the qubit is encoded in magnetically sensitive internal states, as for example in the case of the states $|D_{5/2}, m_j = 3/2\rangle$ and $|S_{1/2ss}, m_j = 1/2\rangle$ utilized in the $^{40}Ca^+$ ion [33, 34]. Under these conditions, magnetic-dephasing noise emerges as the main source of decoherence, since phonon heating and qubit decay are expected to produce an impact in a much longer time scale, ~ 100 ms and ~ 1 s, respectively [50]. In the same manner, trapped-ion experiments involving magnetic-field gradients are also affected by magnetic-field fluctuations, leading to similar coherence times, as in the case of the microwave-driven $^{171}Yb^+$ ion [22]. See Table 1.1 in [51] for a comparison of coherence times between different ion species. Therefore, no further sources of noise have been included in our analysis since the examined total evolution time is, at least, one order of magnitude shorter than their typical rates, and thus, their effect is expected to be negligible. In the following we first present how to account for magnetic-field fluctuations in the trapped-ion Hamiltonian, how to model them, and then, how to apply CCD methods to overcome this noise while at the same time realizing a QRM.

5.2.2.1 Magnetic-Field Fluctuations

To exemplify the suitability of CCD methods, we consider a relatively short, yet realistic, coherence time $T_2 = 3$ ms as a consequence of magnetic-field fluctuations. Besides the well-controlled trapped-ion Hamiltonian, H_{TI} given in Eq. (5.1), noisy terms which can be enclosed in a stochastic Hamiltonian description adding terms of the form $H_{n,} = \delta_m^\alpha(t)\sigma_\alpha/2$ for $\alpha = x$, y and z, where $\delta_m^\alpha(t)$ stands for a stochastic time-dependent fluctuation. Note however that while transverse noise, $\alpha = x, y$, produces transitions between internal levels and is related to qubit decay (T_1), its parallel component gives account of pure dephasing noise, or equivalently, T_2. Since for trapped-ion setups $T_2 \ll T_1$, and we constrain ourselves to evolution times $\sim T_2$, we only scrutinize the role of the parallel noise component (pure dephasing), and therefore, the noisy trapped-ion Hamiltonian becomes (by simplicity we denote $\delta_m(t) \equiv \delta_m^z(t)$) $\tilde{H}_{TI} = \delta_m(t)\sigma_z/2 + H_{TI}$, that is,

$$\tilde{H}_{TI} = \frac{\delta_m(t)}{2}\sigma_z + \frac{\omega_I}{2}\sigma_z + \nu a^\dagger a + \sum_j \frac{\Omega_j}{2}\sigma_x \left[e^{i(\eta(a+a^\dagger)-\omega_j t-\phi_j)} + \text{H.c.}\right]. \quad (5.14)$$

These magnetic-field fluctuations can be modeled as an OU noise [46–49]. While this simple Gaussian noise model allows to gain theoretical insight, as its time-evolution is known exactly, it provides a finite-width spectral density and correctly reproduces experimental observations [43]. Hence, it has become a good description of realistic noise in various setups, as in trapped-ions or nitrogen-vacancy centers in diamond [43]. We leave the details of the OU process for the Appendix F, and quote here the main useful results for our discussion.

Let $\delta_m(t)$ be a stochastic variable that follows an OU process. Then, since OU noise is Gaussian, $\delta_m(t)$ is fully determined by its first two moments, i.e., by two parameters.

These parameters are commonly known as relaxation time τ and diffusion constant c. The exact update formula for $\delta_m(t)$ reads [48, 49]

$$\delta_m(t) = \delta_m(0)e^{-t/\tau} + \left[\frac{c\tau}{2}\left(1 - e^{-2t/\tau}\right)\right]^{1/2} N(t) \tag{5.15}$$

such that $N(t)$ stands for a normal-distributed random variable, that is, $\overline{N(t)} = 0$ and $\overline{N(t)N(t')} = \delta(t - t')$. Note that the overline stands for the stochastic ensemble average. Hence, $\delta_m(t)$ experiences fluctuations around its zero-mean value, $\overline{\delta_m(t)} = 0$, as depicted in Fig. 5.5a for two different parameters τ and c. The spectral density reads

$$S(f) = \frac{2c\tau^2}{1 + 4\pi^2\tau^2 f^2}, \quad f \geq 0, \tag{5.16}$$

whose form has been plotted in Fig. 5.5c. For latter developments, it is insightful defining a characteristic frequency $f_{\text{cr}} = 1/(2\pi\tau)$ at which the spectral density becomes half of its zero-frequency value, i.e., $S(f_{\text{cr}}) = S(0)/2$.

The relaxation time of the noise, τ, sets the width of the spectral density, while the diffusion constant gives account of the intensity of the noise and will be fixed to correctly reproduce the coherence time T_2. Indeed, in Ref. [21], $\tau \approx 100$ μs is proposed as a good estimate for the correlation time of the magnetic noise. For the results presented here we take $\tau = 50$ μs. Then, the diffusion constant c is chosen such that the definition of coherence time is satisfied, $\langle\sigma_{x,y}(T_2)\rangle = e^{-1}\langle\sigma_{x,y}(0)\rangle$. Considering simply an initial state $|\uparrow\rangle_x = (|e\rangle + |g\rangle)/\sqrt{2}$ evolving under $H_{n,z}$ for a single run of the stochastic fluctuation leads to

$$\langle\sigma_x(t)\rangle = \cos\left(\Xi(t)\right), \quad \text{with} \quad \Xi(t) = \int_0^t dt'\, \delta_m(t'). \tag{5.17}$$

and hence, $\overline{\langle\sigma_x(T_2)\rangle} = e^{-1/2\overline{\Xi^2(t)}}$. Since $\delta_m(t)$ is a Gaussian noise, it is possible to obtain an analytic expression for $\overline{\Xi^2}(t)$, allowing us to establish the required relation between τ, c and T_2 (see Appendix F)

$$c = \frac{2}{\tau^2\left[T_2 - \frac{\tau}{2}\left(3 - 4e^{-T_2/\tau} + e^{-2T_2/\tau}\right)\right]}. \tag{5.18}$$

Therefore, knowing τ and choosing a specific coherence time T_2, the diffusion constant follows from the previous expression. For $\tau \ll T_2$, the expression further simplifies to $c \approx 2/(\tau^2 T_2)$, which conveys an exponential decay of $\langle\sigma_x(t)\rangle$. This is indeed the typical situation in trapped-ion setups, as observed experimentally [32], since $T_2 \sim$ ms and $\tau \sim$ μs. On the other limit, $\tau \gtrsim T_2$, we find a Gaussian decay of the coherence. This is illustrated in Fig. 5.5b, where we show the analytic result $\langle\sigma_x(t)\rangle$

and the one obtained numerically after averaging 1000 stochastic trajectories for two cases, $\tau = 50\,\mu s$ and $\tau = 5$ ms, both featuring $T_2 = 3$ ms.

Note that the Fourier components $\tilde{\delta}_m(\omega)$ follow from the spectral density. The basic operating principle of the continuous dynamical decoupling is exemplified in Fig. 5.5c and d. In the former the spectral density is plotted, highlighting three different frequencies Ω_D, which correspond to $\overline{\langle \sigma_x(t) \rangle}$ in (d). For larger frequencies $S(\Omega_D/(2\pi)) \ll 1$, qubit coherence is preserved during longer times, prolonging T_2. Having explained continuous dynamical decoupling and how to model dynamical decoupling, we proceed now to implement finally this method for obtaining a QRM in a trapped-ion setup.

5.2.3 Concatenated Continuous Dynamical Decoupling

Our goal here consists in applying the idea developed in Sect. 5.2.1 for the elimination of magnetic-field fluctuations from Eq. (5.14). In particular, under same the procedure followed in Sect. 5.1, the noisy trapped-ion Hamiltonian $\tilde{H}_{TI}$ would still realize a QRM, but with an explicit noise contribution, $\tilde{H}_{QRM} = \frac{\delta_m(t)}{2}\sigma_z + H_{QRM}$, which jeopardizes its correct and faithful realization. Note that this simple result follows because the considered approximations and rotating frames commute with σ_z. We will refer to this unprotected method (the strategy presented in Sect. 5.1) as zeroth-layer.

5.2.3.1 First Layer

In order to eliminate magnetic-field fluctuations, a driving at resonant frequency $\omega_{D1} = \omega_I$ and amplitude $\Omega_{D1} > f_{cr}$ is imperative. Evidently, the Rabi frequency is much smaller than the qubit transition frequency, $\Omega_{D1} \ll \omega_I$, and thus, $\tilde{H}_{TI}$ in a rotating frame with respect to $H_{TI,0} = \omega_I\sigma_z/2 + \nu a^\dagger a$, after the optical RWA reads

$$\tilde{H}^I_{TI,1} \equiv \mathcal{U}^\dagger_{TI,0}\tilde{H}_{TI,1}\mathcal{U}_{TI,0} \approx \frac{\delta_m(t)}{2}\sigma_z + \frac{\Omega_{D1}}{2}\left[\sigma^+ e^{i\eta_{D1}(ae^{-i\nu t}+a^\dagger e^{i\nu t})} + \text{H.c.}\right]$$
$$+ \sum_j \frac{\Omega_j}{2}\left[\sigma^+ e^{i\eta_j(ae^{-i\nu t}+a^\dagger e^{i\nu t})}e^{i((\omega_I-\omega_j)t-\phi_j)} + \text{H.c.}\right],$$

$$(5.19)$$

where we have already chosen $\phi_{D1} = 0$. In this manner, we can encode now the qubit in the dressed basis defined by $\Omega_{D1}\sigma_x/2$, in which the effect of $\delta_m(t)\sigma_z/2$ is suppressed. We remark that the Lamb-Dicke parameter η_{D1} can adopt an arbitrary small value, although not necessarily zero. Moreover, within the Lamb-Dicke regime and applying the vibration RWA, the Jaynes-Cummings interaction terms are achieved by driving red- and blue-sidebands, $\omega_{r,b} = \omega_I \mp \nu - \delta_{r,b}$. Indeed, choosing

Table 5.1 Trapped-ion parameters to simulate the QRM using CCD scheme. Zeroth layer corresponds to the scheme presented in Sect. 5.1. Qubit basis and interaction stand for the spin axis in which the terms in QRM appear. Recall that we have considered $\Omega_{r,b} \equiv \Omega$ and $\eta_{r,b} \equiv \eta$.

	Zeroth layer	First layer	Second layer
ω_r	$\omega_I - \nu - \delta_r$	$\omega_I - \nu - \delta_r$	$\omega_I - \nu - \delta_r$
ω_b	$\omega_I + \nu - \delta_b$	$\omega_I + \nu - \delta_b$	—
ω_{D1}	—	ω_I	ω_I
ω_{D2}	—	—	ω_I
$\phi_{r,b}$	$3\pi/2$	$3\pi/2$	$3\pi/2$
ϕ_{D1}	—	0	0
ϕ_{D2}	—	—	$\pi/2$
Qubit basis	σ_z	σ_x	σ_y
Interaction	σ_x	σ_y	σ_x
$\tilde{\Omega}$	$\frac{1}{2}(\delta_b + \delta_r)$	Ω_{D1}	Ω_{D2}
$\tilde{\omega}_0$	$\frac{1}{2}(\delta_b - \delta_r)$	$-\delta_r$	$-\delta_r$
$\tilde{\lambda}$	$\frac{\eta\Omega}{2}$	$\frac{\eta\Omega}{2}$	$\frac{\eta\Omega}{4}$

$\phi_{r,b} = 0$ and $\delta_b = -\delta_r = \delta$, the previous Hamiltonian adopts a more recognizable form ($\Omega_{r,b} \equiv \Omega$ and $\eta_{r,b} \equiv \eta$)

$$\tilde{H}^I_{\mathrm{TI},1} \approx \frac{\Omega_{D1}}{2}\sigma_x - \frac{\eta\Omega}{2}\sigma_y\left(ae^{-i\delta t} + a^\dagger e^{i\delta t}\right). \tag{5.20}$$

Indeed, the previous Hamiltonian corresponds to a QRM in a rotating frame of $\delta a^\dagger a$. That is, $\mathcal{U}_{\mathrm{QRM}} \approx \mathcal{U}_E \mathcal{U}_{\mathrm{TI}}$ where now $\mathcal{U}_E = e^{-it(-\omega_I \sigma_z/2 + (\delta - \nu)a^\dagger a)}$. The realized QRM has a rotated spin basis compared with the zeroth-layer method, and it is protected against magnetic-field fluctuations as long as $\Omega_{D1} > f_{\mathrm{cr}}$. This latter requirement constraints the parameter regime of the realized QRM since Ω_{D1} directly gives the qubit frequency of the QRM, i.e., $\tilde{\Omega} = \Omega_{D1}$. Nevertheless, one can still rely on different strategies to achieve small values of $\tilde{\Omega}$ while at the same time protecting the setup against dephasing noise [27]. Recall that we have utilized three radiation sources, driving a carrier excitation, and detuned red- and blue-sidebands. In Table 5.1 we collect and present the used trapped-ion parameters as well as how they relate to those of the resulting QRM, while numerical results are discussed after introducing an additional layer of protection.

5.2.3.2 Second Layer

The main idea behind the CCD scheme consists in applying consecutively radiation sources that average out further noise sources. Here we demonstrate the applicability of such a concatenated scheme to realize a QRM, shielded not only against magnetic-field fluctuations, as in the previous case (first layer), but also against laser amplitude fluctuations, another relevant noise source [1]. Let consider that the Rabi frequencies are not completely stable but fluctuate around Ω_j as $\Omega_j(t) = \Omega_j(1 + \delta_{\Omega_j}(t))$. Moreover, once magnetic-field fluctuations have been overcome, laser-amplitude fluctuations may still spoil quantum coherence within the considered time scale. For that, we characterize $\delta_{\Omega_j}(t)$ again as an OU process, although with a slower noise, $\tau_\Omega = 1$ ms following [1], while taking a relative strength of 0.1% ($p = 0.001$), that is, $c_\Omega = 2p^2/\tau_\Omega$. In this manner, the main source of noise resulting from the first layer appears to be $\Omega_{D1}\delta_{\Omega_{D1}}(t)\sigma_x/2$. Note that the latter causes pure dephasing noise in the realized QRM by the first layer scheme. Therefore, it can be handled in the same manner as we did to eliminate $\delta_m(t)\sigma_z/2$ from Eq. (5.14). The main complication however resides in that an additional driving needs to be introduced such that, after the first interaction picture, the same procedure as explained in Sect. 5.2.1 can be applied. In particular, the trapped-ion Hamiltonian $\tilde{H}^I_{\text{TI},1}$ after the optical RWA (as well as assuming $\eta_{Dk} \ll 1$ and $\Omega_{Dk}\eta_k \ll \nu$ for $k = 1, 2$) becomes

$$\tilde{H}^I_{\text{TI},1} \approx \frac{\delta_m(t)}{2}\sigma_z + \frac{\Omega_{D1}}{2}\sigma_x + \frac{\Omega_{D1}\delta_{\Omega_{D1}}}{2}\sigma_x + \Omega_{D2}\cos(\Omega_{D1}t)\left[\sigma^+ e^{-i\phi_{D2}} + \text{H.c.}\right]$$
$$+ \sum_j \frac{\Omega_j}{2}\left[\sigma^+ e^{i\eta_j(ae^{-i\nu t}+a^\dagger e^{i\nu t})}e^{i((\omega_I-\omega_j)t-\phi_j)} + \text{H.c.}\right], \quad (5.21)$$

where we have considered a modulated Rabi frequency $\Omega_{D2}\cos(\Omega_{D1}t)$ for the an auxiliary carrier driving, $\omega_{D2} = \omega_I$.[1] In addition, for simplicity the laser-amplitude fluctuation is only explicitly written for the Ω_{D1} term. Recall that in order to eliminate noise in the σ_x basis, we must provide an orthogonal driving, and thus we choose $\phi_{D2} = \pi/2$. Then, we move to an additional rotating frame with respect to $\Omega_{D1}\sigma_x/2$. Assuming that $\Omega_{D1} > f_{\text{cr}}$, $\delta_m(t)$ is averaged out, while at the same time we require $\Omega_{D2} \ll \Omega_{D1}$ to further simplify the Hamiltonian neglecting terms rotating at frequencies Ω_{D1}. Therefore, we finally obtain

$$\tilde{H}^{II}_{\text{TI},1} \approx \frac{\Omega_{D2}}{2}\sigma_y + \sum_j \frac{\Omega_j}{4}\left[\sigma_x e^{i\eta_j(ae^{-i\nu t}+a^\dagger e^{i\nu t})}e^{i((\omega_I-\omega_j)t-\phi_j)} + \text{H.c.}\right]. \quad (5.22)$$

The noise $\Omega_{D1}\delta_{\Omega_{D1}}\sigma_x/2$ is canceled if $\Omega_{D2} > 1/(2\pi\tau_\Omega)$, which together with the condition $\Omega_{D1} > f_{\text{cr}} = 1/(2\pi\tau)$ leads to $\tau_\Omega \ll \tau$. The latter requirement can be though of as a necessary condition to successfully eliminate this noise by means of CCD scheme. From previous developments is now evident that Eq. (5.22) corresponds to a QRM driving a single sideband, say $\omega_r = \omega_I - \nu - \delta_r$. In Table 5.1 we show the used trapped-ion parameters as well as how they relate to those of the resulting QRM. Finally, we comment that the realization of a QRM in a parameter

---

[1]Note that a time-independent Rabi frequency could have been used instead. For that, one would have to tune the frequency $\omega_{D2} = \omega_I + \Omega_{D1}$ in order to provide the required resonant terms with $\Omega_{D1}\sigma_x/2$.

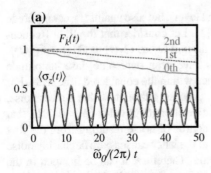
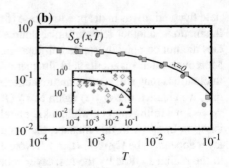

Fig. 5.6 Realization of the QRM characteristic dynamics in two interesting regimes. In **a** we show $\langle \sigma_z(t) \rangle$ (oscillating curve) and the fidelity of each of the considered trapped-ion scheme with the ideal and noiseless QRM, for a resonant QRM and $\tilde{g} = 1/4$. Red, blue and green lines correspond to zeroth, first and second layer, while solid-black lines to the ideal QRM results. The initial state is $|\psi(0)\rangle = |0\rangle (|e\rangle + |g\rangle)/\sqrt{2}$. Recall that for the first and second layer, the spin basis rotated. In **b** we show the universal scaling function $S_{\sigma_z}(x, T)$ and the data points obtained considering a first layer and $\tilde{\Omega}/\tilde{\omega}_0 = 50$ (red circles) and 100 (blue squares). In the inset the results for the zeroth and second layer. Trapped-ion results have been obtained after averaging 200 (**a**) and 100 (**b**) stochastic trajectories. See main text for further details

regime in which both $\tilde{\Omega} \gg \tilde{\omega}_0$ and $\tilde{\lambda} \gg \tilde{\omega}_0$ is expected to fail because $\tilde{\Omega} = \Omega_{D2}$, while $\Omega_{D2} \ll \Omega_{D1}$ to satisfy the RWA. Hence, although an inspection of the critical dynamics is hindered, this scheme may be useful to obtain a faithful realization of the QRM in other appealing regimes.

5.2.3.3 Numerical Simulations

We show the suitability of the CCD scheme to obtain a faithful realization of the QRM in presence of magnetic-field fluctuations as well as laser-amplitude noise. For that, using a trapped-ion Hamiltonian with their corresponding number of drivings and parameters (see Table 5.1), we realize the QRM in two parameter regimes. First, in the resonant case, $\tilde{\Omega} = \tilde{\omega}_0$ and coupling $\tilde{g} = 1/4$, whose main and emblematic hallmark resides in the Jaynes-Cummings oscillations. Second, in the limit of $\tilde{\Omega}/\tilde{\omega}_0 \gg 1$ under a nearly adiabatic protocol to reconstruct the universal scaling function $S_{\sigma_z}(x, T)$, as discussed in Chap. 4 and in Sect. 5.1 of this chapter. In the following paragraph we provide a list of the parameters used for the numerical simulations that correspond to the results plotted in Fig. 5.6.

We take as $\nu = 2\pi \times 1.36$ MHz as the trap frequency, while the Lamb-Dicke parameters $\eta_{r,b} = \eta = 0.06$ and $\eta_{D1,D2} = 0.01$ for the considered light fields. As aforementioned, we take a coherence time of $T_2 = 3$ ms, and model the magnetic-field fluctuations as an OU process with $\tau = 50$ μs, while laser-amplitude noise features $\tau_\Omega = 100$ ms and a $p = 0.001$ of relative strength. Then, to realize a resonant QRM with $\tilde{\omega}_0 = \tilde{\Omega} = 2\pi \times 5$ kHz and $\tilde{g} = 1/4$ (recall $\tilde{g} = 2\tilde{\lambda}/\sqrt{\tilde{\omega}_0 \tilde{\Omega}}$), we select: (zeroth layer) $\delta_b = 2\pi \times 10$ kHz, $\delta_r = 0$, and $\Omega_{r,b} = 2\pi \times 20.83$ kHz, (first layer) $\delta_r = -\delta_b = 2\pi \times 5$ kHz, $\Omega_{D1} = 2\pi \times 5$ kHz and $\Omega_{r,b} = 2\pi \times 20.83$ kHz, and (second layer) $\delta_r = 2\pi \times 5$ kHz, $\Omega_{D2} = 2\pi \times 5$ kHz and $\Omega_{D1} = 40\Omega_{D2} = 2\pi \times 200$ kHz, $\Omega_r = 2\pi \times 41.67$ kHz. For the inspection of the dynamics of the QRM involving the QPT, we consider $\tilde{\omega}_0/\tilde{\Omega} = 50$ and 100, and adiabatically quenching the ground state towards the critical point, $\tilde{g} = 1$. The considered parameters of the QRM are $\tilde{\omega}_0 = 2\pi \times 1$ kHz for the zeroth and first layer, while $\tilde{\omega}_0 = 2\pi \times 400$Hz for the second layer, where the qubit frequency is simply given by $\tilde{\Omega}/\tilde{\omega}_0 = 50$ and 100. The final Rabi frequency, required to reach the critical coupling parameter in a time $0.02 \leq \tau_Q\tilde{\omega}_0/(2\pi) \leq 8.6$, amounts to $2\pi \times 117.8$ ($2\pi \times 166.7$) kHz for $\tilde{\Omega}/\tilde{\omega}_0 = 50$ (100) and zeroth and first layers, while for the second it results in $2\pi \times 94.3$ ($2\pi \times 133.3$) kHz. Moreover, in the second layer we set a large amplitude, $\Omega_{D2} = 2\pi \times 200$ kHz, for which the failure of the required RWA to attain the QRM (Eq. (5.22)) is evident since the ratio Ω_{D1}/Ω_{D2} becomes too small, namely 10 and 5 for the two considered values $\tilde{\Omega}/\tilde{\omega}_0$. Decreasing further Ω_{D2} leads however to long evolution times, and at the same time, reducing the performance on suppressing laser-amplitude noise.

In Fig. 5.6 we plot the results of the numerical simulations, for the three schemes using a trapped-ion Hamiltonian together with the aimed ideal and noiseless QRM dynamics. We remark that, as for the results presented in Sect. 5.1, we assume solely the optical RWA. In addition, magnetic-field and laser-amplitude fluctuations have been included in the three schemes. In (a) we show the representative Rabi oscillations of the QRM in a resonant case, considering an initial state $|\psi(0)\rangle = |0\rangle (|e\rangle + |g\rangle)/\sqrt{2}$, and plotting $\langle\sigma_z(t)\rangle$ as well as the state fidelity $F_k(t) = |\langle\psi_{\mathrm{TI},k}(t)| \psi_{\mathrm{QRM}}(t)\rangle|$ where $|\psi_{\mathrm{TI},k}(t)\rangle$ represents the trapped-ion evolved wave function making use of the kth layer and $|\psi_{\mathrm{QRM}}(t)\rangle$ the targeted quantum state evolved under the ideal H_{QRM}. In this particular case, the performance of the each subsequent layer improves the previous, achieving $F_2 \sim 0.99$ at the end of the evolution, in contrast to the poor $F_0 \sim 0.65$.

Nevertheless, as the CCD scheme relies on the performance of the RWA, different initial states behave in a diverse fashion. This becomes clear dealing with states for which the initially fought noise generates just a global phase. In this case, this happens for states like $|n\rangle |e\rangle$ or $|n\rangle |g\rangle$ evolving under a weakly coupled QRM in the zeroth layer. For these very cases, which have not been plotted here, introducing layers of protection becomes disadvantageous as in the dressed basis the noise may cause relevant transitions. Finally, we inspect the critical dynamics of the QRM by reconstructing $S_{\sigma_z}(x, T)$, as shown in Fig. 5.6b for $x = 0$ and using the first layer scheme, while in the inset the results using zeroth and second layer are plotted. The remarkable improvement of the first compared to the zeroth layer contrasts with the failure of the second layer. Note that, as discussed after attaining Eq. (5.22), the latter scheme is based on a RWA that breaks down in this parameter regime. Therefore, depending on the specific setup conditions, parameter regime and aimed states, CDD may be optimized to accomplish enhanced performances.

5.3 Conclusion and Outlook

In this chapter we have shown that the dynamics of the QRM in the extreme parameter regime required to observe precursors of the QPT can be realized in a trapped-ion experiment. In this setup, the QRM is realized by means of two light fields driving near-resonance red- and blue-sidebands, and thus, coherently coupling the internal electronic states of the ion (encoding the qubit) with the motion along one direction of the trap [28, 29]. However, the parameter regime in which the QPT emerges challenges its correct realization using a trapped-ion Hamiltonian. Recall that since the latter adopts the form of a QRM after a number of approximations, it is not evident that critical traits of the QRM can be observed under realistic conditions.

In order to probe the critical dynamics of the QRM, we rely on the finite-frequency scaling functions, widely discussed in previous Chaps. 3 and 4. Although we mainly focused on the spin population, σ_z, because it is experimentally more accessible than phonon observables, the latter would disclose also rich behavior, as shown in Chap. 4. Pushing the limit of the accessible $\tilde{\Omega}/\tilde{\omega}_0 \gg 1$ values, we find that carrier excitations coming from detuned sidebands strongly deteriorate the scaling functions, obscuring the universal behavior. We solve this limitation by using a standing-wave disposition of the light fields, which despite of being experimentally demanding, will definitely allow us to accomplish larger $\tilde{\Omega}/\tilde{\omega}_0$ ratios. Moreover, by analyzing realistic noise sources we find that nonequilibrium finite-frequency scaling function is more robust against noise than its equilibrium counterpart since the former can be attained by faster drivings, i.e., shorter quench times than those required to adiabatically prepare the ground state. We also consider the applicability of continuous dynamical methods to implement a noise resilient, yet tunable, QRM in a trapped-ion setup, and the followed procedure to cope with magnetic-field fluctuations is detailed. Moreover, we corroborate that the recently proposed *concatenated continuous dynamical decoupling* in the realm of nitrogen-vacancy centers in diamond [43], can be also applied to trapped ions, where further sources of noise are tackled.

This chapter completes and complements previous chapters involving the QRM, finally undertaking here its possible experimental implementation. As a matter of fact, we have corroborated that a single trapped-ion experiment can explore the dynamics of a superradiant QPT, typically appearing in the thermodynamic limit of infinitely many particles. In this respect, and despite the great advances in trapped-ion technologies, scaling up the number of ions to observe critical phenomena, while preserving quantum coherence, controllability and efficient state reconstruction still constitutes a daunting task [10, 11]. In this manner, the QRM emerges as an excellent candidate where quantum critical features can be examined, without the need of scaling up the system and thus, without affecting quantum control and coherence of a single quantum register.

Needless to say, multiple issues regarding possible experimental conditions remain to be analyzed and determined, as well as potential schemes to preserve quantum coherence. Indeed, while the present chapter has been intended to be sufficiently general, the implementation of the proposed schemes need to be optimized depending on particular conditions. Moreover, since the QRM is an ubiquitous model in nature, its realization is not constrained to trapped ions, but quite the opposite. Indeed, another platforms relevant for quantum technologies, such as circuit and cavity QED [3, 4], cold atoms [6, 7] or even in spin-mechanical systems [8], may enable the exploration of distinct features of the universal dynamics of a superradiant QPT with the most fundamental quantum entities, a qubit and a single mode.

References

1. H.Häffner, C.F. Roos, R. Blatt, Quantum computing with trapped ions. Phys. Rep. **469**, 155 (2008). https://doi.org/10.1016/j.physrep.2008.09.003
2. D. Leibfried, R. Blatt, C. Monroe, D. Wineland, Quantum dynamics of single trapped ions. Rev. Mod. Phys. **75**, 281 (2003). https://doi.org/10.1103/RevModPhys.75.281
3. S. Haroche, J.-M. Raimond, Exploring the quantum: atoms, cavities, and photons. Oxford University Press, Oxford (2006)
4. M.H. Devoret, R.J. Schoelkopf, Superconducting circuits for quantum information: an outlook. Science **339**, 1169 (2013). https://doi.org/10.1126/science.1231930
5. M. Aspelmeyer, T.J. Kippenberg, F. Marquardt, Cavity optomechanics. Rev. Mod. Phys. **86**, 1391 (2014). https://doi.org/10.1103/RevModPhys.86.1391
6. S. Felicetti, E. Rico, C. Sabin, T. Ockenfels, J. Koch, M. Leder, C. Grossert, M. Weitz, E. Solano, Quantum Rabi model in the Brillouin zone with ultracold atoms. Phys. Rev. A **95**, 013827 (2017). https://doi.org/10.1103/PhysRevA.95.013827
7. P. Schneeweiss, A. Dareau, C. Sayrin, Cold-atom based implementation of the quantum Rabi model (2017). arxiv.org/abs/1706.07781
8. M. Abdi, M.-J. Hwang, M. Aghtar, M.B. Plenio, Spin-mechanical scheme with color centers in hexagonal boron nitride membranes. Phys. Rev. Lett. **119**, 233602 (2017). https://doi.org/10.1103/PhysRevLett.119.233602
9. D. Porras, J.I. Cirac, Effective quantum spin systems with trapped ions. Phys. Rev. Lett. **92**, 207901 (2004). https://doi.org/10.1103/PhysRevLett.92.207901
10. K. Kim, M.-S. Chang, S. Korenblit, R. Islam, E.E. Edwards, J.K. Freericks, G.-D. Lin, L.-M. Duan, C. Monroe, Quantum simulation of frustrated Ising spins with trapped ions. Nature **465**, 590 (2010). https://doi.org/10.1038/nature09071
11. R. Islam, E.E. Edwards, K. Kim, S. Korenblit, C. Noh, H. Carmichael, G.-D. Lin, L.-M. Duan, C.-C. Joseph Wang, J.K. Freericks, C. Monroe, Onset of a quantum phase transition with a trapped ion quantum simulator. Nat. Commun. **2**, 377 (2011). https://doi.org/10.1038/ncomms1374
12. K. Pyka, J. Keller, H.L. Partner, R. Nigmatullin, T. Burgermeister, D.M. Meier, K. Kuhlmann, A. Retzker, M.B. Plenio, W.H. Zurek, A. del Campo, T.E. Mehlst"aubler, Topological defect formation and spontaneous symmetry breaking in ion Coulomb crystals. Nat. Commun. **4**, 2291 (2013). https://doi.org/10.1038/ncomms3291
13. S. Ulm, J. Ronagel, G. Jacob, C. Degnther, S.T. Dawkins, U.G. Poschinger, R. Nigmatullin, A. Retzker, M.B. Plenio, F. Schmidt-Kaler, K. Singer, Observation of the Kibble-Zurek scaling law for defect formation in ion crystals. Nat. Commun. **4**, 2290 (2013). https://doi.org/10.1038/ncomms3290
14. J. Zhang, G. Pagano, P.W. Hess, A. Kyprianidis, P. Becker, H. Kaplan, A.V. Gorshkov, Z.-X. Gong, C. Monroe, Observation of a many-body dynamical phase transition with a 53-qubit quantum simulator. Nature **551**, 601 (2017). https://doi.org/10.1038/nature24654
15. P. Jurcevic, H. Shen, P. Hauke, C. Maier, T. Brydges, C. Hempel, B.P. Lanyon, M. Heyl, R. Blatt, C.F. Roos, Direct observation of dynamical quantum phase transitions in an interacting many-body system. Phys. Rev. Lett. **119**, 080501 (2017). https://doi.org/10.1103/PhysRevLett.119.080501

16. A.M. Souza, G.A. 'Alvarez, D. Suter, Robust dynamical decoupling. Phil. Trans. R. Soc. A **370**, 4748 (2012). https://doi.org/10.1098/rsta.2011.0355
17. L. Viola, S. Lloyd, Dynamical suppression of decoherence in two-state quantum systems. Phys. Rev. A **58**, 2733 (1998). https://doi.org/10.1103/PhysRevA.58.2733
18. A. Bermudez, P.O. Schmidt, M.B. Plenio, A. Retzker, Robust trapped-ion quantum logic gates by continuous dynamical decoupling. Phys. Rev. A **85**, 040302 (2012). https://doi.org/10.1103/PhysRevA.85.040302
19. A. Lemmer, A. Bermudez, M.B. Plenio, Driven geometric phase gates with trapped ions. New J. Phys. **15**, 083001 (2013). http://stacks.iop.org/1367-2630/15/i=8/a=083001
20. I. Cohen, P. Richerme, Z.-X. Gong, C. Monroe, A. Retzker, Simulating the Haldane phase in trapped-ion spins using optical fields. Phys. Rev. A **92**, 012334 (2015). https://doi.org/10.1103/PhysRevA.92.012334
21. G. Mikelsons, I. Cohen, A. Retzker, M.B. Plenio, Universal set of gates for microwave dressed-state quantum computing. New J. Phys. **17**, 053032 (2015). http://stacks.iop.org/1367-2630/17/i=5/a=053032
22. N. Timoney, I. Baumgart, M. Johanning, A.F. Varon, M.B. Plenio, A. Retzker, C. Wunderlich, Quantum gates and memory using microwave-dressed states. Nature **476**, 185 (2011). https://doi.org/10.1038/nature10319
23. T.R. Tan, J.P. Gaebler, R. Bowler, Y. Lin, J.D. Jost, D. Leibfried, D.J. Wineland, Demonstration of a dressed-state phase gate for trapped ions. Phys. Rev. Lett. **110**, 263002 (2013). https://doi.org/10.1103/PhysRevLett.110.263002
24. R. Puebla, J. Casanova, M.B. Plenio, A robust scheme for the implementation of the quantum Rabi model in trapped ions. New. J. Phys. **18**, 113039 (2016). http://stacks.iop.org/1367-2630/18/i=11/a=113039
25. R. Puebla, M.-J. Hwang, J. Casanova, M.B. Plenio, Probing the dynamics of a superradiant quantum phase transition with a single trapped ion. Phys. Rev. Lett. **118**, 073001 (2017a). https://doi.org/10.1103/PhysRevLett.118.073001
26. R. Puebla, M.-J. Hwang, J. Casanova, M.B. Plenio, Protected ultrastrong coupling regime of the two-photon quantum Rabi model with trapped ions. Phys. Rev. A **95**, 063844 (2017b). https://doi.org/10.1103/PhysRevA.95.063844
27. R. Puebla, J. Casanova, M.B. Plenio, Magnetic-field fluctuations analysis for the ion trap implementation of the quantum Rabi model in the deep strong coupling regime. J. Mod. Opt. **65**, 745 (2018). https://doi.org/10.1080/09500340.2017.1404651
28. J.S. Pedernales, I. Lizuain, S. Felicetti, G. Romero, L. Lamata, E. Solano, Quantum Rabi model with trapped ions. Sci. Rep. **5**, 15472 (2015). https://doi.org/10.1038/srep15472
29. J.I. Cirac, A.S. Parkins, R. Blatt, P. Zoller, "Dark" squeezed states of the motion of a trapped ion. Phys. Rev. Lett. **70**, 556 (1993). https://doi.org/10.1103/PhysRevLett.70.556
30. L.S. Brown, G. Gabrielse, Geonium theory: physics of a single electron or ion in a Penning trap. Rev. Mod. Phys. **58**, 233 (1986). https://doi.org/10.1103/RevModPhys.58.233
31. W. Paul, Electromagnetic traps for charged and neutral particles. Rev. Mod. Phys. **62**, 531 (1990). https://doi.org/10.1103/RevModPhys.62.531
32. D.J. Wineland, C. Monroe, W.M. Itano, D. Leibfried, B.E. King, D.M. Meekhof, Experimental issues in coherent quantum-state manipulation of trapped atomic ions. J. Res. Natl. Inst. Stand. Technol. **103**, 259 (1998). https://doi.org/10.6028/jres.103.019
33. R. Gerritsma, G. Kirchmair, F. Zahringer, E. Solano, R. Blatt, C.F. Roos, Quantum simulation of the Dirac equation. Nature **463**, 68 (2010). https://doi.org/10.1038/nature08688
34. R. Gerritsma, B.P. Lanyon, G. Kirchmair, F. Z"ahringer, C. Hempel, J. Casanova, J.J. Garc'ia-Ripoll, E. Solano, R. Blatt, C.F. Roos, Quantum simulation of the Klein paradox with trapped ions. Phys. Rev. Lett. **106**, 060503 (2011). https://doi.org/10.1103/PhysRevLett.106.060503
35. S. Olmschenk, K.C. Younge, D.L. Moehring, D.N. Matsukevich, P. Maunz, C. Monroe, Manipulation and detection of a trapped Yb$^+$ hyperfine qubit. Phys. Rev. A **76**, 052314 (2007). https://doi.org/10.1103/PhysRevA.76.052314
36. J. Casanova, G. Romero, I. Lizuain, J.J. Garc'ia-Ripoll, E. Solano, Deep strong coupling regime of the Jaynes-Cummings model. Phys. Rev. Lett. **105**, 263603 (2010a). https://doi.org/10.1103/PhysRevLett.105.263603

37. S. Haroche, Nobel lecture: controlling photons in a box and exploring the quantum to classical boundary. Rev. Mod. Phys. **85**, 1083 (2013). https://doi.org/10.1103/RevModPhys.85.1083
38. S. Felicetti, J.S. Pedernales, I.L. Egusquiza, G. Romero, L. Lamata, D. Braak, E. Solano, Spectral collapse via two-phonon interactions in trapped ions. Phys. Rev. A **92**, 033817 (2015). https://doi.org/10.1103/PhysRevA.92.033817
39. J.I. Cirac, R. Blatt, P. Zoller, W.D. Phillips, Laser cooling of trapped ions in a standing wave. Phys. Rev. A **46**, 2668 (1992). https://doi.org/10.1103/PhysRevA.46.2668
40. T.E. deLaubenfels, K.A. Burkhardt, G. Vittorini, J.T. Merrill, K.R. Brown, J.M. Amini, Modulating carrier and sideband coupling strengths in a standing-wave gate beam. Phys. Rev. A **92**, 061402 (2015). https://doi.org/10.1103/PhysRevA.92.061402
41. H.-P. Breuer, F. Pretuccione, The theory of open quantum systems. Oxford University Press, Oxford (2002)
42. D.A. Lidar, Review of decoherence-free subspaces, noiseless subsystems, and dynamical decoupling, in *Quantum information and computation for chemistry* (Wiley, 2014), pp. 295–354. https://doi.org/10.1002/9781118742631.ch11
43. J.-M. Cai, B. Naydenov, R. Pfeiffer, L.P. McGuinness, K.D. Jahnke, F. Jelezko, M.B. Plenio, A. Retzker, Robust dynamical decoupling with concatenated continuous driving. New J. Phys. **14**, 113023 (2012). http://stacks.iop.org/1367-2630/14/i=11/a=113023
44. L. Lamata, J. Le'on, T. Sch"atz, E. Solano, *Dirac equation and quantum relativistic effects in a single trapped ion*, https://doi.org/10.1103/PhysRevLett.98.253005 Phys. Rev. Lett. **98**, 253005 (2007)
45. J. Casanova, J.J. Garc'ia-Ripoll, R. Gerritsma, C.F. Roos, E. Solano, Klein tunneling and Dirac potentials in trapped ions. Phys. Rev. A **82**, 020101 (2010b). https://doi.org/10.1103/PhysRevA.82.020101
46. G.E. Uhlenbeck, L.S. Ornstein, On the theory of the Brownian motion. Phys. Rev. **36**, 823 (1930). https://doi.org/10.1103/PhysRev.36.823
47. M.C. Wang, G.E. Uhlenbeck, On the theory of the Brownian motion II. Rev. Mod. Phys. **17**, 323 (1945). https://doi.org/10.1103/RevModPhys.17.323
48. D.T. Gillespie, Exact numerical simulation of the Ornstein-Uhlenbeck process and its integral. Phys. Rev. E **54**, 2084 (1996a). https://doi.org/10.1103/PhysRevE.54.2084
49. D.T. Gillespie, The mathematics of Brownian motion and Johnson noise. Am. J. Phys. **64**, 225 (1996b). https://doi.org/10.1119/1.18210
50. F. Schmidt-Kaler, S. Gulde, M. Riebe, T. Deuschle, A. Kreuter, G. Lancaster, C. Becher, J. Eschner, H. Hffner, R. Blatt, The coherence of qubits based on single Ca^+ ions. J. Phys. B At. Mol. Opt. Phys. **36**, 623 (2003). http://stacks.iop.org/0953-4075/36/i=3/a=319
51. A.H. Burrell, High fidelity readout of trapped ion qubits, Ph.D dissertation, School University of Oxford (2010)

Chapter 6
Quantum Kibble–Zurek Mechanism

The Kibble–Zurek (KZ) mechanism, the paradigmatic theory addressing nonequilibrium dynamics involving continuous phase transitions, has played a key role throughout this thesis. The main ideas of this mechanism have been explained in the Introduction and examined in the classical realm in Chap. 2. In addition, we have extended its arguments to a zero-dimensional system exhibiting a quantum phase transition (QPT), as shown in Chap. 4, which precisely explains the emergence of universal power-law relations in terms of equilibrium critical exponents. As a matter of fact, the seminal works published in 2005 [1–4] scrutinized the validity of KZ arguments when traversing a quantum critical point, and found that KZ mechanism also succeeds at predicting the witnessed power-law scaling of defect formation, as it does in classical phase transitions. In the quantum realm, although quasiparticle excitations play the role of long-lived topological defects of the traditional KZ, same *freeze-out* arguments can be still applied, which was dubbed quantum Kibble–Zurek (QKZ) mechanism. See Introduction, Sect. 1.3 for a detailed yet general derivation, and Chap. 4 for its application to the QRM. The aim of this Chapter consists in elucidating nonequilibrium dynamics of a truly and experimentally feasible quantum many-body system traversing a QPT, and whether its nonequilibrium aspects agree with QKZ predictions.

However, despite the pertinence and generality of these theoretical works [1–4], an experimental confirmation of QKZ predictions has long remained elusive. Indeed, it has only been very recently when QKZ scaling laws have been finally observed in Bose-Einstein condensates [5, 6]. The dearth of experiments in this field stems from the fact that QKZ inquiry requires bringing an entire and large quantum many-body system through a QPT in a controllable manner, while at the same time safeguarding the system from decoherence processes—certainly a tremendous and challenging task. Note in addition the experiments reported in [7, 8] where QKZ arguments were tested relying explicitly on a Landau-Zener problem.

On the other hand, since the introduction of a ferromagnetism toy model by E. Ising in 1925 [9], that now bears his name, the Ising model has became a

© Springer Nature Switzerland AG 2018

R. Puebla, *Equilibrium and Nonequilibrium Aspects of Phase Transitions in Quantum Physics*, Springer Theses, https://doi.org/10.1007/978-3-030-00653-2_6

cornerstone in statistical physics, with relevance in many and disparate areas of science. In its standard form, the Ising model comprises only nearest-neighbors interactions. In particular, its relevance does not only stem from being one of the simplest models showing classical and QPTs, but also because in certain cases it admits an exact solution [10–12]—assuredly the holy grail of statistical physics and a rare exception in the realm of many-body systems. Therefore, it is not surprising why the Ising model is regarded as a testbed for novel methods and theories in statistical mechanics. In particular, in the quantum realm, the one-dimensional Ising model with transverse field features a QPT in the thermodynamic limit of infinitely many spins $N \to \infty$ [11, 12]. Moreover, the extension of KZ mechanism to the quantum realm was exemplified first in this paradigmatic model [2–4]. We refer to the Introduction for further details regarding the Ising model, and its critical traits.

Following a recent theoretical proposal [13], it has been experimentally demonstrated that a platform of N trapped ions can be adequately engineered to reproduce the physics of an Ising model [14–20]. In particular, the resulting spin-spin interaction approximately decays as a power-law with the distance, $1/r^\alpha$, where the exponent α can be tuned. Motivated by these ground-breaking experiments in which dozens of ions have been trapped while preserving coherent interactions, the accessible number of spins might be already large enough to attain sufficiently prominent QPT precursors, and thus, posing the question of whether critical features may be examined in such a system, as QKZ mechanism. This is precisely the goal of the present Chapter, although it is worth mentioning that a similar study has been carried out recently in Ref. [21]. We remark that the results presented are solely intended to provide indications on that QKZ scaling may be readily verified in a truly quantum many-body system with state-of-the-art ion trap technology. Therefore, we leave for future work an in-depth inspection of the critical properties, both in- and out-of-equilibrium, of the long-range transverse field Ising model by means of approximate methods well suited for one-dimensional models [22] that would allow us to delve into large system sizes, otherwise unattainable.

6.1 Long-Range Transverse Field Ising Model

The long-range transverse field Ising model describes N spins interacting at long distances, unlike the nearest-neighbors interaction contemplated in its standard form. In particular, we focus on a one-dimensional system where the interaction between the ith and jth spin decays as $J_{i,j} = J_0 |i - j|^{-\alpha}$ for $i \neq j$ with $\alpha > 0$. In addition, there is a homogeneous magnetic field g orienting the spins to a transverse direction. The Hamiltonian of such a system can be written as

$$H_{\text{LRI}} = \sum_{i<j} J_{i,j} \sigma_i^x \sigma_j^x + g \sum_i \sigma_i^z, \tag{6.1}$$

which is sketched in Fig. 6.1a. Evidently, in the $\alpha \to \infty$ limit, one recovers the standard nearest-neighbors interaction. Indeed, the interplay between the three parameters of H_{LRI}, namely, J_0, α and g conceals rich physics, as we sketch in the following.

Firstly, we note that H_{LRI} is symmetric under the transformation $\sigma_i^x \to -\sigma_i^x$, $\sigma_i^z \to \sigma_i^z$. This Z_2 symmetry does not break for finite N, and allows us to split the Hilbert space as $\mathcal{H} = \mathcal{H}_+ \oplus \mathcal{H}_-$ where the $+(-)$ sign corresponds to even (odd) number of excited spins $|\uparrow\rangle$. Indeed, since $\mathcal{H}$ can be spanned by the states of the form $|\gamma_1, \gamma_2, \ldots, \gamma_N\rangle$ where γ can be either $\uparrow$ or $\downarrow$, and because $H_{LRI}|\gamma_1, \gamma_2, \ldots, \gamma_N\rangle$ does not change the parity of the total number of excited spins, $\Pi = e^{i\pi/2(\sum_i \sigma_i^z + 1)}$ is a conserved quantity. That is, the states are labeled as $\Pi|\gamma_1, \gamma_2, \ldots, \gamma_N\rangle = \pm|\gamma_1, \gamma_2, \ldots, \gamma_N\rangle$, where $+(-)$ indicates its parity symmetry.

For zero external magnetic field, $g = 0$, it is straightforward to notice that the ground state depends on the sign of J_0, which defines the ferro or anti-ferromagnetic character of the system. For $J_0 > 0$, the anti-ferromagnetic interaction provokes a staggered spin alignment, and the ground state becomes doubly degenerate, namely, a superposition of the Néel states $|\rightarrow \leftarrow \rightarrow \ldots \leftarrow\rangle$ and $|\leftarrow \rightarrow \leftarrow \ldots \rightarrow\rangle$ such that $\sigma^x|\leftarrow\rangle = +|\leftarrow\rangle$ and $\sigma^x|\rightarrow\rangle = -|\rightarrow\rangle$. The emergence of a Néel state, $|\psi_{\text{Neel}}\rangle = \frac{1}{\sqrt{2}}(|\leftarrow \rightarrow \leftarrow \ldots \rightarrow\rangle \pm |\rightarrow \leftarrow \rightarrow \ldots \leftarrow\rangle)$, or equivalently, the magnetic frustration, embodies the main trait of antiferro-magnetic interaction. On the contrary, for ferromagnetic interaction $J_0 < 0$, the spins align in absence of magnetic field resulting as well in two degenerated ground states, $|\rightarrow \rightarrow \ldots \rightarrow\rangle$ and $|\leftarrow \leftarrow \ldots \leftarrow\rangle$, which we denote by $|\psi_F\rangle$. On the other hand, for $g \gg |J_0|$, the system becomes fully polarized in the z-direction $|\psi_P\rangle = |\downarrow\downarrow \ldots \downarrow\rangle$, that is, paramagnetic in the x direction. We recall that, since $g \geq 0$, and because we will consider an even number of spins, we restrict our calculations to the even parity subspace, $\mathcal{H}_+$, as it encloses $|\psi_P\rangle$ the ground state for $g \gg |J_0|$.

In the $N \to \infty$, the Z_2 symmetry spontaneously breaks in the ferro- or anti-ferromagnetic phase $g < g_c(\alpha)$, where $g_c(\alpha)$ denotes the quantum critical point as a function of α. In addition, $g_c(\alpha)$ depends on the sign of J_0. In the thermo-

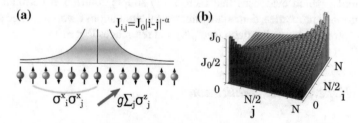

Fig. 6.1 Schematic illustration of the one-dimensional long-range Ising model with a transverse field, whose Hamiltonian is given in Eq. (6.1). The spins, which interact through $\sigma_i^x \sigma_j^x$ with a coupling $J_{i,j} = J_0|i - j|^{-\alpha}$ for $i \neq j$ (**a**), are subject to a uniform magnetic field g in the z direction. This long-range interacting model can be realized experimentally, where α can be tuned between 0 and 3 (see for further details). Indeed, in **b** we show a possible experimental coupling matrix $J_{i,j}$ for $N = 24$, where $\alpha \approx 0.5$, obtained from Eq. (6.2)

dynamic limit, and for $g < g_c(\alpha)$, the system acquires a non-zero magnetization $\langle m_F \rangle = \langle 1/N \sum_i \sigma_i^x \rangle \neq 0$ and $\langle m_{AF} \rangle = \langle 1/N \sum_i (-1)^i \sigma_i^x \rangle \neq 0$, for ferro- and anti-ferromagnetic couplings, while $m_{F,AF} \equiv 0$ in the paramagnetic phase. Indeed, $\langle m_{F,AF} \rangle$ is a good order parameter of the phase transition. Recall that for the nearest-neighbor counterpart, $\langle m \rangle \propto |g - g_c|^\beta$ with $\beta = 1/8$ its critical exponent. However, this Z_2 symmetry holds for any finite N compelling the eigenstates to $\langle m \rangle \equiv 0$. For that reason, it is convenient to compute $\langle m_{F,AF}^2 \rangle$ instead. We will discuss more in detail the ground-state properties as well as nonequilibrium dynamics in the following sections.

For $\alpha = 0$, the interaction $J_{i,j}$ looses its spatial dependence as all the spins inter-act equally strongly with the rest, and hence, the system becomes fully connected. Introducing the total spin operators $S_\beta = \sum_i \sigma_i^\beta$, it leads to $S_x^2 = \sum_{i \neq j} \sigma_i^x \sigma_j^x + N$, and therefore, H_{LRI} adopts the form of a Lipkin-Meshkov-Glick model, $H_{LMG} = J_0 S_x^2 + g S_z$ [23].

In the classical realm, the one-dimensional long-range Ising model has been sub-ject to an intense research during the last decades, aiming to determine the full phase diagram, their corresponding critical exponents as well as the connection between short-range and long-range behavior. Recall that the Hamiltonian of a classical Ising model is attained when spin operators commute, and it is typically studied in the absence of an external magnetic field, $g = 0$. As we have already mentioned in the Introduction, the standard one-dimensional Ising model does not undergo a phase transition at any finite temperature $T > 0$. This situation dramatically changes when the interaction becomes long ranged, as demonstrated by F. Dyson in [24], who proved that there is indeed a transition for ferromagnetic couplings when $1 < \alpha < 2$, while for $\alpha \leq 1$ the energy is not longer a extensive quantity. Furthermore, the phase dia-gram becomes more exotic, if possible. Indeed, for $\alpha = 2$ there is a phase transition but of a Kosterlitz–Thouless type [25] as suggested by analytical [26] and numer-ical results [27], while for $\alpha > 2$ the system recovers its short-range nature and hence, there is no phase transition at any finite temperature. The situation becomes clearly more complicated in higher dimensions (but still below the upper critical dimension[1]), which evidences that there is no simple connection between short- and long-range properties, demanding numerical simulations involving large system sizes to correctly determine them (see [28] and references therein).

6.1.1 Experimental Realization

As aforementioned, the model described in Eq. (6.1) can be implemented experimen-tally, where the range of the interactions, determined by α, can be tuned. This was first proposed in [13] and first realized in [14] involving few spins or qubits (see also

---

[1] Recall that the upper critical dimension is defined as the spatial dimension of a system after which the mean field description becomes exact. For the nearest-neighbors Ising model $d_U = 4$.

Refs. [15–18]). Nevertheless, the tremendous advances in trapped-ion technologies have made it possible to explore much richer physics of a truly quantum many-body system using the H_{LRI} as a testbed, as dozens of trapped ions can be now coherently controlled [29–31]. Needless to say, these experiments have reached a level of complexity that challenges the classical capabilities for an exact computation of spin dynamics.

We can differentiate two manners of implementing H_{LRI} in a trapped-ion setup (i) either manipulating optical transitions of $^{40}\text{Ca}^+$ ions, as carried out in the group of R. Blatt in Innsbruck [19, 31, 32], or (ii) using $^{171}\text{Yb}^+$ ions, where the qubit is encoded using hyperfine clock states [33], as performed in the laboratory led by C. Monroe [15–18, 20, 29, 30]. In the following we briefly explain the latter scheme.

Considering N atomic ions of $^{171}\text{Yb}^+$, trapped in a linear Paul trap, the qubits are encoded in the hyperfine clock states $|F = 1, m_F = 0\rangle$ and $|F = 0, m_F = 0\rangle$ of the state $^2S_{1/2}$ of the valence electron and spin-$\frac{1}{2}$ nucleus. Here F and m_F correspond to the total atomic angular momentum and its projection along the axis given by a weak magnetic field ~ 4 Gauss. These two hyperfine states encode the spin states, $|\uparrow\rangle$ and $|\downarrow\rangle$ and are separated by $\omega_{\text{HF}} = 2\pi \times 12.642$ GHz [33]. Applying two off-resonant Raman beams homogeneously onto the ions, with wave-number difference $\Delta\mathbf{k}$, and each of them driving red and blue sidebands of the transverse trapped-ion motion, $\omega_0 \pm \mu$, with μ comparable to the transverse motional frequencies, an effective Ising-type interaction is obtained

$$J_{i,j} = \Omega_i \Omega_j \frac{(\Delta k)^2}{2M} \sum_m \frac{b_{i,m} b_{j,m}}{\mu^2 - \omega_m^2} \tag{6.2}$$

where $b_{i,m}$ stands for the transformation of the ith ion with the mth normal mode of frequency ω_m [34], Ω_i represents the Rabi frequency on the ith ion, and M the ion mass, which is valid within the Lamb-Dicke regime and $|\mu - \omega_m| \gg \eta_{i,m}\Omega_i$. Note that $\eta_{i,m}$ stands for the Lamb-Dicke parameter $\eta_{i,m} = b_{i,m}\Delta k(2M\omega_m)^{-1/2}$. Hence, one can control the effective range of spin-spin interactions by tuning μ, namely, from infinitely ranged when $\mu \sim \omega_{\text{com}}$ (center of mass mode frequency) to a dipole-dipole type of interaction if $\mu \gg \omega_{\text{com}}$. In this manner, spin-spin interaction $J_{i,j}$ approximately adopts a power-law decaying form, $J_{i,j} = J_0|i - j|^{-\alpha}$ where $0 < \alpha < 3$. See Fig. 6.1b for a realistic coupling matrix $J_{i,j}$ with $\alpha \approx 0.5$. Finally, choosing the lasers with a beatnote detuning μ asymmetrically around the carrier by a value g, $\omega_0 \pm \mu + g$, a global Stark shift is produced, which has the same effect as a uniform transverse magnetic field $g \sum_i \sigma_i^z$. Note that the frequencies J_0 and g lie in the range of kHz [18, 20, 30]. Moreover, we stress that high-fidelity measurements can be performed in this platform, as well as high-fidelity state preparation. For further details, we refer to [13, 15, 16, 20].

6.1.2   Ground-State Properties

In the following we tackle the ground-state properties of a one-dimensional long-range quantum Ising model, H_{LRI}. As commented previously, when the exponent $\alpha < 2$, H_{LRI} undergoes a classical phase transition at sufficiently low temperature, but still $T > 0$, while for $\alpha > 2$, the behavior of the paradigmatic nearest-neighbors is retrieved and thus only a phase transition occurs at $T = 0$. Besides the existence of classical phase transitions, H_{LRI} features distinct QPTs in the thermodynamic limit $N \to \infty$ depending on the value of $\alpha \geq 0$. Yet, how the critical traits of H_{LRI} vary as a function of α remains to be disclosed, although big efforts have been devoted to it recently [21, 35, 36]. In particular, although it is well established that the QPT falls into the universality class of the nearest-neighbors type when $\alpha > 2$, how the critical exponents modify for smaller α values needs to be determined. In the following we sketch the main theoretical ingredients aiming to ascertain such a dependence, as well as to locate the value of the critical magnetic-field $g_c(\alpha)$.

As discussed at the beginning of this Section, we expect to observe a crossover between paramagnetic states, for $g \gg |J_0|$, where all the spins are oriented along the z-axis, to a ferromagnetic or anti-ferromagnetic ordering for $g \ll |J_0|$, depending of the sign of J_0. The properties of the ground state can be quantified resorting to the two-point correlation function,

$$C_{i,j} = \langle \sigma_i^x \sigma_j^x \rangle - \langle \sigma_i^x \rangle \langle \sigma_j^x \rangle. \tag{6.3}$$

In particular, when $g \gg |J_0|$ the ground state becomes $|\psi_P\rangle$ and thus $C_{i,j} \approx 0$ is expected. On the other hand, when $g < g_c(\alpha)$, an anti-ferromagnetic ground state emerges $|\psi_{Neel}\rangle$, or $|\psi_F\rangle$ for the ferromagnetic case. Hence, the two-point correlation function $C_{i,j}$ acquires a utterly different shape: $C_{i,j} = (-1)^{i-j}$ for anti-ferromagnetic and $C_{i,j} = 1$ for a ferromagnetic case. Loosely speaking, the critical point $g_c(\alpha)$ finds itself along the crossover between these two behaviors. In Fig. 6.2 we plot $C_{N/2,N/2+k}$ for $N = 20$ spins, $\alpha = 2.5$, and three different magnetic field values, which reveal the aforementioned crossover.

In addition, ground-state properties also disclose such crossover, which in the $N \to \infty$ are separated by a QPT. For example, the energy gap between the ground and first excited state Δ becomes minimal in the vicinity of the critical point, ultimately leading to $\Delta \propto |g - g_c|^{z\nu}$ in the $N \to \infty$ limit [11]. Moreover, in the same spirit of the two-point correlation, the overlap between the actual ground state of H_{LRI} and the states $|\psi_{Neel}\rangle$, $|\psi_F\rangle$ and $|\psi_P\rangle$ provides a good estimate of the crossover between (anti-)ferromagnetic and paramagnetic phases. Finally, it is convenient to introduce the so-called Binder cumulant [37],

$$B(N, g) = \frac{1}{2} \left(3 - \frac{\langle m^4 \rangle}{\langle m^2 \rangle^2} \right), \tag{6.4}$$

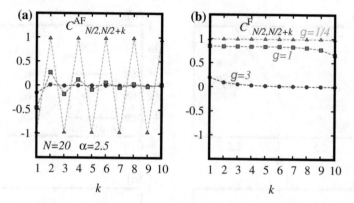

Fig. 6.2 Two-point correlation functions $C_{i,j} = \left\langle \sigma_i^x \sigma_j^x \right\rangle - \left\langle \sigma_i^x \right\rangle \left\langle \sigma_j^x \right\rangle$ on the ground state for anti-ferromagnetic ($J_0 > 0$) and ferromagnetic couplings ($J_0 < 0$), shown in **a** and **b**, respectively, and for different values of the external magnetic field g, $N = 20$ spins and an interaction decaying as $J = \pm|i - j|^{-\alpha}$ with $\alpha = 2.5$ (that is $|J_0| = 1$). For large magnetic field $g_0 \gg |J_0|$, the system shows little correlation along x direction, as it is polarized along the z axis, and thus $C_{i,j}$ adopts small values. Reducing g, the system enters in the (anti-)ferromagnetic phase, and $C_{i,j}$ reveals the predominance of a (Néel) ferromagnetic state. In **a** $C_{i,j}$ shows alternating sign due to the staggered Néel state, $|\varphi_{\text{Neel}}\rangle$. To the contrary, for ferromagnetic couplings, $C_{i,j}$ indicates a collective alignment of the spins along the x axis, as shown in **b**

where the dependence on N and g stems from $\langle m^n \rangle$, with $m^{\text{F,AF}} = 1/N \sum_i (\pm 1)^i \sigma_i^x$ the order parameter. In the $N \to \infty$, the Binder cumulant features a sharp transition at the critical point. Indeed, $B(N, g)$ is particularly helpful to locate the critical point $g_c(\alpha)$, as we comment later. These quantities are plotted in Fig. 6.3 for anti-ferromagnetic couplings. Needless to say, although $C_{i,j}$ helps to identify the nature of the ground state, it does not provide an accurate location of the critical point, which can be accomplished based on finite-size scaling theory and will be explained in the following.

6.1.2.1 Locating the Critical Point and Critical Exponents

The typical procedure to determine the position of the critical point of a phase transition and its associated critical exponents relies on finite-size scaling theory. In particular, the order parameter displays a behavior $\langle m \rangle \propto |g - g_c|^\beta$ close to the critical point g_c. As discussed in the Introduction, according to the finite-size scaling theory [38, 39], any quantity $\mathcal{A}$ behaving as $\mathcal{A} = \mathcal{A}_0 |g - g_c|^{\gamma_\mathcal{A}}$ in the thermodynamic limit admits the following form for finite N

$$\mathcal{A}(N, g) = |g - g_c|^{\gamma_\mathcal{A}} F_\mathcal{A}(|g - g_c|^\nu N), \qquad (6.5)$$

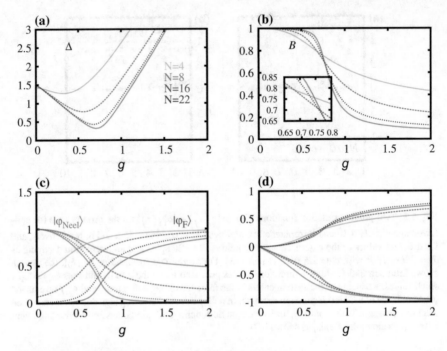

Fig. 6.3 Precursors of the QPT in a H_{LRI} with anti-ferromagnetic couplings, $J_0 > 0$, for $\alpha = 2.5$ and considering four different system sizes, $N = 4, 8, 16$ and 22. As the system size increases, ground state properties become sharper close to the critical point, as one can observe in **a** for the energy gap between the ground and second-excited state. In **b** we plot the Binder cumulant $B(N, g)$, whose intersections between different system sizes indicate an estimate of the critical point $g_c(\alpha)$. As a consequence of higher-order corrections to the finite-size scaling hypothesis, $B(N, g^*) = B(N', g^*)$ with $g^* \to g_c(\alpha)$ for $NN' \to \infty$. See the inset for a zoom close to the region where distinct $B(N, g)$ cross. The overlap of the ground state of H_{LRI}, $\left|\varphi_0^{\text{AF}}(g, \alpha)\right|$, with the expected states at $g \ll J_0$ (Néel state) and $g \gg J_0$ (fully polarized state), denoted by $|\psi_{\text{Neel}}\rangle$ and $|\psi_{\text{zP}}\rangle$, respectively, is plotted in **c**. The overlap $|\langle\varphi_0^{\text{AF}}(g, \alpha)|\varphi_{\text{Neel}}\rangle|$ is represented by red lines, while $|\langle\varphi_0^{\text{AF}}(g, \alpha)|\psi_P\rangle|$ is shown with blue lines. Finally, in **d**, we plot $\sum_i \langle\sigma_i^z\rangle/N$ (red lines) and $\sum_i^{N-1}\langle 1 + \sigma_i^x\sigma_{i+1}^x\rangle/N$ (blue lines), quantities that will be essential for addressing defect formation. See main text for details

where higher-order corrections are neglected, and with $F_A(x)$ the universal scaling function, satisfying $\lim_{x\to\infty} F_A(x) = A_0$ and $\lim_{x\to 0} F_A(x) \sim x^{-\gamma_A/\nu}$. Hence, it follows that $A(N, g = g_c) \propto N^{-\gamma_A/\nu}$. Nevertheless, the previous assumption works for large N, and thus, including the first higher-order correction, we expect $A(N, g = g_c) \propto N^{-\gamma_A/\nu}(1 + a_A N^{-\omega_A})$ where a_A is simply a constant and $\omega_A \geq 0$ a scaling exponent, which accounts for higher-order corrections and becomes sub-leading for $N \gg 1$.

At this stage, the usefulness of the Binder cumulant $B(N, g)$ becomes visible, Eq. (6.4). In particular, upon introducing Eq. (6.5) and considering $g = g_c$, $B(N, g_c)$ adopts a size-independent form. This property is extremely helpful to locate the critical point: computing $B(N, g)$ for different system sizes N and N', they intersect at one particular g value, which would correspond precisely to the critical point g_c. Nevertheless, due to higher-order finite-size corrections, this is not exact, and $B(N, g_c) \neq B(N', g_c)$. Instead, the Binder cumulant for two different N intersects at a certain value g^*, that is, $B(N, g^*) = B(N', g^*)$, such that $g^* \to g_c$ when $NN' \to \infty$. See Fig. 6.3b for an illustration of $B(N, g)$ for anti-ferromagnetic couplings and different system sizes. Indeed, the critical point follows from $g_c^*(x) = g_c(1 + bx^{-\omega})$ with b and $\omega > 0$ constants, while $x = NN'$ as done in [40]. Note as well the similar procedures carried out in [28, 41].

In order to determine the critical exponent ν we first recast the scaling function as $F_{m_n}(\tilde{x}) = N^{-n\beta/\nu}\phi_{m_n}(\tilde{x})$ with $\tilde{x} = x^{1/\nu} = |g - g_c|N^{1/\nu}$, where now $\phi_{m_n}(\tilde{x}) = a_{m,0}$ for $\tilde{x} \to 0$. Then, one can show

$$\ln \frac{\partial}{\partial g} \langle m^n \rangle \bigg|_{g=g_c} = \frac{1 - n\beta}{\nu} \ln N + \ln \frac{\partial \phi_{m_n}(\tilde{x})}{\partial \tilde{x}} \bigg|_{\tilde{x}=0}, \qquad (6.6)$$

where the last term produces just a constant shift. In this manner, by computing the derivatives of $\langle m^2 \rangle$ and $\langle m^4 \rangle$ at the critical point, ν can be determined without a prior knowledge of any critical exponent, as it follows from

$$2 \ln \frac{\partial}{\partial g} \langle m^2 \rangle \bigg|_{g=g_c} - \ln \frac{\partial}{\partial g} \langle m^4 \rangle \bigg|_{g=g_c} = \frac{1}{\nu} \ln N + K, \qquad (6.7)$$

where K stands for a constant. Finally, once ν is determined one can rely on standard finite-size scaling methods to attain different critical exponents, either from the scaling function itself $F_{\mathcal{A}}(x)$ or from scaling at the critical point, $\langle \mathcal{A} \rangle (N, g_c) \propto N^{-\gamma_{\mathcal{A}}/\nu}$. In particular, the dynamical critical exponent z follows from the excitation energy, whose behavior in the thermodynamic limit is given by $\Delta(g) \propto |g - g_c|^{z\nu}$ [11], and therefore $\Delta(N, g_c) \propto N^{-z}$ for large N.

We remark that, in order to accurately determine the critical traits of H_{LRI}, larger system sizes are required, $N \gg 1$, as we rely on power-law relations on N to obtain g_c and critical exponents (see Eq. (6.7)). Nevertheless, since the dimension of the Hilbert space grows exponentially with N (as 2^{N-1} if one restricts to one parity subspace), exact numerical methods soon become unfeasible, and suitable approximate techniques must be employed [21, 35]. We leave however this task for future work. Instead, in order to gain insight into H_{LRI}, we show the energy gap Δ as a function of g and α, which allows us to provide a rough estimate of $g_c(\alpha)$, plotted in Fig. 6.4.

6.1.3 Nonequilibrium Dynamics and Scaling Laws

As explained in the Introduction, nonequilibrium dynamics involving phase transitions is an exciting problem and of interest in several fields. In this thesis we have

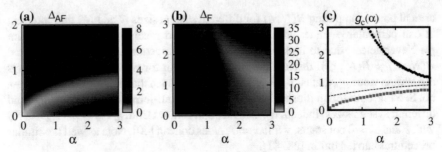

Fig. 6.4 Color map, in gray scale, of the energy gap Δ as a function of the exponent α of the interaction and the external magnetic field g, for anti-ferromagnetic (**a**) and ferromagnetic couplings (**b**). The results have been computed for $N = 18$ spins. Note the different scale in g and of Δ (encoded by the color map) for both cases. A rough estimate of the critical point as function of α can be obtained from $\tilde{g}_c(\alpha) = \min_g \Delta(g, \alpha)$. This is precisely plotted in **c**, where light-gray squares and black circles correspond to anti-ferromagnetic ($g_c^{AF}(\alpha) \leq 1$) and ferromagnetic case ($g_c^F(\alpha) \geq 1$), respectively. The dashed lines represent a theoretical approximation of $g_c^{AF,F}(\alpha)$ derived in [21] relying on a truncated Jordan-Wigner approach, namely, $g_c^{AF}(\alpha) = \zeta(\alpha)$ and $g_c^F(\alpha) = (1 - 2^{1-\alpha})\zeta(\alpha)$ where $\zeta(x) = \sum_{n=1}^{\infty} n^{-x}$ represents the Riemann zeta function. It is worth recalling that $g_c = 1$ (dotted line) for nearest-neighbors couplings, which is qualitatively well reproduced, $g_c(\alpha \gg 1) \to 1$

focused mainly on the universal fingerprints inherited by nonequilibrium states when the singular behavior characteristic of a continuous phase transition participates in the dynamics. In this respect, the celebrated KZ mechanism predicts a set power-law scaling in terms of the equilibrium critical exponents and the finite rate at which the critical point is traversed (see Introduction, Sect. 1.3). In this thesis we have tested these KZ predictions both in a classical and a QPT (see Chaps. 2 for Ginzburg-Landau and 4 for the QPT in the quantum Rabi model), and in addition, we have exploited the finite-size scaling relations to attain its nonequilibrium counterpart that encompasses the KZ scaling as a limiting case. In the following we examine the KZ problem in the one-dimensional long-range transverse field Ising model, H_{LRI}. However, we emphasize that the results serve simply as a proof-of-principle on that the KZ mechanism could be readily tested in a truly quantum many-body system, while we leave for future work the exhaustive inspection of different parameter regimes and nonequilibrium scaling functions.

As studied in the seminal works given in Refs. [1–4], the KZ mechanism can be successfully translated to the quantum realm, where quantum excitations play the role of topological defects in the standard KZ mechanism formulation. This quantum counterpart of the KZ mechanism (QKZ) appropriately describes the power-law behavior exhibited by the density of defects $n_d \sim \tau_Q^{-d\nu/(z\nu+1)}$ when the magnetic field is quenched at a rate $g(t) \propto \tau_Q^{-1}$ crossing the QPT of the paradigmatic one-dimensional nearest-neighbors Ising model with transverse field (see Introduction for a further discussion on this model) [2–4]. In this manner, at the end of the quench, the state consists of a quantum superposition of different states with different number of defects. In particular, for ferromagnetic couplings, these possible states possess local order within domains, namely,

$$|\dots \leftarrow\leftarrow\leftarrow\leftarrow\rightarrow\rightarrow\rightarrow\rightarrow\rightarrow\leftarrow\leftarrow\leftarrow\leftarrow\leftarrow\leftarrow\rightarrow\rightarrow\rightarrow \dots\rangle . \tag{6.8}$$

As an example, the previous state explicitly shows four domains that are separated by three defects. We remark that, although the situation is similar to the classical KZ mechanism, the underlying mechanism is distinct as now the final quantum state comprises a superposition of different number of domains and defects. In addition, it is worth emphasizing that the QKZ mechanism applies even to a system under unitary dynamics which preserves the parity symmetry (this symmetry is only spontaneously broken in the $N \to \infty$ limit). Nevertheless, quantum excitations are promoted as a consequence of the finite-rate ramp across the critical point and the number of defects is expected to follow a KZ scaling.

The protocol we consider here is similar and experimentally feasible [17, 18]. The magnetic field varies linearly in time $g(t) = g_0 + (g_1 - g_0)t/\tau_Q$. For simplicity, we choose $g_1 = 0$ ($g(t) = g_0(1 - t/\tau_Q)$) and consider $g_0 \gg g_c$ such that a fully z-polarized initial state $|\psi(0)\rangle = |\psi_P\rangle |\downarrow\downarrow \dots \downarrow\rangle$ is a suitable description, to a very good approximation, of the actual ground state of H_{LRI}, $|\langle\varphi_0^{\mathrm{AF,F}}|\psi_P\rangle| \approx 1$. Then, the number of defects can be obtained measuring $N - 1$ times a two-site observable, that is, $\sum_i^{N-1} \sigma_i^x \sigma_{i+1}^x$. In particular, depending on the couplings, we quantify the density of defects by

$$n_d^{\mathrm{AF}} = \frac{1}{2(N-1)} \sum_i^{N-1} \left(1 + \sigma_i^x \sigma_{i+1}^x\right), \tag{6.9}$$

$$n_d^{\mathrm{F}} = \frac{1}{2(N-1)} \sum_i^{N-1} \left(1 - \sigma_i^x \sigma_{i+1}^x\right). \tag{6.10}$$

Note that the sum runs up to $N - 1$ due to open boundary conditions and the difference of sign depending on $J_0 > 0$ (AF) and $J_0 < 0$ (F). For a strictly adiabatic passage ($\tau_Q \to \infty$), one retrieves the ground state at $g = 0$, and therefore $n_d^{\mathrm{AF,F}} \equiv 0$ since $\sum_i^{N-1} \langle\sigma_i^x \sigma_{i+1}^x\rangle = -N + 1$ for anti-ferromagnetic and $N - 1$ for ferromagnetic couplings. Moreover, for $g = 0$, the excited states exhibit $\sum_i^{N-1} \langle\sigma_i^x \sigma_{i+1}^x\rangle = -N + 1 + k$ or $N - 1 - k$, with k a positive integer number, that corresponds to the number of kinks. A wrong spin alignment in the ordered states represents a defect or kink, as illustrated previously. Since we consider an even number of spins, and we have constrained ourselves to the even parity subspace, only even number of kinks can be formed. In particular, the first excited state for anti-ferromagnetic couplings possesses $\sum_i^{N-1} \langle\sigma_i^x \sigma_{i+1}^x\rangle = -N + 3$ for $g = 0$.

In this manner, since the time-evolved state $|\psi(\tau_Q)\rangle$ can be written as

$$|\psi(\tau_Q)\rangle = \sum_k c_k(\tau_Q) \, |\varphi_{2k}(g=0)\rangle \qquad (6.11)$$

where $|\varphi_{2k}(g=0)\rangle$ is the $2k$th eigenstate of H_{LRI}, and thus, the density of defects follows from $n_d = \sum_k 2k|c_k(\tau_Q)|^2$ which is then expected to adopt a universal form if a QPT is involved in the dynamics $n_d \sim \tau_Q^{-d\nu/(z\nu+1)}$ and as long as $d\nu < 2(z\nu+1)$, as otherwise QKZ scaling becomes sub-leading and the standard τ_Q^{-2} dominates [12, 42, 43]. Recall that, applying adiabatic perturbation theory (APT) we could argue that $|c_k(\tau_Q)|^2 \sim \tau_Q^{-d\nu/(z\nu+1)}$ (see Appendix E). Finally, if α is sufficiently large, ν and z are expected to adopt its short-range values, namely $\nu = z = 1$, which together with $d = 1$ provides the QKZ predicted scaling, $n_d \sim \tau_Q^{-1/2}$.

We obtain $|\psi(t)\rangle$ solving the time evolution of $|\psi(0)\rangle = |\psi_P\rangle$ under H_{LRI} with realistic couplings $J_{i,j}$, Eq. (6.2), and subject to different quench rates across the critical point $g_c(\alpha)$. Then, we compute the number of defects from Eq. (6.9). We remark that we only focus on an anti-ferromagnetic scenario. Although QKZ mechanism can be examined in both cases, in the ferromagnetic scenario $g_c(\alpha)$ largely increases as α approaches 1, and thus one would need to prepare an initial state at $g_0 \gg |J_0|$. Furthermore, for $\alpha < 1$ one would need to normalize $J_{i,j}$ to properly achieve a thermodynamic limit [44]. Note that an initial value $g_0 = 2|J_0|$ together with $0.5 \lesssim \alpha \lesssim 1.5$ can be currently achieved [20]. The main results are collected in Fig. 6.5, where we have considered $N = 16$ and $N = 24$ spins and the spin-spin interactions approximately fall off with an exponent $\alpha \approx 1.41$, 1.16 and 0.76. As expected, for a too fast evolution across the critical point, $J_0\tau_Q \ll 1$, the number of defects saturates, while it decreases as the ramp is performed slower, eventually decaying in a power-law fashion τ_Q^{μ}. According to the QKZ mechanism, $\mu = -d\nu/(z\nu+1)$, which assuming short-range critical exponents leads to $\mu = -1/2$. For $N = 16$ and $\alpha \approx 1.41$, a fit in the region $5 \leq J_0\tau_Q \leq 50$ provides an exponent $\mu = -0.48(1)$, while for $N = 24$ and $\alpha \approx 1.16$, $\mu = -0.49(1)$ for $5 \leq J_0\tau_Q \leq 30$, in agreement with QKZ prediction for short-range interactions, as well as with the results reported in [21]. However, for $\alpha \approx 0.76$, we find an exponent $\mu = -0.20(1)$. Finally, since $J_0 \approx 2\pi \times 1$ kHz [18, 20, 30] and because the effect of decoherence processes may be relevant for $\tau_Q \gtrsim 3$ ms [18, 20], we should consider evolution times $5 \leq J_0\tau_Q \leq 20$. A fit within this region for $N = 24$ and $\alpha = 1.16$ still provides a good agreement with QKZ prediction. However, a closer inspection to the critical exponents for $\alpha \approx 0.76$ is required to address this regime, and to verify whether QKZ still applies.

We have performed a primary inspection of the nonequilibrium dynamics in a one-dimensional long-range Ising model with transverse field, which suggests that QKZ mechanism could be corroborated with state-of-the-art ion trap technology. The number of qubits and required α exponents are indeed available within current technology, as well as the possibility of performing ramps varying the external magnetic field [17, 18, 20]. Moreover, the effect of decoherence processes may be disregarded for evolution times $\tau_Q \lesssim 3$ ms [18, 20], which for a typical value

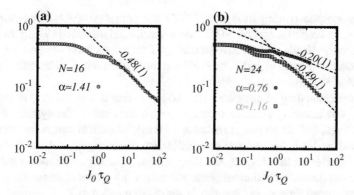

Fig. 6.5 Density of formed defects in a long-range Ising model, H_{LRI}, as a function of the quench time τ_Q, for $g(t) = g_0(1 - t/\tau_Q)$ and $g_0 = 2J_0$, and for different interaction profiles obtained from Eq. (6.2), which approximately follows $J_{i,j} \approx J_0|i - j|^{-\alpha}$, with α indicated in the plots. We consider an initial state fully polarized state in the z direction, $|\psi(0)\rangle = |\varphi_F\rangle = |\downarrow\downarrow \dots \downarrow\rangle$, as it corresponds to the ground state of H_{LRI} for $g \gg |J_0|$. In addition, $J_0 > 0$, and thus, at $g(\tau_Q) = 0$ one can measure the amount of kinks or defects resorting to n_d^{AF} (see Eq. (6.9)). Since $g_0 = 2J_0 > g_c(\alpha)$ for anti-ferromagnetic couplings, the critical point is traversed and defect formation may follow a power-law in a universal fashion as predicted by QKZ, $n_d \sim \tau_Q^{-d\nu/(z\nu+1)}$. Assuming short-range critical exponents, $d = z = \nu = 1$, it results in $\tau_Q^{-1/2}$. In **a** we show the results for $N = 16$ spins and $\alpha \approx 1.41$, which suggest that QKZ scaling is indeed attained (a fit provides an exponent $-0.48(1)$). The same applies to a larger system, $N = 24$ with $\alpha \approx 1.16$, as plotted in **b**, resulting in an exponent $-0.49(1)$. For an interaction much long ranged, as $\alpha \approx 0.76$, the QKZ prediction based on short-range critical exponents breaks down. See main text for details

$J_0 \approx 2\pi \times 1$ kHz [18, 20, 30] would still enable the observation of QKZ scaling. Distinct experimental imperfections may be yet relevant and a more detailed analysis of the potential impact of noise sources is pertinent. Finally, we comment that, as we have shown in the QRM, one can resort to nonequilibrium scaling functions, requiring shorter evolution times to witness its universal behavior (see Chap. 4).

6.2 Conclusion and Outlook

In this Chapter we have analyzed an experimentally feasible model where the QKZ mechanism could be tested. Thanks to the formidable advance in trapped-ion technologies, quantum simulation of systems with sufficiently large number of particles required to observe first indications of critical phenomena are now experimentally accessible. In particular, one can engineer the effective interactions undergone by N trapped ions, which adopt a Ising-like form [13, 45]. The resulting interaction decays approximately as a power-law with the distance, whose exponent α can be tuned. The realization of such an interacting quantum many-body system paves the way for the inspection of novel critical phenomena in the quantum realm, which encompasses an eventual confirmation of the QKZ among other prospects. For example, the recent exploration of a dynamical QPT deserves special mention [19, 20]. Briefly, in contrast to the standard QPT, a dynamical QPT manifests itself as non-analyticities in the Loschmidt echo at particular evolution times of an initial state evolving under a Hamiltonian [46].

The understanding of complex many-body systems lies at the core of statistical physics. The interplay between quantum correlations within the system and from surroundings, distinct phases of matter and nonequilibrium effects is far from being elucidated, and it is the ultimate responsible for the emergence of exotic quantum states of matter. In this respect, the Ising model, a paradigmatic model exhibiting classical and quantum continuous phase transitions, has served as an excellent testbed for diverse novel aspects and theories in statistical mechanics.

The recent work [21] deserves special mention, as it pursues the same goal as the results presented in this Chapter, and where matrix-product states have been utilized to extrapolate critical points, and to estimate nonequilibrium power-law exponents as a function of α, that is, of the interaction range. Although the mentioned Ref. [21] contains valuable results, a concise determination of critical exponents as well as finite-size scaling functions, both in- and out-of-equilibrium, remain to be elucidated. An interesting open question resides in whether QKZ successfully applies when long-range interaction dominates. The inspection of these issues would allow to disclose the universality class of the QPTs in such long-range interacting system featuring an exotic phase diagram. Finally, we stress that we leave a detailed investigation of large system sizes relying on approximate methods for future work.

References

1. B. Damski, The simplest quantum model supporting the Kibble-Zurek mechanism of topological defect production: Landau-Zener transitions from a new perspective. Phys. Rev. Lett. **95**, 035701 (2005). https://doi.org/10.1103/PhysRevLett.95.035701
2. W.H. Zurek, U. Dorner, P. Zoller, Dynamics of a quantum phase transition. Phys. Rev. Lett. **95**, 105701 (2005). https://doi.org/10.1103/PhysRevLett.95.105701
3. J. Dziarmaga, Dynamics of a quantum phase transition: exact solution of the quantum Ising model. Phys. Rev. Lett. **95**, 245701 (2005). https://doi.org/10.1103/PhysRevLett.95.245701
4. A. Polkovnikov, Universal adiabatic dynamics in the vicinity of a quantum critical point. Phys. Rev. B **72**, 161201 (2005). https://doi.org/10.1103/PhysRevB.72.161201
5. M. Anquez, B.A. Robbins, H.M. Bharath, M. Boguslawski, T.M. Hoang, M.S. Chapman, Quantum Kibble-Zurek mechanism in a spin-1 Bose-Einstein condensate. Phys. Rev. Lett. **116**, 155301 (2016). https://doi.org/10.1103/PhysRevLett.116.155301
6. L.W. Clark, L. Feng, C. Chin, Universal space-time scaling symmetry in the dynamics of bosons across a quantum phase transition. Science **354**, 606 (2016). https://doi.org/10.1126/science.aaf9657
7. J.-M. Cui, Y.-F. Huang, Z. Wang, D.-Y. Cao, J. Wang, W.-M. Lv, L. Luo, A. del Campo, Y.-J. Han, C.-F. Li, G.-C. Guo, Experimental trapped-ion quantum simulation of the Kibble-Zurek dynamics in momentum space. Sci. Rep. **6**, 33381 (2016). https://doi.org/10.1038/srep33381
8. X.-Y. Xu, Y.-J. Han, K. Sun, J.-S. Xu, J.-S. Tang, C.-F. Li, G.-C. Guo, Quantum simulation of Landau-Zener model dynamics supporting the Kibble-Zurek mechanism. Phys. Rev. Lett. **112**, 035701 (2014). https://doi.org/10.1103/PhysRevLett.112.035701

9. E. Ising, Beitrag zur Theorie des Ferromagnetismus. Z. Phys. **31**, 253 (1925). https://doi.org/10.1007/BF02980577
10. R.J. Baxter, *Exactly Solved Models in Statistical Mechanics* (Academic Press, London, 1989)
11. S. Sachdev, *Quantum Phase Transitions*, 2nd edn. (Cambridge University Press, Cambridge, UK, 2011)
12. A. Dutta, G. Aeppli, B.K. Chakrabarti, U. Divakaran, T.F. Rosenbaum, D. Sen, *Quantum Phase Transitions in Transverse Field Spin Models: From Statistical Physics to Quantum Information* (Cambridge University Press, Cambridge, 2015)
13. D. Porras, J.I. Cirac, Effective quantum spin systems with trapped ions. Phys. Rev. Lett. **92**, 207901 (2004). https://doi.org/10.1103/PhysRevLett.92.207901
14. A. Friedenauer, H. Schmitz, J.T. Glueckert, D. Porras, T. Schaetz, Simulating a quantum magnet with trapped ions. Nat. Phys. **4**, 757 (2008). https://doi.org/10.1038/nphys1032
15. K. Kim, M.-S. Chang, R. Islam, S. Korenblit, L.-M. Duan, C. Monroe, Entanglement and tunable spin-spin couplings between trapped ions using multiple transverse modes. Phys. Rev. Lett. **103**, 120502 (2009). https://doi.org/10.1103/PhysRevLett.103.120502
16. K. Kim, M.-S. Chang, S. Korenblit, R. Islam, E.E. Edwards, J.K. Freericks, G.-D. Lin, L.-M. Duan, C. Monroe, Quantum simulation of frustrated Ising spins with trapped ions. Nature **465**, 590 (2010). https://doi.org/10.1038/nature09071
17. R. Islam, E.E. Edwards, K. Kim, S. Korenblit, C. Noh, H. Carmichael, G.-D. Lin, L.-M. Duan, C.-C. Joseph Wang, J.K. Freericks, C. Monroe, Onset of a quantum phase transition with a trapped ion quantum simulator. Nat. Commun. **2**, 377 (2011). https://doi.org/10.1038/ncomms1374
18. R. Islam, C. Senko, W.C. Campbell, S. Korenblit, J. Smith, A. Lee, E.E. Edwards, C.-C.J. Wang, J.K. Freericks, C. Monroe, Emergence and frustration of magnetism with variable-range interactions in a quantum simulator. Science **340**, 583 (2013). https://doi.org/10.1126/science.1232296
19. P. Jurcevic, H. Shen, P. Hauke, C. Maier, T. Brydges, C. Hempel, B.P. Lanyon, M. Heyl, R. Blatt, C.F. Roos, Direct observation of dynamical quantum phase transitions in an interacting many-body system. Phys. Rev. Lett. **119**, 080501 (2017). https://doi.org/10.1103/PhysRevLett.119.080501
20. J. Zhang, G. Pagano, P.W. Hess, A. Kyprianidis, P. Becker, H. Kaplan, A.V. Gorshkov, Z.-X. Gong, C. Monroe, Observation of a many-body dynamical phase transition with a 53-qubit quantum simulator. Nature **551**, 601 (2017). https://doi.org/10.1038/nature24654
21. D. Jaschke, K. Maeda, J.D. Whalen, M.L. Wall, L.D. Carr, Critical phenomena and Kibble-Zurek scaling in the long-range quantum Ising chain. New J. Phys. **19**, 033032 (2017), http://stacks.iop.org/1367-2630/19/i=3/a=033032
22. U. Schollwöck, The density-matrix renormalization group in the age of matrix product states. Ann. Phys. (N. Y.) **326**, 96 (2011). https://doi.org/10.1016/j.aop.2010.09.012
23. H. Lipkin, N. Meshkov, A. Glick, Validity of many-body approximation methods for a solvable model. Nucl. Phys. **62**, 188 (1965). https://doi.org/10.1016/0029-5582(65)90862-X
24. F.J. Dyson, Existence of a phase-transition in a one-dimensional Ising ferromagnet. Commun. Math. Phys. **12**, 91 (1969). https://doi.org/10.1007/BF01645907
25. J.M. Kosterlitz, D.J. Thouless, Ordering, metastability and phase transitions in two-dimensional systems. J. Phys. C **6**, 1181 (1973), http://stacks.iop.org/0022-3719/6/i=7/a=010
26. J. Fröhlich, T. Spencer, The phase transition in the one-dimensional Ising model with $1/r^2$ interaction energy. Comm. Math. Phys. **84**, 87 (1982), https://projecteuclid.org:443/euclid.cmp/1103921047
27. E. Luijten, H. Meßingfeld, Criticality in one dimension with inverse square-law potentials. Phys. Rev. Lett. **86**, 5305 (2001). https://doi.org/10.1103/PhysRevLett.86.5305
28. M.C. Angelini, G. Parisi, F. Ricci-Tersenghi, Relations between short-range and long-range Ising models. Phys. Rev. E **89**, 062120 (2014). https://doi.org/10.1103/PhysRevE.89.062120
29. P. Richerme, Z.-X. Gong, A. Lee, C. Senko, J. Smith, M. Foss-Feig, S. Michalakis, A.V. Gorshkov, C. Monroe, Non-local propagation of correlations in quantum systems with long-range interactions. Nature **511**, 198 (2014). https://doi.org/10.1038/nature13450

30. J. Smith, A. Lee, P. Richerme, B. Neyenhuis, P.W. Hess, P. Hauke, M. Heyl, D.A. Huse, C. Monroe, Many-body localization in a quantum simulator with programmable random disorder. Nat. Phys. **12**, 907 (2016). https://doi.org/10.1038/nphys3783
31. B.P. Lanyon, C. Maier, M. Holzäpfel, T. Baumgratz, C. Hempel, P. Jurcevic, I. Dhand, A.S. Buyskikh, A.J. Daley, M. Cramer, M.B. Plenio, R. Blatt, C.F. Roos, Efficient tomography of a quantum many-body system. Nat. Phys. EP (2017). https://doi.org/10.1038/nphys4244
32. P. Jurcevic, B.P. Lanyon, P. Hauke, C. Hempel, P. Zoller, R. Blatt, C.F. Roos, Quasiparticle engineering and entanglement propagation in a quantum many-body system. Nature **511**, 202EP (2014). https://doi.org/10.1038/nature13461
33. S. Olmschenk, K.C. Younge, D.L. Moehring, D.N. Matsukevich, P. Maunz, C. Monroe, Manipulation and detection of a trapped Yb^+ hyperfine qubit. Phys. Rev. A **76**, 052314 (2007). https://doi.org/10.1103/PhysRevA.76.052314
34. D. James, Quantum dynamics of cold trapped ions with application to quantum computation. App. Phys. B **66**, 181 (1998). https://doi.org/10.1007/s003400050373
35. T. Koffel, M. Lewenstein, L. Tagliacozzo, Entanglement entropy for the long-range Ising chain in a transverse field. Phys. Rev. Lett. **109**, 267203 (2012). https://doi.org/10.1103/PhysRevLett.109.267203
36. P. Hauke, L. Tagliacozzo, Spread of correlations in long-range interacting quantum systems. Phys. Rev. Lett. **111**, 207202 (2013). https://doi.org/10.1103/PhysRevLett.111.207202
37. K. Binder, Finite size scaling analysis of Ising model block distribution functions. Z. Phys. B **43**, 119 (1981). https://doi.org/10.1007/BF01293604
38. M.E. Fisher, M.N. Barber, Scaling theory for finite-size effects in the critical region. Phys. Rev. Lett. **28**, 1516 (1972). https://doi.org/10.1103/PhysRevLett.28.1516
39. J.G. Brankov, *Introduction to Finite-size Scaling* (Leuven University Press, Leuven, 1996)
40. S. Mukherjee, A. Rajak, B.K. Chakrabarti, Classical-to-quantum crossover in the critical behavior of the transverse-field Sherrington-Kirkpatrick spin glass model. Phys. Rev. E **92**, 042107 (2015). https://doi.org/10.1103/PhysRevE.92.042107
41. E. Luijten, H.W.J. Blöte, Classical critical behavior of spin models with long-range interactions. Phys. Rev. B **56**, 8945 (1997). https://doi.org/10.1103/PhysRevB.56.8945
42. A.K. Chandra, A. Das, B.K.C. (Eds.), *Quantum Quenching, Annealing and Computation* (Springer, Berlin, 2010)
43. C. De Grandi, V. Gritsev, A. Polkovnikov, Quench dynamics near a quantum critical point: application to the sine-Gordon model. Phys. Rev. B **81**, 224301 (2010). https://doi.org/10.1103/PhysRevB.81.224301
44. M. Kac, C.J. Thompson, Critical behavior of several lattice models with long-range interaction. J. Math. Phys. **10**, 1373 (1969). https://doi.org/10.1063/1.1664976
45. X.-L. Deng, D. Porras, J.I. Cirac, Effective spin quantum phases in systems of trapped ions. Phys. Rev. A **72**, 063407 (2005). https://doi.org/10.1103/PhysRevA.72.063407
46. M. Heyl, A. Polkovnikov, S. Kehrein, Dynamical quantum phase transitions in the transverse-field Ising model. Phys. Rev. Lett. **110**, 135704 (2013). https://doi.org/10.1103/PhysRevLett.110.135704

Chapter 7
Concluding Remarks and Outlook

In this thesis we have studied equilibrium and nonequilibrium aspects of continuous phase transitions in distinct systems. We have made special emphasis on their nonequilibrium features, a much less understood topic than their static counterparts, aiming to elucidate questions such as to what extent the well-established universal static properties apply to the dynamics. In this regard, we have examined and extended the predicted nonequilibrium scaling laws dictated by Kibble–Zurek (KZ) mechanism in classical and quantum systems, covering standard many-body systems such as a Coulomb crystal [1] and an Ising model. Moreover, we have found that a small quantum system, namely the quantum Rabi model (QRM), turns out to be a suitable model to test and learn from the rich phenomenology of phase transitions [2–5]. In the following we provide a brief summary of the main outcomes of the work presented throughout the thesis, as well as some interesting directions for future research, while we refer the reader to the conclusions of different chapters for a more detailed account of the main results in each of them.

The KZ mechanism explains the unavoidable departure from equilibrium as a consequence of a diverging relaxation time, a hallmark of continuous phase transition at the critical point, and thus promoting defect formation. As a nearly-adiabatic dynamics across the critical point inherits fingerprints of equilibrium divergences, the dynamics is said to be universal. However, KZ predictions are based on the features of a phase transition in the thermodynamic limit, and thus cannot account for finite-size systems. Therefore, in order to elucidate universal properties in nonequilibrium scenarios embracing finite-size effects a more general framework is required. In Chap. 2 we have analyzed a Ginzburg–Landau model [6] and a realistic zigzag phase transition in a Coulomb crystal [7, 8], driving the system at finite speed towards the ordered phase. This forces a local symmetry breaking promoting defect formation, which turns out to be a suitable scenario to examine KZ physics, as indicated in previous theoretical works (see [9–11] for the Ginzburg-Landau model and [12–15] for zigzag phase transition in Coulomb crystals) and experimentally verified in two ion trap experiments [16, 17]. In this part, we have studied this problem using a

© Springer Nature Switzerland AG 2018
R. Puebla, *Equilibrium and Nonequilibrium Aspects of Phase Transitions in Quantum Physics*, Springer Theses, https://doi.org/10.1007/978-3-030-00653-2_7

Fokker–Planck approach, aiming to obtain phase-space probability distributions in a deterministic manner. Under certain approximations, this method is more efficient than computing ensemble averages from individual stochastic trajectories. Moreover, we show as well that non-linear interaction terms are not essential to capture KZ scaling laws, and thus the corresponding physics can be well described by a Gaussian approximation of both Ginzburg–Landau model and zigzag phase transition in a Coulomb crystal, which has been published in [1]. The developed framework may allow to gain valuable insights in systems driven out of equilibrium, not necessarily involving phase transitions. In this regard, it may certainly be interesting to delve into nonequilibrium thermodynamics quantities such as work done, entropy production and fluctuation theorems [18, 19]. Needless to say, their connection with critical phenomena, defect formation and symmetry breaking represents an exciting prospect for future work (note the recent works in these directions [20–22]).

The investigation presented in this thesis leads us to reconsider one of the most common conceptions about phase transitions. This well established notion is concisely summarized by L. P. Kadanoff: *"In particular, phase transitions cannot occur in any finite system; they are solely a property of infinite systems, [...] there cannot be any phase transition in any finite system described by the Ising model or indeed any statistical system with everything in it being finite"* [23]. That is, there can only be a phase transition if there are infinitely many system components. We however show that it is indeed possible to attain critical behavior while keeping finite the number of constituents of a system, thus challenging this idea of phase transitions. We have dubbed finite-component system phase transition this novel manner to achieve critical behavior. In particular, we exemplify this finite-component system phase transition in Chap. 3 considering QRM, a fundamental model in quantum physics that describes a single spin interacting with a single bosonic mode [24]. In this sense, the QRM finds itself far from being in the thermodynamic limit and one would not expect to observe any kind of phase transition. Note however the previous works which pointed out certain footprints of critical behavior [25–29]. Instead of scaling system constituents, the QRM undergoes a quantum phase transition (QPT) in a suitable parameter limit, which involves an infinitely large spin splitting Ω over the bosonic frequency $\Omega/\omega_0 \rightarrow \infty$, and at the same time an infinitely large coupling strength $\lambda/\omega_0 \rightarrow \infty$ such that $g \equiv 2\lambda/\sqrt{\Omega\omega_0}$ remains finite. This finite-component system phase transition exploits the inherent infinitely large Hilbert space provided by the harmonic oscillator, and takes place in a parameter limit in which an infinite part of the Hilbert space becomes relevant.

In Chap. 3 we provide a detailed derivation of the equilibrium or static properties of the QRM in this limit, which discloses all the hallmarks of a QPT. In particular, it bears the same similarities as those of the Dicke model [30]—the multi-spin version of the QRM—and Lipkin–Meshkov–Glick model [31], which undergo a QPT in the standard thermodynamic limit. This type of QPT is known as superradiant QPT [32–37]. As in any other conventional phase transition, we examine the impact of a finite-size system, which in the QRM translates to a finite Ω/ω_0 value, and find that finite-size scaling theory is still applicable (or finite-*frequency* scaling to stress that we are interested in finite values of the parameters, while the system always comprises a spin and a single bosonic mode). This allows us to assert that the QRM falls into the same universality class of that of Dicke and Lipkin–Meshkov–Glick models, as they share critical exponents and finite-size scaling functions. A closer inspection to higher excited states unveils the presence of a logarithmic singularity in the density of states, which is known as excited-state quantum phase transition [38–40]. We then examine its nonequilibrium dynamics in Chap. 4, in the spirit of the KZ problem, that is, driving the system towards the critical point. Remarkably, we find a perfect agreement with the KZ predictions, and thus, extending the application of KZ mechanism to a zero-dimensional system. Note that, prior to our work, the KZ mechanism was considered to fail or not applicable to these systems [41, 42]. As done in the first part of the thesis, we extend the universality to a nonequilibrium scenario, which encompasses KZ scaling laws and finite-frequency systems. The Dicke model and QRM exhibit equivalent nonequilibrium finite-size scaling functions, supporting their equivalence and the universality of the dynamics. Hence, despite the apparent simplicity of the QRM, the presented theoretical work strongly suggests that the QRM is a suitable model to study critical dynamics. Finally, we propose to probe the dynamics of the QPT realizing a QRM in an ion trap, which allows us to explore a superradiant QPT with *only* one single trapped ion, as we explain in Chap. 5. As a trapped-ion setup is prone to experimental imperfections and different noise sources, we rely on nonequilibrium scaling functions and demonstrate that their observation is feasible under realistic parameters and noise. Furthermore, as magnetic field fluctuations are acknowledged as the main source of noise in certain trapped-ion schemes, we propose to apply dynamical decoupling methods to attain a robust and noise-resilient QRM with a trapped ion. Part of the results presented in Chaps. 3, 4 and 5, have been published in Refs. [2–5].

In this manner, the QRM emerges as a good candidate to analyze other interesting scenarios. For example, we have not mentioned possible design of optimized protocols aiming to attain reliable ground states in much shorter evolution times than those required under a naive linear quench [43–45]. These optimized protocols may be of considerable experimental relevance and may open the possibility to test static traits of the QPT by almost-adiabatic ground-state preparation. In addition, the QRM may also exhibit nonequilibrium phase transitions (not to be confused with nonequilibrium traits of a QPT, as we have studied here), which refer to non-trivial stable states when parameters of the system are periodically driven, as shown in the Dicke model [46]. Another interesting prospect resides in the dissipative effect in the dynamics, which may ultimately lead to critical transitions in steady states, i.e., a dissipative phase transition, and recently studied in [47].

However, more work needs to be done to establish a rigorous framework of finite-component phase transitions. In this line, the recent work done in [48] deserves special mention, as it extends this notion to a different system, a Jaynes–Cummings lattice model. Although a similar phenomenology is observed, a continuous $U(1)$ symmetry spontaneously breaks instead, leading to a Goldstone mode. In addition, it is worth noting a recent work which has analyzed the impact of different weights of counter-rotating terms (an anisotropic QRM) on the QPT [49]. Hence, finite-component system phase transitions may be present in a broad range of physical systems. On the other hand, investigation of system-environment interactions is also of considerable interest, not only due to its experimental relevance but also to study various theoretical aspects. For example, it has been reported that different spectral properties of an external bath may modify the critical exponents of the superradiant QPT in the Dicke model [50, 51]. Moreover, environment-induced transitions may play an important role during adiabatic passages in the vicinity of a QPT [52–55]. As a consequence of the ubiquity of spin and boson interaction, the QRM naturally arises in different platforms. Here we have only focused on its trapped-ion realization, however, the QRM may be attained in other setups such as circuit QED [56], cold-atom system [57] or by using color centers in membranes [58]. Nevertheless, depending on the particular platform, it might be challenging to reach the extreme parameter limit required to witness critical fingerprints of the QRM.

In the last part of this thesis we have studied the quantum Kibble–Zurek mechanism (QKZ) considering an experimentally feasible quantum many-body system, namely, a long-range Ising model with transverse field. The QKZ differs from its classical version on that quantum excitations produced during a finite-time, yet unitary, evolution traversing a QPT follow a universal power law, in the same fashion as the number of topological defects does in the KZ mechanism. Theoretically proposed in [59–62], the predicted QKZ scaling laws have been only very recently confirmed in [63, 64]. Chapter 6 intends to provide a proof of concept indicating that QKZ mechanism can be tested and verified in a long-range Ising model within current technology. Thanks to the great experimental progress in trapped-ion setups (high fidelity measurements and state preparation, and high degree of isolation and control [65]), the realization of this model in the laboratory has opened the door to examine striking quantum effects and exotic phases of matter, such as the onset of ferromagnetic QPTs [66–69], quantum time crystal [70] and dynamical QPTs [71, 72]. Hence, the long-range Ising model with transverse field has become an excellent model where novel aspects of quantum many-body physics can be tested. Needless to say, these works also constitute a great step forward in the quest for quantum simulation and computation. We recall that the effective spin-spin interaction decays as a power-law with the distance between them, a feature that crucially modifies the universality of the phase transition. However, for sufficiently short-range interactions the model falls into the universality class of its nearest-neighbors counterpart. We discuss the ground-state properties, and observe that, for a feasible number of spins, $N \sim 20$, and realistic couplings, the QKZ scaling might already be experimentally observed. We however leave for future work an in-depth analysis of the rich and compelling static and dynamical features of this model. For example, in order to answer the question of whether QKZ scaling holds when long-range dominates, one needs first to determine static critical exponents, thus requiring the inspection of large system sizes to obtain reliable estimates based on finite-size scaling. Because the Hilbert space grows exponentially with the number of spins, exact procedures become soon unfeasible, thus demanding suitable approximate techniques such as density matrix renormalization group methods [73], as done in [74, 75].

References

1. R. Puebla, R. Nigmatullin, T.E. Mehlstäubler, M.B. Plenio, Fokker-Planck formalism approach to Kibble-Zurek scaling laws and nonequilibrium dynamics. Phys. Rev. B **95**, 134104 (2017a). https://doi.org/10.1103/PhysRevB.95.134104

2. M.-J. Hwang, R. Puebla, M.B. Plenio, Quantum phase transition and universal dynamics in the Rabi model. Phys. Rev. Lett. **115**, 180404 (2015). https://doi.org/10.1103/PhysRevLett.115.180404

3. R. Puebla, M.-J. Hwang, M.B. Plenio, Excited-state quantum phase transition in the Rabi model. Phys. Rev. A **94**, 023835 (2016a). https://doi.org/10.1103/PhysRevA.94.023835

4. R. Puebla, J. Casanova, M.B. Plenio, A robust scheme for the implementation of the quantum Rabi model in trapped ions. New J. Phys. **18**, 113039 (2016b), http://stacks.iop.org/1367-2630/18/i=11/a=113039

5. R. Puebla, M.-J. Hwang, J. Casanova, M.B. Plenio, Probing the dynamics of a superradiant quantum phase transition with a single trapped ion. Phys. Rev. Lett. **118**, 073001 (2017b). https://doi.org/10.1103/PhysRevLett.118.073001

6. L.D. Landau, E.M. Lifshitz, *Statistical Physics*, 3rd edn. (Butterworth-Heinemann, Oxford, 1980)

7. D.H.E. Dubin, T.M. O'Neil, Trapped nonneutral plasmas, liquids, and crystals (the thermal equilibrium states). Rev. Mod. Phys. **71**, 87 (1999). https://doi.org/10.1103/RevModPhys.71.87

8. R.C. Thompson, Ion Coulomb crystals. Cont. Phys. **56**, 63 (2015). https://doi.org/10.1080/00107514.2014.989715

9. P. Laguna, W.H. Zurek, Density of kinks after a quench: when symmetry breaks, how big are the pieces? Phys. Rev. Lett. **78**, 2519 (1997). https://doi.org/10.1103/PhysRevLett.78.2519

10. P. Laguna, W.H. Zurek, Critical dynamics of symmetry breaking: quenches, dissipation, and cosmology. Phys. Rev. D **58**, 085021 (1998). https://doi.org/10.1103/PhysRevD.58.085021

11. E. Moro, G. Lythe, Dynamics of defect formation. Phys. Rev. E **59**, R1303(R) (1999). https://doi.org/10.1103/PhysRevE.59.R1303

12. G. De Chiara, A. del Campo, G. Morigi, M.B. Plenio, A. Retzker, Spontaneous nucleation of structural defects in inhomogeneous ion chains. New J. Phys. **12**, 115003 (2010), http://stacks.iop.org/1367-2630/12/i=11/a=115003

13. A. del Campo, A. Retzker, M.B. Plenio, The inhomogeneous Kibble-Zurek mechanism: vortex nucleation during Bose-Einstein condensation. New J. Phys. **13**, 083022 (2011), http://stacks.iop.org/1367-2630/13/i=8/a=083022

14. A. del Campo, G. De Chiara, G. Morigi, M.B. Plenio, A. Retzker, Structural defects in ion chains by quenching the external potential: the inhomogeneous Kibble-Zurek mechanism. Phys. Rev. Lett. **105**, 075701 (2010). https://doi.org/10.1103/PhysRevLett.105.075701

15. R. Nigmatullin, A. del Campo, G. De Chiara, G. Morigi, M.B. Plenio, A. Retzker, Formation of helical ion chains. Phys. Rev. B **93**, 014106 (2016). https://doi.org/10.1103/PhysRevB.93.014106

16. K. Pyka, J. Keller, H.L. Partner, R. Nigmatullin, T. Burgermeister, D.M. Meier, K. Kuhlmann, A. Retzker, M.B. Plenio, W.H. Zurek, A. del Campo, T.E. Mehlstäubler, Topological defect formation and spontaneous symmetry breaking in ion Coulomb crystals. Nat. Commun. **4**, 2291 (2013). https://doi.org/10.1038/ncomms3291

17. S. Ulm, J. Roßnagel, G. Jacob, C. Degünther, S.T. Dawkins, U.G. Poschinger, R. Nigmatullin, A. Retzker, M.B. Plenio, F. Schmidt-Kaler, K. Singer, Observation of the Kibble–Zurek scaling law for defect formation in ion crystals, Nat. Commun. **4**, 2290 (2013). https://doi.org/10.1038/ncomms3290

18. C. Jarzynski, Nonequilibrium equality for free energy differences. Phys. Rev. Lett. **78**, 2690 (1997). https://doi.org/10.1103/PhysRevLett.78.2690

19. G.E. Crooks, Entropy production fluctuation theorem and the nonequilibrium work relation for free energy differences. Phys. Rev. E **60**, 2721 (1999). https://doi.org/10.1103/PhysRevE.60.2721

20. T.M. Hoang, H.M. Bharath, M.J. Boguslawski, M. Anquez, B.A. Robbins, M.S. Chapman, Adiabatic quenches and characterization of amplitude excitations in a continuous quantum phase transition. Proc. Natl. Acad. Sci. **113**, 9475 (2016). https://doi.org/10.1073/pnas.1600267113

21. F. Cosco, M. Borrelli, P. Silvi, S. Maniscalco, G. De Chiara, Nonequilibrium quantum thermodynamics in Coulomb crystals. Phys. Rev. A **95**, 063615 (2017). https://doi.org/10.1103/PhysRevA.95.063615

22. S. Deffner, Kibble-Zurek scaling of the irreversible entropy production. Phys. Rev. E **96**, 052125 (2017). https://doi.org/10.1103/PhysRevE.96.052125

23. L.P. Kadanoff, More is the same; phase transitions and mean field theories. J. Stat. Phys. **137**, 777 (2009). https://doi.org/10.1007/s10955-009-9814-1

24. M.O. Scully, M.S. Zubairy, *Quantum Optics* (Cambridge University Press, Cambridge, England, 1997)

25. A.P. Hines, C.M. Dawson, R.H. McKenzie, G.J. Milburn, Entanglement and bifurcations in Jahn-Teller models. Phys. Rev. A **70**, 022303 (2004). https://doi.org/10.1103/PhysRevA.70.022303

26. G. Levine, V.N. Muthukumar, Entanglement of a qubit with a single oscillator mode. Phys. Rev. B **69**, 113203 (2004). https://doi.org/10.1103/PhysRevB.69.113203

27. S. Ashhab, F. Nori, Qubit-oscillator systems in the ultrastrong-coupling regime and their potential for preparing nonclassical states. Phys. Rev. A **81**, 042311 (2010). https://doi.org/10.1103/PhysRevA.81.042311

28. L. Bakemeier, A. Alvermann, H. Fehske, Quantum phase transition in the Dicke model with critical and noncritical entanglement. Phys. Rev. A **85**, 043821 (2012). https://doi.org/10.1103/PhysRevA.85.043821

29. S. Ashhab, Superradiance transition in a system with a single qubit and a single oscillator. Phys. Rev. A **87**, 013826 (2013). https://doi.org/10.1103/PhysRevA.87.013826

30. R.H. Dicke, Coherence in spontaneous radiation processes. Phys. Rev. **93**, 99 (1954). https://doi.org/10.1103/PhysRev.93.99

31. H. Lipkin, N. Meshkov, A. Glick, Validity of many-body approximation methods for a solvable model. Nucl. Phys. **62**, 188 (1965). https://doi.org/10.1016/0029-5582(65)90862-X

32. C. Emary, T. Brandes, Quantum chaos triggered by precursors of a quantum phase transition: the Dicke model. Phys. Rev. Lett **90**, 044101 (2003a). https://doi.org/10.1103/PhysRevLett.90.044101

33. C. Emary, T. Brandes, Chaos and the quantum phase transition in the Dicke model. Phys. Rev. E **67**, 66203 (2003b). https://doi.org/10.1103/PhysRevE.67.066203

34. N. Lambert, C. Emary, T. Brandes, Entanglement and the phase transition in single-mode superradiance. Phys. Rev. Lett. **92**, 073602 (2004). https://doi.org/10.1103/PhysRevLett.92.073602

35. N. Lambert, C. Emary, T. Brandes, Entanglement and entropy in a spin-boson quantum phase transition. Phys. Rev. A **71**, 053804 (2005). https://doi.org/10.1103/PhysRevA.71.053804

36. P. Ribeiro, J. Vidal, R. Mosseri, Thermodynamical limit of the Lipkin-Meshkov-Glick model. Phys. Rev. Lett. **99**, 050402 (2007). https://doi.org/10.1103/PhysRevLett.99.050402

37. P. Ribeiro, J. Vidal, R. Mosseri, Exact spectrum of the Lipkin-Meshkov-Glick model in the thermodynamic limit and finite-size corrections. Phys. Rev. E **78**, 021106 (2008). https://doi.org/10.1103/PhysRevE.78.021106

38. P. Cejnar, M. Macek, S. Heinze, J. Jolie, J. Dobes, Monodromy and excited-state quantum phase transitions in integrable systems: collective vibrations of nuclei. J. Phys. A Math. Theor. **39**, L515 (2006), http://stacks.iop.org/0305-4470/39/i=31/a=L01

39. P. Cejnar, P. Stránský, Impact of quantum phase transitions on excited-level dynamics. Phys. Rev. E **78**, 031130 (2008). https://doi.org/10.1103/PhysRevE.78.031130

40. M. Caprio, P. Cejnar, F. Iachello, Excited state quantum phase transitions in many-body systems. Ann. Phys. (N.Y.) **323**, 1106 (2008). https://doi.org/10.1016/j.aop.2007.06.011

41. T. Caneva, R. Fazio, G.E. Santoro, Adiabatic quantum dynamics of the Lipkin-Meshkov-Glick model. Phys. Rev. B **78**, 104426 (2008). https://doi.org/10.1103/PhysRevB.78.104426

42. O.L. Acevedo, L. Quiroga, F.J. Rodríguez, N.F. Johnson, New dynamical scaling universality for quantum networks across adiabatic quantum phase transitions. Phys. Rev. Lett. **112**, 030403 (2014). https://doi.org/10.1103/PhysRevLett.112.030403
43. S. van Frank, M. Bonneau, J. Schmiedmayer, S. Hild, C. Gross, M. Cheneau, I. Bloch, T. Pichler, A. Negretti, T. Calarco, S. Montangero, Optimal control of complex atomic quantum systems. Sci. Rep. **6**, 34187 (2016). https://doi.org/10.1038/srep34187
44. R. Barankov, A. Polkovnikov, Optimal nonlinear passage through a quantum critical point. Phys. Rev. Lett. **101**, 076801 (2008). https://doi.org/10.1103/PhysRevLett.101.076801
45. E. Torrontegui, S. Ibáñez, S. Martínez-Garaot, M. Modugno, A. del Campo, D. Guéry-Odelin, A. Ruschhaupt, X. Chen, J.G. Muga, in *Advances In Atomic, Molecular, and Optical Physics*, vol. 62, ed. by E. Arimondo, P.R. Berman, C.C. Lin (Academic Press, 2013), pp. 117–169. https://doi.org/10.1016/B978-0-12-408090-4.00002-5
46. V.M. Bastidas, C. Emary, B. Regler, T. Brandes, Nonequilibrium quantum phase transitions in the Dicke model. Phys. Rev. Lett. **108**, 043003 (2012). https://doi.org/10.1103/PhysRevLett.108.043003
47. M.-J. Hwang, P. Rabl, M.B. Plenio, Dissipative phase transition in the open quantum Rabi model. Phys. Rev. A **97**, 013825 (2018). https://doi.org/10.1103/PhysRevA.97.013825
48. M.-J. Hwang, M.B. Plenio, Quantum phase transition in the finite Jaynes-Cummings lattice systems. Phys. Rev. Lett. **117**, 123602 (2016). https://doi.org/10.1103/PhysRevLett.117.123602
49. M. Liu, S. Chesi, Z.-J. Ying, X. Chen, H.-G. Luo, H.-Q. Lin, Universal scaling and critical exponents of the anisotropic quantum Rabi model. Phys. Rev. Lett. **119**, 220601 (2017). https://doi.org/10.1103/PhysRevLett.119.220601
50. D. Nagy, P. Domokos, Nonequilibrium quantum criticality and non-Markovian environment: critical exponent of a quantum phase transition. Phys. Rev. Lett. **115**, 043601 (2015). https://doi.org/10.1103/PhysRevLett.115.043601
51. D. Nagy, P. Domokos, Critical exponent of quantum phase transitions driven by colored noise. Phys. Rev. A **94**, 063862 (2016). https://doi.org/10.1103/PhysRevA.94.063862
52. D. Patanè, A. Silva, L. Amico, R. Fazio, G.E. Santoro, Adiabatic dynamics in open quantum critical many-body systems. Phys. Rev. Lett. **101**, 175701 (2008). https://doi.org/10.1103/PhysRevLett.101.175701
53. D. Patanè, L. Amico, A. Silva, R. Fazio, G.E. Santoro, Adiabatic dynamics of a quantum critical system coupled to an environment: scaling and kinetic equation approaches. Phys. Rev. B **80**, 024302 (2009). https://doi.org/10.1103/PhysRevB.80.024302
54. P. Nalbach, Adiabatic-Markovian bath dynamics at avoided crossings. Phys. Rev. A **90**, 042112 (2014). https://doi.org/10.1103/PhysRevA.90.042112
55. P. Nalbach, S. Vishveshwara, A.A. Clerk, Quantum Kibble-Zurek physics in the presence of spatially correlated dissipation. Phys. Rev. B **92**, 014306 (2015). https://doi.org/10.1103/PhysRevB.92.014306
56. N.K. Langford, R. Sagastizabal, M. Kounalakis, C. Dickel, A. Bruno, F. Luthi, D.J. Thoen, A. Endo, L. DiCarlo, Experimentally simulating the dynamics of quantum light and matter at deep-strong coupling. Nat. Comm. **8**, 1715 (2017). https://doi.org/10.1038/s41467-017-01061-x
57. P. Schneeweis, A. Dareau, C. Sayrin, Cold-atom based implementation of the quantum Rabi model (2017). arXiv:1706.07781
58. M. Abdi, M.-J. Hwang, M. Aghtar, M.B. Plenio, Spin-mechanical scheme with color centers in hexagonal boron nitride membranes. Phys. Rev. Lett. **119**, 233602 (2017). https://doi.org/10.1103/PhysRevLett.119.233602
59. B. Damski, The simplest quantum model supporting the Kibble-Zurek mechanism of topological defect production: Landau-Zener transitions from a new perspective. Phys. Rev. Lett. **95**, 035701 (2005). https://doi.org/10.1103/PhysRevLett.95.035701
60. W.H. Zurek, U. Dorner, P. Zoller, Dynamics of a quantum phase transition. Phys. Rev. Lett. **95**, 105701 (2005). https://doi.org/10.1103/PhysRevLett.95.105701
61. J. Dziarmaga, Dynamics of a quantum phase transition: exact solution of the quantum Ising model. Phys. Rev. Lett. **95**, 245701 (2005). https://doi.org/10.1103/PhysRevLett.95.245701

62. A. Polkovnikov, Universal adiabatic dynamics in the vicinity of a quantum critical point. Phys. Rev. B **72**, 161201 (2005). https://doi.org/10.1103/PhysRevB.72.161201
63. L.W. Clark, L. Feng, C. Chin, Universal space-time scaling symmetry in the dynamics of bosons across a quantum phase transition. Science **354**, 606 (2016). https://doi.org/10.1126/science.aaf9657
64. M. Anquez, B.A. Robbins, H.M. Bharath, M. Boguslawski, T.M. Hoang, M.S. Chapman, Quantum Kibble-Zurek mechanism in a spin-1 Bose-Einstein condensate. Phys. Rev. Lett. **116**, 155301 (2016). https://doi.org/10.1103/PhysRevLett.116.155301
65. D. Leibfried, R. Blatt, C. Monroe, D. Wineland, Quantum dynamics of single trapped ions. Rev. Mod. Phys. **75**, 281 (2003). https://doi.org/10.1103/RevModPhys.75.281
66. A. Friedenauer, H. Schmitz, J.T. Glueckert, D. Porras, T. Schaetz, Simulating a quantum magnet with trapped ions. Nat. Phys. **4**, 757 (2008). https://doi.org/10.1038/nphys1032
67. K. Kim, M.-S. Chang, S. Korenblit, R. Islam, E.E. Edwards, J.K. Freericks, G.-D. Lin, L.-M. Duan, C. Monroe, Quantum simulation of frustrated Ising spins with trapped ions. Nature **465**, 590 (2010). https://doi.org/10.1038/nature09071
68. R. Islam, E.E. Edwards, K. Kim, S. Korenblit, C. Noh, H. Carmichael, G.-D. Lin, L.-M. Duan, C.-C. Joseph Wang, J.K. Freericks, C. Monroe, Onset of a quantum phase transition with a trapped ion quantum simulator. Nat. Commun. **2**, 377 (2011). https://doi.org/10.1038/ncomms1374
69. R. Islam, C. Senko, W.C. Campbell, S. Korenblit, J. Smith, A. Lee, E.E. Edwards, C.-C.J. Wang, J.K. Freericks, C. Monroe, Emergence and frustration of magnetism with variable-range interactions in a quantum simulator. Science **340**, 583 (2013). https://doi.org/10.1126/science.1232296
70. J. Zhang, P.W. Hess, A. Kyprianidis, P. Becker, A. Lee, J. Smith, G. Pagano, I.-D. Potirniche, A.C. Potter, A. Vishwanath, N.Y. Yao, C. Monroe, Observation of a discrete time crystal. Nature **543**, 217 (2017a). https://doi.org/10.1038/nature21413
71. P. Jurcevic, H. Shen, P. Hauke, C. Maier, T. Brydges, C. Hempel, B.P. Lanyon, M. Heyl, R. Blatt, C.F. Roos, Direct observation of dynamical quantum phase transitions in an interacting many-body system. Phys. Rev. Lett. **119**, 080501 (2017). https://doi.org/10.1103/PhysRevLett.119.080501
72. J. Zhang, G. Pagano, P.W. Hess, A. Kyprianidis, P. Becker, H. Kaplan, A.V. Gorshkov, Z.-X. Gong, C. Monroe, Observation of a many-body dynamical phase transition with a 53-qubit quantum simulator. Nature **551**, 601 (2017b). https://doi.org/10.1038/nature24654
73. U. Schollwöck, The density-matrix renormalization group in the age of matrix product states. Ann. Phys. (N.Y.) **326**, 96 (2011). https://doi.org/10.1016/j.aop.2010.09.012
74. D. Jaschke, K. Maeda, J.D. Whalen, M.L. Wall, L.D. Carr, Critical phenomena and Kibble-Zurek scaling in the long-range quantum Ising chain. New J. Phys. **19**, 033032 (2017), http://stacks.iop.org/1367-2630/19/i=3/a=033032
75. T. Koffel, M. Lewenstein, L. Tagliacozzo, Entanglement entropy for the long-range Ising chain in a transverse field. Phys. Rev. Lett. **109**, 267203 (2012). https://doi.org/10.1103/PhysRevLett.109.267203

Appendix A
Survey on Stochastic Processes

In this Appendix, we present a very brief revision of stochastic processes, starting from the Langevin equation, to later introduce the Fokker-Planck equation [1], which has been used in Chap. 2. These equations dictate the time evolution of the probability distributions, both in the high friction limit and in the general case known as Smoluchowski and Kramers equations, respectively. In addition, we discuss the fluctuation-dissipation theorem.

A.1 Brownian Motion and Langevin Equation

Let us consider a particle of mass m subject to a force $f(x) = -dV(x)/dx$ with $V(x)$ the potential, as well as to a stochastic force $\zeta(t)$ that effectively describes the interaction of the particle with the infinitely many degrees of freedom of the environment. Then, the equation of motion of this simple system reads

$$m\ddot{x}(t) + \gamma\dot{x}(t) = f(x) + \zeta(t), \tag{A.1}$$

where γ represents the friction coefficient, as its surrounding fluid opposes to the movement. This expression is known as Langevin equation, which for the typical Brownian motion the noise adopts a Gaussian form, namely,

$$\langle \zeta(t) \rangle = 0 \tag{A.2}$$
$$\langle \zeta(t)\zeta(t') \rangle = \sigma^2 \delta(t - t'), \tag{A.3}$$

where σ represents the amplitude of the thermal fluctuations. Intuitively, thermal fluctuations must be related to the dissipated energy through the friction coefficient, an intuition that becomes clear in the fluctuation-dissipation theorem [2], as we present in the following.

© Springer Nature Switzerland AG 2018
R. Puebla, *Equilibrium and Nonequilibrium Aspects of Phase Transitions
in Quantum Physics*, Springer Theses, https://doi.org/10.1007/978-3-030-00653-2

A.2   Fluctuation-Dissipation Theorem

In order to derive the fluctuation-dissipation theorem we follow a standard procedure. We start considering the Langevin equation, Eq. (A.1), which can be formally integrated as

$$\dot{x}(t) = \dot{x}(0)e^{-\frac{\gamma}{m}t} + \frac{1}{m}\int_0^t dt'\, e^{-\frac{\gamma}{m}(t-t')}\zeta(t'), \qquad (A.4)$$

where we have considered that the particle moves freely, i.e., $f(x) \equiv 0$. Then, we shall analyze how the particle diffuses and thermal equilibrium is attained, and specifically, how it relates to the noise amplitude, σ. For that reason, we first take the limit $t \to \infty$, multiply by $\dot{x}(t)$ and take stochastic average,

$$\lim_{t\to\infty}\langle \dot{x}^2(t)\rangle = \frac{1}{m^2}\int_0^\infty dt'' \int_0^\infty dt'\, e^{-\frac{\gamma}{m}(2t-t'-t'')}\langle \zeta(t')\zeta(t'')\rangle. \qquad (A.5)$$

Assuming $\langle \zeta(t')\zeta(t'')\rangle \equiv \langle \zeta(0)\zeta(t''-t')\rangle$, i.e., the noise correlation is independent of the specific time and just depends on time differences, and defining $r = t - t''$ and $s = t'' - t'$, it follows that

$$\lim_{t\to\infty}\langle \dot{x}^2(t)\rangle = \frac{1}{m^2}\int_0^\infty dr \int_{-\infty}^\infty ds\, e^{-\frac{\gamma}{m}(2r+s)}\langle \zeta(0)\zeta(s)\rangle \qquad (A.6)$$

$$= \frac{1}{2m\gamma}\int_{-\infty}^\infty ds\, e^{-\frac{\gamma}{m}s}\langle \zeta(0)\zeta(s)\rangle. \qquad (A.7)$$

Since the $t \to \infty$ limit ensures that the particle reaches thermal equilibrium, and because of the equipartition theorem, the kinetic energy takes half of the thermal energy, $m\langle \dot{x}^2\rangle/2 = (2\beta)^{-1}$, or simply $\langle \dot{x}^2\rangle = 1/(m\beta)$ with $\beta \equiv (k_B T)^{-1}$. Finally, this leads to

$$\int_{-\infty}^\infty ds\, e^{-\frac{\gamma}{m}s}\langle \zeta(0)\zeta(s)\rangle = \frac{2\gamma}{\beta}, \qquad (A.8)$$

which can be further simplified under the assumption of a short correlated noise, i.e. if $\langle \zeta(0)\zeta(t)\rangle$ decays in a time typical time scale τ such that $\tau \ll \gamma/m$, then $\int_{-\infty}^\infty ds\, \langle \zeta(0)\zeta(s)\rangle = 2\frac{\gamma}{\beta}$. This result is known as fluctuation-dissipation theorem as it relates the amplitude of thermal fluctuations with the dissipated by friction. Finally, for white noise, $\langle \zeta(t)\zeta(t')\rangle = \sigma^2\delta(t-t')$ the relation is exact

$$\sigma^2 = \frac{2\gamma}{\beta}. \qquad (A.9)$$

A.3 Fokker-Planck Equation

We derive here the Fokker-Planck counterpart of the Langevin equation, given here in Eq. (A.1), which is known as Kramers equation. For that reason, it is convenient to rewrite the Langevin equation in a general manner as

$$dV = \alpha(X, V)dt + \sqrt{\beta(X, V)dt}\,N_t \tag{A.10}$$

$$dX = Vdt, \tag{A.11}$$

where we have introduced the general functions $\alpha(X, V)$ and $\beta(X, V)$. Note that N_t stands for a random number following a normal distribution, i.e., zero mean and variance unity. In particular, for the case discussed previously, $\beta(x, v) = \sigma^2/m^2$, and $\alpha(x, v) = -\gamma v/m + F(x)$. Then, using

$$\iint dx\, dv\, f(x, v)\frac{\partial p(x, v, t)}{\partial t} = \left\langle \frac{df(X, V)}{dt} \right\rangle \tag{A.12}$$

where $p(x, v, t)$ corresponds to the probability distribution, and since $f(X, V)$ depends on t through X and V, it can be expanded as (keeping only terms up to dt)

$$df = \frac{\partial f}{\partial X}dX + \frac{\partial f}{\partial V}dV + \frac{\partial^2 f}{\partial V^2}\frac{(dV)^2}{2} + \mathcal{O}(dX^2, dV^3) \tag{A.13}$$

$$= \frac{\partial f}{\partial X}Vdt + \frac{\partial f}{\partial V}\left[\alpha(X, V)dt + \sqrt{\beta(X, V)dt}\,N_t\right] + \frac{\partial^2 f}{\partial V^2}\frac{\beta(X, V)N_t^2}{2}dt + \mathcal{O}(dt^2). \tag{A.14}$$

Then, taking the average, noting that $\langle N_t \rangle = 0$ and $\langle N_t^2 \rangle = 1$ by definition, and integrating

$$\iint dx\, dv\, f(x, v)\frac{\partial p(x, v, t)}{\partial t} =$$
$$\iint dx\, dv\, p(x, v, t)\left[v\frac{\partial f}{\partial x} + \alpha(x, v)\frac{\partial f}{\partial v} + \frac{\beta(x, v)}{2}\frac{\partial^2 f}{\partial v^2}\right], \tag{A.15}$$

which upon part-integration of the r.h.s. and assuming that the probability distribution vanishes on the boundaries (neglecting the surface terms), we obtain the Kramers equation,

$$\frac{\partial p(x, v, t)}{\partial t} = -v\frac{\partial p(x, v, t)}{\partial x} - \frac{\partial}{\partial v}[\alpha(x, v)p(x, v, t)] + \frac{1}{2}\frac{\partial^2}{\partial v^2}(\beta(x, v)p(x, v, t)), \tag{A.16}$$

which corresponds to the Eq. (2.17) in the Chap. 2. In a straightforward manner one can derive the Smoluchowski equation, see (2.21), which represents the high friction limit or overdamped dynamics, and it is attained simply considering $dX = \alpha(X)dt + \sqrt{\beta(X)dt}\,N_t$, and thus,

$$\frac{\partial p(x, t)}{\partial t} = -\frac{\partial}{\partial x}[\alpha(x)p(x, t)] + \frac{1}{2}\frac{\partial}{\partial x}[\beta(x)p(x, t)]. \tag{A.17}$$

References

1. H. Risken, *The Fokker-Planck Equation: Methods of Solution and Applications*, 2nd edn. (Springer, New York, 1984)
2. R. Kubo, The fluctuation-dissipation theorem. Rep. Prog. Phys. **29**, 255 (1966). http://stacks.iop.org/0034-4885/29/i=1/a=306

Appendix B
Rotating Wave Approximation

In this Appendix we provide a detailed description of the rotating wave approximation (RWA), which is essential to understand the developments carried out in Chap. 5. Note that this customary approximation the realm of quantum optics [1, 2, 3, 4] can be applied to a quantum Rabi model to neglect counter-rotating terms, as commented in Chap. 3. The main task consist in analyzing under what conditions the following Hamiltonian

$$H(t) = \frac{\omega}{2}\sigma_z + \frac{\Omega}{2}\left[\sigma^+ e^{i\delta t} + \sigma^- e^{-i\delta t}\right] \tag{B.1}$$

can be well approximated by $H_0 = \omega\sigma_z/2$. For that, it is insightful to move to an interaction picture with respect to H_0,

$$H_1^I(t) \equiv e^{-itH_0}(H(t) - H_0)e^{itH_0} = \frac{\Omega}{2}\left[\sigma^+ e^{i(\delta+\omega)t} + \sigma^- e^{-i(\delta+\omega)t}\right]. \tag{B.2}$$

If the Hamiltonian given in Eq. (B.1) is well approximated by H_0, the time-evolution propagator of the previous Hamiltonian must be close to the unity, $\mathcal{U}_1^I(t) \equiv Te^{-i\int_0^t dt' H_1^I(t')} \approx \mathbb{1}$ with T the time-ordering operator. Upon a Magnus expansion, $\mathcal{U}_1^I(t) = e^{\sum_k^\infty \Delta_k(t)}$ where $\Delta_k(t)$ contains the k-order time integrals, whose first three terms read

$$\Delta_1(t) = -i\int_0^t dt_1 H_1^I(t_1) \tag{B.3}$$

$$\Delta_2(t) = -\frac{1}{2}\int_0^t dt_1 \int_0^{t'} dt_2 \left[H_1^I(t_1), H_1^I(t_2)\right] \tag{B.4}$$

$$\Delta_3(t) = i\frac{1}{6}\int_0^t dt_1 \int_0^{t_1} dt_2 \int_0^{t_2} dt_3 \left(\left[H_1^I(t_1), \left[H_1^I(t_2), H_1^I(t_3)\right]\right] + \left[H_1^I(t_3), \left[H_1^I(t_2), H_1^I(t_1)\right]\right]\right),$$

we find that $\Delta_1(t) = i\Omega/(2(\delta+\omega))[-\sigma^+(e^{it(\delta+\omega)} - 1) + \sigma^-(e^{-it(\delta+\omega)} - 1)]$ and $\Delta_2(t) = -i\Omega^2/(4(\delta+\omega)^2)\sigma_z[(\delta+\omega)t - \sin((\delta+\omega)t)]$. Note that, there are terms

© Springer Nature Switzerland AG 2018
R. Puebla, *Equilibrium and Nonequilibrium Aspects of Phase Transitions in Quantum Physics*, Springer Theses, https://doi.org/10.1007/978-3-030-00653-2

which do not give an accumulated effect in time, and therefore, provided $\Omega \ll (\delta + \omega)$, the time-evolution propagator, up to $\mathcal{O}(\Omega^3 t/(\delta + \omega)^2)$, reads

$$\mathcal{U}_1^I(t) \approx e^{-it\frac{\Omega^2}{4(\delta+\omega)}\sigma_z}. \tag{B.5}$$

Hence, the approximation $H(t) \approx H_0$ is valid if $\Omega/(\delta + \omega) \ll 1$, or more specifically, $\Omega^2 t \ll (\delta + \omega)$. Note that the same result follows considering instead a Dyson expansion of $\mathcal{U}_1^I(t)$.

Finally, we comment that the performance of a RWA can be enhanced by including the effective time-evolution obtained in Eq. (B.5). Indeed, since $\mathcal{U}(t) = \mathcal{U}_0(t)\mathcal{U}_1^I(t)$, we notice that Eq. (B.1) effectively translates to a shift in the qubit frequency as $\omega' = \omega + \Omega^2/(2(\delta + \omega))$, the so-called Bloch-Siegert shift [4],

$$H(t) \approx H' = \frac{\omega'}{2}\sigma_z = \frac{\omega}{2}\left(1 + \frac{\Omega^2}{2\omega(\delta + \omega)}\right)\sigma_z. \tag{B.6}$$

References

1. D.F. Walls, G.J. Milburn, *Quantum Optics*, 2nd edn. (Springer, Berlin Heidelberg, 2008)
2. M.O. Scully, M.S. Zubairy, *Quantum Optics* (Cambridge University Press, Cambridge, England, 1997)
3. C. Gerry, P. Knight, *Introductory Quantum Optics* (Cambridge University Press, Cambridge, England, 2004)
4. A.B. Klimov, S.M. Chumakov, *A Group-Theoretical Approach to Quantum Optics* (Wiley-VCH, 2009)

Appendix C
Effective Hamiltonians of the Quantum Rabi Model

As explained in Chap. 3, Sect. 3.2, we are interested in obtaining the low-energy effective Hamiltonians of the QRM in the $\Omega/\omega_0 \to \infty$ limit, together with $\lambda/\omega_0 \to \infty$, while keeping $g \equiv 2\lambda/\sqrt{\Omega\omega_0}$ finite. In this Appendix we discuss the detailed procedure to attain the an effective Hamiltonian, together with its first-order perturbative correction in ω_0/Ω. In addition, we also show the standard procedure to diagonalize them, Sect. C.2, and the derivation of the entanglement entropy Sect. C.3. In this manner the material presented here complements the mathematical aspects of the developments shown in Chap. 3.

C.1 First-Order Correction for the Normal Phase

Here we provide the mathematical derivation to attain the first-order correction Hamiltonian for the normal phase, $0 \leq g \leq 1$, denoted by H_{np}^{Ω}, which reduces to H_{np} in the $\Omega/\omega_0 \to \infty$ limit. We start considering the QRM Hamiltonian, H_{QRM}, which can be written as

$$H_{\mathrm{QRM}} = H_0 - \lambda V \qquad (C.1)$$

where

$$H_0 = \omega a^\dagger a + \frac{\Omega}{2}\sigma_z, \quad V = (a + a^\dagger)\sigma_x. \qquad (C.2)$$

The unperturbed Hamiltonian H_0 has decoupled spin subspaces $\mathcal{H}_\downarrow$ and $\mathcal{H}_\uparrow$. For a ratio $\Omega/\omega_0 \gg 1$, the low-energy eigenstates of H_0 are confined in the subspace $\mathcal{H}_\downarrow$, and they are simply those of a simple harmonic oscillator. However, the interaction Hamiltonian V introduces coupling between the two subspaces and therefore, the structure of these low-energy eigenstates will change as soon as the interaction term is included. As commented in Sect. 3.2, our strategy consists in finding a unitary

© Springer Nature Switzerland AG 2018
R. Puebla, *Equilibrium and Nonequilibrium Aspects of Phase Transitions in Quantum Physics*, Springer Theses, https://doi.org/10.1007/978-3-030-00653-2

transformation U such that $U^\dagger H_{\text{QRM}} U$ has decoupled spin subspaces, i.e., it is free of coupling between $|\uparrow\rangle$ and $|\downarrow\rangle$, which is similar to the Schrieffer-Wolff transformation [1, 2]. In particular, the unitary transformation can be expressed as $U = e^S$ where S is an anti-Hermitian and block-off-diagonal with respect to the spin subspaces. The transformed Hamiltonian then reads

$$H' = e^{-S} H_{\text{QRM}} e^S = \sum_{k=0}^{\infty} \frac{1}{k!} \left[H_{\text{QRM}}, S\right]^{(k)} \tag{C.3}$$

where $[H, S]^{(k)} \equiv \left[[H, S]^{(k-1)}, S\right]$ denotes the k-order nested commutator, and $[H, S]^{(0)} \equiv H$. As we aim to obtain a transformed H' free of coupling between spin subspaces, it is convenient to split H' in diagonal and off-diagonal parts making use of the block-off-diagonal S and block-diagonal V, i.e.,

$$H_d' = \sum_{k=0}^{\infty} \frac{1}{(2k)!} [H_0, S]^{(2k)} - \sum_{k=0}^{\infty} \frac{1}{(2k+1)!} [\lambda V, S]^{(2k+1)}, \tag{C.4}$$

$$H_{od}' = \sum_{k=0}^{\infty} \frac{1}{(2k+1)!} [H_0, S]^{(2k+1)} - \sum_{k=0}^{\infty} \frac{1}{(2k)!} [\lambda V, S]^{(2k)}. \tag{C.5}$$

Then, we need to look for S such that H_{od}' vanishes. However, since we are interested in the $\Omega/\omega_0 \gg 1$ limit, we require that H_{od}' vanishes up a certain order of λ. Indeed, in Sect. 3.2 we have seen that for the $\Omega/\omega_0 \to \infty$ limit, only up to the second order in λ is needed, as higher-order terms vanish. Nevertheless, in order to accomplish a H_{np}^Ω, we must go up to fourth order, i.e., the terms of H_{od}' up to λ^4 must vanish. Denoting the generator $S \equiv \lambda S_1 + \lambda^3 S_3$, H_{od}' up to λ^4 can be written as

$$H_{od}' = [H_0, S] + \frac{1}{6} [H_0, S]^{(3)} - \lambda V - \frac{1}{2} [\lambda V, S]^{(2)} + \mathcal{O}(\lambda^5) \tag{C.6}$$

$$= \lambda [H_0, S_1] + \lambda^3 [H_0, S_3] + \frac{\lambda^3}{6} [[[H_0, S_1], S_1], S_1] - \lambda V - \frac{\lambda^3}{2} [[V, S_1], S_1] + \mathcal{O}(\lambda^5),$$

and collecting terms in λ and λ^3, $H_{od}' = 0$ leads to

$$[H_0, S_1] = V \tag{C.7}$$

$$[H_0, S_3] = \frac{1}{3} [[V, S_1], S_1]. \tag{C.8}$$

A generator S that fulfills the previous conditions is

$$S_1 = \frac{1}{\Omega} (a + a^\dagger)(\sigma_+ - \sigma_-) + \mathcal{O}\left(\frac{\omega_0}{\Omega^2}\right) \tag{C.9}$$

$$S_3 = -\frac{4}{3\Omega^3} (a + a^\dagger)^3 (\sigma_+ - \sigma_-) + \mathcal{O}\left(\frac{\omega_0}{\Omega^4}\right). \tag{C.10}$$

Now, making use of the Eq. (C.4),

$$H' = H_0 + \frac{1}{2}[H_0, S]^{(2)} + \frac{1}{24}[H_0, S]^{(4)} - [\lambda V, S] - \frac{1}{6}[\lambda V, S]^{(3)} + \mathcal{O}(\lambda^6) \qquad (C.11)$$

$$= H_0 - \frac{\lambda^2}{2}[V, S_1] - \frac{\lambda^4}{2}[V, S_3] + \frac{\lambda^4}{24}[[[V, S_1], S_1], S_1] + \mathcal{O}(\lambda^6) \qquad (C.12)$$

$$= \omega_0 a^\dagger a + \frac{\Omega}{2}\sigma_z + \frac{\omega_0^2 g^2}{4\Omega} + \frac{\omega_0 g^2}{4}(a + a^\dagger)^2 \sigma_z - \frac{\omega_0^2 g^4}{16\Omega}(a + a^\dagger)^4 \sigma_z + \mathcal{O}\left(\frac{g^2 \omega_0^3}{\Omega^2}\right),$$

where we have introduced the dimensionless coupling constant $g = 2\lambda/\sqrt{\Omega\omega_0}$. The previous Hamiltonian contains terms up to first order in ω_0/Ω without coupling between the spin subspaces. Hence, after projecting onto $\mathcal{H}_\downarrow$, we finally obtain

$$H_{np}^\Omega \equiv \langle\downarrow| H' |\downarrow\rangle = \omega_0 a^\dagger a - \frac{\omega_0 g^2}{4}(a + a^\dagger)^2 + \frac{\omega_0^2 g^4}{16\Omega}(a + a^\dagger)^4 - \frac{\Omega}{2} + \frac{g^2 \omega_0^2}{4\Omega},$$

which corresponds to the Hamiltonian given in Eq. (3.23).

C.2 Diagonalization

Here, we briefly show how these effective low-energy Hamiltonians, H_{np} and H_{sp}, which present a quadratic term, $(a + a^\dagger)^2$, can be diagonalized by means of squeezing operators. These Hamiltonians can be written in a compact manner as

$$H_q = c_1 a^\dagger a + c_2(a + a^\dagger)^2 + c_0, \qquad (C.13)$$

where $c_{1,2,3}$ represent the different parameters. Then, we transform H_q via the squeezing operator $S[r] = e^{r/2(a^{\dagger 2} - a^2)}$ considering $r \in \mathbb{R}$,

$$S^\dagger[r]H_q S[r] = (c_1 + 2c_2)\left(\cosh(2r)a^\dagger a + \frac{1}{2}\sinh(2r)(a^2 + (a^\dagger)^2) + \sinh^2 r\right)$$

$$+ c_2\left(2\sinh(2r)a^\dagger a + \cosh(2r)(a^2 + (a^\dagger)^2) + \sinh(2r) + 1\right) + c_0$$

$$= a^\dagger a\left((c_1 + 2c_2)\cosh(2r) + 2c_2\sinh(2r)\right)$$

$$+ (a^2 + (a^\dagger)^2)\left(c_2\cosh(2r) + \frac{(c_1 + 2c_2)}{2}\sinh(2r)\right)$$

$$+ (c_1 + 2c_2)\sinh^2 r + c_2(\sinh(2r) + 1) + c_0 \qquad (C.14)$$

Note that with the definition of $S[r]$, the bosonic operators transform as $S^\dagger[r]aS[r] = a\cosh r + a^\dagger \sinh r$ and $S^\dagger[r]a^\dagger S[r] = a^\dagger \cosh r + a \sinh r$. Then, the previous transformation adopts a diagonal form when the coefficient accompanying the term $a^2 + (a^\dagger)^2$ vanishes. Thus, the squeezing parameter that diagonalizes H_q follows from

$$c_2 \cosh(2r_s) + \frac{(c_1 + 2c_2)}{2} \sinh(2r_s) = 0 \qquad (C.15)$$

that is,

$$r_s = -\frac{1}{4} \log \left(\frac{4c_2}{c_1} + 1 \right) \qquad (C.16)$$

For H_{np}, the parameters are $c_1 = \omega_0$, $c_2 = -\omega_0 g^2/4$ and $c_0 = -\Omega/2$. Thus, the squeezing parameter is $r_{np} = -1/4 \log(1 - g^2)$. For the superradiant phase, the parameters c_2 and c_0 are different, with $c_2 = -\omega_0/(4g^4)$ and $c_0 = -\Omega/4(g^2 + g^{-2})$, which lead to $r_{sp} = -1/4 \log(1 - g^{-4})$. In general, under the choice of r_s, Eq. (C.16), the transformed Hamiltonian takes the following form

$$S^\dagger[r_s]H_q S[r_s] = \sqrt{c_1(c_1 + 4c_2)}a^\dagger a + \frac{c_1}{2} \left(\sqrt{1 + \frac{4c_2}{c_1}} - 1 \right) + c_0, \qquad (C.17)$$

which after substituting the parameters for H_{np}, we recover the expression given in Chap. 3, Eq. (3.14),

$$S^\dagger[r_{np}]H_{np} S[r_{np}] = \omega_0\sqrt{1 - g^2}a^\dagger a + \frac{\omega_0}{2} \left(\sqrt{1 - g^2} - 1 \right) - \frac{\Omega}{2} \qquad (C.18)$$

$$= \epsilon_{np}(g)a^\dagger a + E_{G,np}(g), \qquad (C.19)$$

where we have introduced the excitation energy $\epsilon_{np}(g) = \omega_0\sqrt{1 - g^2}$ and the ground-state energy $E_{G,np}(g) = (\epsilon_{np}(g) - \omega_0)/2 - \Omega/2$. In a straightforward manner, we obtain the diagonalized H_{sp},

$$S^\dagger[r_{sp}]H_{sp} S[r_{sp}] = \omega_0\sqrt{1 - g^{-4}}a^\dagger a + \frac{\omega_0}{2} \left(\sqrt{1 - g^{-4}} - 1 \right) - \frac{\Omega}{4}(g^2 + g^{-2})$$

$$= \epsilon_{sp}(g)a^\dagger a + E_{G,sp}(g), \qquad (C.20)$$

with $\epsilon_{sp}(g) = \omega_0\sqrt{1 - g^{-4}}$ and ground-state energy $E_{G,sp}(g) = (\epsilon_{sp}(g) - \omega_0)/2 - \Omega(g^2 + g^{-2})/4$.

C.3 Entanglement Entropy

Here we derive the expression of the von Neumann entropy across the QPT given in the Chap. 3. The von Neumann entanglement entropy S_{vN} can be written as

$$S_{vN} = -\text{Tr}\left[\rho_s \log_2 \rho_s\right], \tag{C.21}$$

where ρ_s here denotes the reduced spin density matrix. Since we are interested in the ground-state properties, $\rho = \left|\varphi_{np}^0(g)\right\rangle\left\langle\varphi_{np}^0(g)\right|$ for $0 < g < 1$ and $\left|\varphi_{sp}^0(g)\right\rangle$ for the superradiant phase. Certainly, for the normal phase it is straightforward to show that $S_{vN} = 0$ since $\rho_s = \left|\downarrow\right\rangle\left\langle\downarrow\right|$. To the contrary, for the superradiant phase we consider the symmetrized ground state, that is,

$$\left|\varphi_{sp}^0(g)\right\rangle_S = \frac{1}{\sqrt{2}}\left(\left|\varphi_{sp}^0(g)\right\rangle_+ \pm \left|\varphi_{sp}^0(g)\right\rangle_-\right), \tag{C.22}$$

with

$$\left|\varphi_{sp}^n(g)\right\rangle_\pm = \mathcal{D}[\pm\alpha_g(g)]\mathcal{S}[r_{sg}(g)]\left|n\right\rangle\left|\downarrow^\pm\right\rangle, \tag{C.23}$$

where the spin states read

$$\left|\downarrow^\pm\right\rangle = \mp\sqrt{\frac{1-g^{-2}}{2}}\left|\uparrow\right\rangle + \sqrt{\frac{1+g^{-2}}{2}}\left|\downarrow\right\rangle. \tag{C.24}$$

Then, from the full density matrix, $\rho = \left|\varphi_{sp}^0(g)\right\rangle_S\left\langle\varphi_{sp}^0(g)\right|_S$, we obtain the reduced spin density matrix as

$$\rho_s = \frac{1}{\pi}\int d^2\beta\ \left|\beta\right\rangle\left\langle\beta\right|\rho \tag{C.25}$$

$$= \frac{1}{2\pi}\int d^2\beta\ \left(\left|\downarrow^+\right\rangle\left\langle\downarrow^+\right|\left\langle\beta|\alpha_g\right\rangle\left\langle\alpha_g|\beta\right\rangle + \left|\downarrow^-\right\rangle\left\langle\downarrow^-\right|\left\langle\beta|-\alpha_g\right\rangle\left\langle-\alpha_g|\beta\right\rangle \right.$$
$$\left. - \left|\downarrow^-\right\rangle\left\langle\downarrow^+\right|\left\langle\beta|+\alpha_g\right\rangle\left\langle-\alpha_g|\beta\right\rangle - \left|\downarrow^+\right\rangle\left\langle\downarrow^-\right|\left\langle\beta|-\alpha_g\right\rangle\left\langle+\alpha_g|\beta\right\rangle\right) \tag{C.26}$$

where $\left|\beta\right\rangle = \mathcal{D}(\beta)\left|0\right\rangle$ denotes a coherent state and $d^2\beta = |\beta|d|\beta|d\theta$ since $\beta = |\beta|e^{i\theta}$. Then, considering that the overlap $\left\langle\alpha_g|\beta\right\rangle = \delta_{\alpha_g,\beta}$. Note that, in general $\left\langle\alpha|\beta\right\rangle = e^{-(|\alpha|^2+|\beta|^2)/2-\alpha^*\beta}$, however, since the displacement $\alpha_g \rightarrow \infty$ in the $\Omega/\omega_0 \rightarrow \infty$, the overlap vanishes exponentially. Recall that $\alpha_g = \sqrt{\Omega/(4g^2\omega_0)}\sqrt{g^4 - 1}$. Thus, in this limit ρ_s results in

$$\rho_s = \frac{1}{2}\left(|\downarrow^+\rangle\langle\downarrow^+| + |\downarrow^-\rangle\langle\downarrow^-|\right) \tag{C.27}$$

$$= \frac{1}{2}\left((1 - g^{-2})\,|\uparrow\rangle\,\langle\uparrow| + (1 + g^{-2})\,|\downarrow\rangle\,\langle\downarrow|\right). \tag{C.28}$$

Note that ρ_s is already diagonal in the spin basis, and therefore S_{vN} simply reads

$$S_{\mathrm{vN}} = -\frac{1 - g^{-2}}{2}\log_2\left(\frac{1 - g^{-2}}{2}\right) - \frac{1 + g^{-2}}{2}\log_2\left(\frac{1 + g^{-2}}{2}\right), \tag{C.29}$$

which corresponds to the Eq. (3.38).

References

1. J.R. Schrieffer, P.A. Wolff, Relation between the Anderson and Kondo Hamiltonians. Phys. Rev. **149**, 491 (1966). https://doi.org/10.1103/PhysRev.149.491
2. S. Bravyi, D.P. DiVincenzo, D. Loss, Schrieffer–Wolff transformation for quantum many-body systems. Ann. Phys. (N. Y.) **326**, 2793 (2011). https://doi.org/10.1016/j.aop.2011.06.004

Appendix D
Semiclassical Approach of the Quantum Rabi Model

In this Appendix we present a detailed analysis of the semiclassical density of states of the QRM that allowed us to unveil the existence of an excited-state quantum phase transition (ESQPT) in the $\Omega/\omega_0 \to \infty$ limit, as well as relate the singular density of states to the mean-field observables. For that purpose, we first show how to obtain the semiclassical density of states, and then, the subsequent and necessary derivations to attain analytical expressions that disclose its divergence at a certain critical excitation energy, given in Sect. D.1. Finally, in Sect. D.2, we show how to obtain Eq. (3.84) by means of the Hellmann-Feynman theorem.

D.1 Semiclassical Density of States

The semiclassical density of states of the QRM plays a key role in the Chap. 3.84, as it exhibits singular behavior at a certain non-zero excitation energy. The semiclassical density of states, $\nu_{\mathrm{scl}}(\varepsilon, g)$ can be written as

$$\nu_{\mathrm{scl}}(\varepsilon, g) = \frac{1}{\omega_0 \pi} \int dx dp \, \delta \left[\varepsilon - 2H_{\mathrm{scl}}^-(x, p, g)/\Omega \right] \tag{D.1}$$

$$= \frac{1}{\omega_0 \pi} \int dx dp \, \delta \left[\varepsilon - p^2 - x^2 + \sqrt{1 + 2g^2 x^2} \right] \tag{D.2}$$

$$= \frac{1}{\omega_0 \pi} \int dx \left(\frac{\delta[p - p_+]}{|\partial_p 2H_{\mathrm{scl}}(x, p, g)/\Omega|_{p=p_+}} + \frac{\delta[p - p_-]}{|\partial_p 2H_{\mathrm{scl}}(x, p, g)/\Omega|_{p=p_-}} \right)$$

$$= \frac{2}{\omega_0 \pi} \int_{x_1}^{x_2} dx \, \frac{1}{\sqrt{\varepsilon - x^2 + \sqrt{1 + 2g^2 x^2}}} \tag{D.3}$$

© Springer Nature Switzerland AG 2018
R. Puebla, *Equilibrium and Nonequilibrium Aspects of Phase Transitions in Quantum Physics*, Springer Theses, https://doi.org/10.1007/978-3-030-00653-2

where $p_{\pm} = \pm\sqrt{\varepsilon - x^2 + \sqrt{1 + 2g^2x^2}}$ are the positive and negative roots of $\varepsilon - 2H_{\text{scl}}^{-}(x, p, g)/\Omega = 0$. Note that we only take into account $H_{\text{scl}}^{-}(\varepsilon, g)$, as the positive branch is only relevant at energies $\varepsilon > 1$. The integration limits are given by

$$x_1 = \sqrt{\varepsilon + g^2 - \sqrt{g^4 + 2\varepsilon g^2 + 1}} \,\Theta\left[\varepsilon_c - \varepsilon\right] \tag{D.4}$$

$$x_2 = \sqrt{\varepsilon + g^2 + \sqrt{g^4 + 2\varepsilon g^2 + 1}}. \tag{D.5}$$

In this manner, we are able to compute $\nu_{\text{scl}}(\varepsilon, g)$. However, as shown in Chap. 3, we can provide analytical expressions for some relevant and interesting cases, namely $\varepsilon_c = -1$ for $g = 1$ and $g > 1$. We start considering $g = 1$, for which the ground-state energy is simply $\varepsilon_c = -1$. Then, for an energy $\varepsilon = \varepsilon_c + \delta\varepsilon$ with $0 < \delta\varepsilon \ll 1$, we can expand the integration limits as well as the Eq. (D.3) such that

$$\nu_{\text{sc}}(\varepsilon_c + \delta\varepsilon, g = 1) \approx \frac{2}{\omega_0\pi} \int_0^{(2\delta\varepsilon)^{1/4}} dx \left(\frac{1}{\sqrt{\delta\varepsilon - \frac{x^4}{2}}} + \mathcal{O}(x^6)\right) = \frac{\Gamma(5/4)}{\Gamma(3/4)} \frac{2^{5/4}}{\omega_0\sqrt{\pi}} (\delta\varepsilon)^{-1/4}.$$

Therefore, the semiclassical density of states diverges at the critical point of the QPT as $\nu_{\text{scl}}(\varepsilon, g = 1) \propto (\varepsilon - \varepsilon_c)^{-1/4}$ for $\varepsilon - \varepsilon_c \ll 1$.

On the other hand, for $g > 1$ we first consider $0 < \delta\varepsilon \ll 1$. In this case, the integration limits result in $x_2 = \sqrt{2(g^2 - 1)} + \mathcal{O}(\delta\varepsilon)$ and $x_1 = 0$. Then, we split the integration interval as $\int_{x_1}^{x_2} dx = \int_{x_1}^{x_m} dx + \int_{x_m}^{x_2} dx$ and the arbitrary intermediate value x_m is chosen such that $0 < x_m \ll 1$ but still greater than $\delta\varepsilon$. Then,

$$\nu_{\text{sc}}(\varepsilon_c + \delta\varepsilon, g > 1) \approx \frac{2}{\omega_0\pi} \int_0^{x_m} dx \left(\frac{1}{\sqrt{\delta\varepsilon + (g^2 - 1)x^2}} + \mathcal{O}(x^4)\right) + K \tag{D.6}$$

$$\approx \frac{1}{\omega_0\pi\sqrt{g^2 - 1}} \ln\left(\frac{2x_m^2(g^2 - 1)}{\delta\varepsilon}\right) + K, \tag{D.7}$$

where K denotes the result of the integral from x_m to x_2, which only produces a constant shift, as we demonstrate in the following. Certainly, the denominator of Eq. (D.3) vanishes for $x_2 = \sqrt{2(g^2 - 1)}$ as $\sqrt{\delta\varepsilon}$ and therefore it may lead to a divergence in $\nu_{\text{scl}}(\varepsilon_c, g > 1)$. However, defining yet another intermediate integration limit x_n such that $\int_{x_m}^{x_2} dx = \int_{x_m}^{x_n} dx + \int_{x_n}^{x_2} dx$ with $x_m < x_n < x_2$ and $0 < x_2 - x_n \ll 1$, we can denote the result of the integral between x_m and x_n as $\tilde{K}$ as it does not involve any singularity and thus, it just contributes to $\nu_{\text{scl}}(\varepsilon_c, g > 1)$ with a constant value. Then, the shift K can be expressed as

$$K = \frac{2}{\omega_0 \pi} \int_{x_m}^{x_2} \frac{dx}{\sqrt{\delta \varepsilon - 1 - x^2 + \sqrt{1 + 2g^2 x^2}}}$$

$$= \tilde{K} + \frac{2}{\omega_0 \pi} \int_{x_n}^{x_2} \frac{dx}{\sqrt{\delta \varepsilon - 1 - x^2 + \sqrt{1 + 2g^2 x^2}}}$$

$$= \tilde{K} + \frac{2}{\omega_0 \pi} \int_{x_n}^{x_2} dx \left(\frac{1}{\sqrt{\delta \varepsilon + \frac{2\sqrt{2}(g^2-1)^{3/2}}{2g^2-1}(x - x_2)}} + \mathcal{O}((x - x_2)^2) \right)$$

$$\approx \tilde{K} + \frac{2}{\omega_0 \pi} \frac{(2g^2 - 1)}{\sqrt{2}(g^2 - 1)^{3/2}} \left(\sqrt{\delta \varepsilon + \frac{4(g^2-1)^2 - 2\sqrt{2}(g^2-1)^{3/2}}{2g^2-1}} - \sqrt{\delta \varepsilon} \right),$$

where $x_n = x_2 - \delta x$. Indeed, if we select $\delta \varepsilon = 0$, the previous expression adopts a form

$$K \approx \tilde{K} + \frac{2}{\omega_0 \pi} \frac{2^{1/4}\sqrt{\delta x (2g^2 - 1)}}{(g^2 - 1)^{3/4}}, \tag{D.8}$$

which is analytic for any $g > 1$. Moreover, we now analyze the limit in which the critical energy is approached from the opposite side, i.e., $\varepsilon = \varepsilon_c - \delta \varepsilon$. Then, the integration limits can be expanded as

$$x_1 = \sqrt{\frac{\delta \varepsilon}{g^2 - 1}} + \mathcal{O}(\delta \varepsilon), \tag{D.9}$$

$$x_2 = \sqrt{2(g^2 - 1)} + \mathcal{O}(\delta \varepsilon). \tag{D.10}$$

In a straightforward manner, we obtain

$$\nu_{\text{scl}}(\varepsilon, g) = \frac{2}{\omega_0 \pi} \int_{x_1}^{x_2} dx \left(\frac{1}{\sqrt{(g^2 - 1)x^2 - \delta \varepsilon}} + \mathcal{O}(x^4) \right) \tag{D.11}$$

$$\approx \frac{1}{\omega_0 \pi \sqrt{g^2 - 1}} \left(\ln \left(\frac{8(g^2 - 1)^2}{\delta \varepsilon} \right) \right), \tag{D.12}$$

and therefore, we can conclude that $\nu_{\text{scl}}(\varepsilon, g > 1)$ diverges logarithmically for $|\varepsilon - \varepsilon_c| \ll 1$, as given in Eq. (3.82).

D.2   Hellman-Feynman Theorem

In Chap. 3, we have made use of the Hellmann-Feynman theorem to connect expectation values with derivatives of the accumulated number of states $N(\varepsilon, g)$ (see Eq. 3.84). Briefly, the Hellmann-Feynman theorem indicates [1]

$$\frac{\partial}{\partial \beta} E(\beta) = \langle \varphi(\beta) | \frac{\partial}{\partial \beta} H(\beta) | \varphi(\beta) \rangle , \tag{D.13}$$

where β denotes a real parameter and $|\varphi(\beta)\rangle$ an eigenvector of the Hamiltonian $H(\beta)$. Then, since $H(\beta) |\varphi(\beta)\rangle = E(\beta) |\varphi(\beta)\rangle$, and $\langle \varphi(\beta) | \varphi(\beta) \rangle = 1$, one obtains that

$$\frac{\partial}{\partial \beta} E(\beta) = \frac{\partial}{\partial \beta} \langle \varphi(\beta) | H(\beta) | \varphi(\beta) \rangle \tag{D.14}$$

$$= \langle \varphi(\beta) | \frac{\partial}{\partial \beta} H(\beta) | \varphi(\beta) \rangle + E(\beta) \left\{ \left[\frac{\partial}{\partial \beta} \langle \varphi(\beta) | \right] | \varphi(\beta) \rangle + \langle \varphi(\beta) | \left[\frac{\partial}{\partial \beta} | \varphi(\beta) \rangle \right] \right\} ,$$

which reduces to the Eq. (D.13) since the last two terms correspond to $\partial_\beta (\langle \varphi(\beta) | \varphi(\beta) \rangle) = 0$. Therefore, taking Eq. (D.13) and noting that if an observable $\mathcal{A}$ appears in the Hamiltonian with a linear dependence on a parameter β, i.e., $\mathcal{A} = \partial_\beta H(\beta)$, then

$$\frac{\partial}{\partial \beta} E(\beta) = \langle \varphi(\beta) | \mathcal{A} | \varphi(\beta) \rangle = \langle \mathcal{A} \rangle . \tag{D.15}$$

Therefore, to attain Eq. (3.84), we shall start with Eq. (3.83),

$$\langle \mathcal{A} \rangle (\varepsilon, g) = \frac{1}{\nu_q(\varepsilon, g)} \sum_{k=0, \alpha=\pm} \langle \varphi_k^\alpha | \mathcal{A} | \varphi_k^\alpha \rangle \delta \left[\varepsilon - \varepsilon_k^\alpha \right] \tag{D.16}$$

$$= \frac{1}{\nu_q(\varepsilon, g)} \sum_{k=0, \alpha=\pm} \langle \varphi_k^\alpha | \frac{\partial}{\partial \beta} H(\beta) | \varphi_k^\alpha \rangle \delta \left[\varepsilon - \varepsilon_k^\alpha \right] \tag{D.17}$$

$$= \frac{1}{\nu_q(\varepsilon, g)} \sum_{k=0, \alpha=\pm} \frac{\partial}{\partial \beta} \varepsilon_k^\alpha(\beta) \delta \left[\varepsilon - \varepsilon_k^\alpha \right] . \tag{D.18}$$

On the other hand, the derivative of the accumulated number of states $\partial_\beta N_q(\varepsilon, g)$ reads,

$$\frac{\partial}{\partial\beta}N_q(\varepsilon, g) = \frac{\partial}{\partial\beta}\int_{-\infty}^{\varepsilon}d\varepsilon' \nu_q(\varepsilon', g) \tag{D.19}$$

$$= \nu_q(\varepsilon, g)\frac{\partial\varepsilon}{\partial\beta} - \lim_{x\to-\infty}\nu_q(x, g)\frac{\partial x}{\partial\beta} + \int_{-\infty}^{\varepsilon}d\varepsilon'\frac{\partial}{\partial\beta}\nu_q(\varepsilon, g) \tag{D.20}$$

$$= \int_{-\infty}^{\varepsilon}d\varepsilon'\frac{\partial}{\partial\beta}\sum_{k=0,\alpha=\pm}\delta\left[\varepsilon' - \varepsilon_k^\alpha\right] = \int_{-\infty}^{\varepsilon}d\varepsilon'\sum_{k=0,\alpha=\pm}\frac{\partial\varepsilon'}{\partial\beta}\frac{\partial}{\partial\varepsilon'}\delta\left[\varepsilon' - \varepsilon_k^\alpha\right] \tag{D.21}$$

$$= -\sum_{k=0,\alpha=\pm}\int_{-\infty}^{\varepsilon}d\varepsilon'\frac{\partial}{\partial\varepsilon'}\frac{\partial\varepsilon'}{\partial\beta}\delta\left[\varepsilon' - \varepsilon_k^\alpha\right] = -\sum_{k=0,\alpha=\pm}\frac{\partial}{\partial\beta}\varepsilon_k^\alpha\,\delta\left[\varepsilon - \varepsilon_k^\alpha\right]. \tag{D.22}$$

Note that the energy ε is an external parameter and does not depend on β, while ε_k^α are the energy eigenvalues of $H(\beta)$. Hence, combining Eqs. (D.18) and (D.22), we finally obtain the expression given in Eq. (3.84),

$$\langle\mathcal{A}\rangle(\varepsilon, g) = -\frac{1}{\nu_q(\varepsilon, g)}\frac{\partial}{\partial\beta}N_q(\varepsilon, g), \tag{D.23}$$

whose semiclassical approximation is achieved replacing $\nu_q(\varepsilon, g)$ and $N_q(\varepsilon, g)$ by $\nu_{scl}(\varepsilon, g)$ and $N_{scl}(\varepsilon, g)$, respectively.

Reference

1. C. Cohen-Tannoudji, B. Diu, F. Laloë, *Quantum Mechanics* (Wiley-VCH, Berlin Heidelberg, 2005)

Appendix E
Adiabatic Perturbation Theory and Critical Dynamics

In this Appendix we explain first, in Sect. E.1, the general adiabatic perturbation theory (APT) [1, 2] and how to apply it to the quantum Rabi model (QRM). Then, in Sect. E.2, we show how to compute the time-evolved wave function when the QRM exhibits a quantum phase transition, and in particular, how to calculate the fidelity given in Eq. (4.8) of Chap. 4. In addition, in Sect. E.3 we provide the derivation of the equation of motion given in Eqs. (4.17) and (4.18), (4.31) as well as the required information for Sects. 4.1.3 and 4.1.4.

E.1 Adiabatic Perturbation Theory

The APT aims to obtain a correction to the an ideal adiabatic evolution. In particular, considering a Hamiltonian $H(t)$ that varies in a characteristic time τ_Q, the parameter τ_Q^{-1} is treated perturbatively ($\tau_Q \to \infty$ corresponds to the adiabatic limit) and leading order corrections in τ_Q^{-1} are sought. In the following lines we closely follow the derivation given in [3]. In general, let $H(t)$ denote a time-dependent Hamiltonian such that $H(t) = \sum_n \epsilon_n(t) |\phi_n(t)\rangle$ where $\epsilon_n(t)$ is the energy of the nth instantaneous eigenstate $|\phi_n(t)\rangle$, and considering that H depends on t through the variation of a parameter $g(t) = g_f t / \tau_Q$. Furthermore, we express the wave function at time t in terms of $|\phi_n(t)\rangle$, that is, $|\psi(t)\rangle = \sum_n \alpha_n(t) e^{-i\Theta_n(t)} |\phi_n(t)\rangle$ with $\Theta_n(t) = \int_0^t \epsilon_n(t') dt'$. Then, the Schrödinger equation for the coefficients $\alpha_n(t)$ reads

$$\dot{\alpha}_n(t) = -\sum_m \alpha_m(t) \langle \phi_n(t)| \, \partial_t \, |\phi_m(t)\rangle \, e^{i(\Theta_n(t) - \Theta_m(t))}, \tag{E.1}$$

whose formal solution is given by

$$\alpha_n(g) = -\sum_m \int_0^g dg' \alpha_m(g') \langle \phi_n(g')| \, \partial_{g'} \, |\phi_m(g')\rangle \, e^{i(\Theta_n(g') - \Theta_m(g'))}, \tag{E.2}$$

© Springer Nature Switzerland AG 2018
R. Puebla, *Equilibrium and Nonequilibrium Aspects of Phase Transitions in Quantum Physics*, Springer Theses, https://doi.org/10.1007/978-3-030-00653-2

where we have changed the variables assuming a linear quench $g = \dot{g}t$. Since we are interested in nearly-adiabatic dynamics, the quench rate is small, $\dot{g} \ll 1$, and in addition, we take as initial condition the ground state, $\alpha_0(0) = 1$ and $\alpha_n(0) = 0$ for $n \geq 1$. Then, in the leading order of $\dot{g}$, Eq. (E.2) becomes

$$\alpha_n(g) \approx - \int_0^g dg' \langle \phi_n(g') | \partial_{g'} | \phi_0(g') \rangle e^{i(\Theta_n(g') - \Theta_0(g'))}. \tag{E.3}$$

Moreover, since $\Theta_n(g) = \frac{1}{\dot{g}} \int_0^g \epsilon_n(g')dg'$, the phase term $e^{i(\Theta_n(g) - \Theta_0(g))}$ rapidly oscillates. Using the standard evaluation of a fast oscillating integral, $\int_{x_1}^{x_2} f(x)e^{i\eta g(x)}dx = \frac{1}{i\eta} \frac{f(x)}{g'(x)} e^{iag(x)} \Big|_{x_1}^{x_2} + \mathcal{O}(\eta^{-2})$, we obtain

$$\alpha_n(g) \approx i\dot{g} \frac{\langle \phi_n(g) | \partial_g | \phi_0(g) \rangle}{\epsilon_n(g) - \epsilon_0(g)} e^{i(\Theta_n(g) - \Theta_0(g))} \Big|_0^g + \mathcal{O}(\dot{g}^2). \tag{E.4}$$

Hence, knowing $\langle \phi_n(g) | \partial_g | \phi_0(g) \rangle$ and the energy spectrum, $\epsilon_n(g)$ one obtains the coefficients $\alpha_n(g)$ within the APT, thus valid for $\dot{g} \ll 1$. In addition, it is worth noting that if the Hamiltonian becomes gapless, $\epsilon_n(g) = \epsilon_0(g)$ for a certain n and g, the previous theory does not apply.

In the case a QRM in the $\Omega/\omega_0 \to \infty$ limit, we have have all the ingredients needed to compute $\alpha_n(g)$. As we consider $g(t) = g_f t/\tau_Q$ with $0 \leq g_f \leq g_c = 1$ we remain in the normal phase, and thus $|\phi_n(g)\rangle = |\varphi_{np}^n(g)\rangle = S_{np}[r_{np}(g)] |n\rangle |\downarrow\rangle$ with $\epsilon_n(g) - \epsilon_0(g) = n\omega_0(1 - g^2)^{z\nu}$ with $z\nu = 1/2$. Note that we have explicitly written the dependence on the critical exponents. In addition, since the squeezing parameter reads $r_{np}(g) = -1/4 \ln(1 - g^2)$,

$$\langle \varphi_{np}^n(g)|\partial_g|\varphi_{np}^m(g)\rangle = \frac{1}{2} \frac{\partial r_{np}(g)}{\partial g} \langle \varphi_{np}^m(g) | (a^{\dagger 2} - a^2)| \varphi_{np}^n(g) \rangle = -\frac{\sqrt{2}g}{4(1 - g^2)} \delta_{m,2}.$$

Therefore, at leading order in τ_Q^{-1}, the only non-vanishing correction to an adiabatic evolution reads

$$\alpha_2(g) \approx -i \frac{g_f}{\tau_Q} \frac{ge^{i(\Theta_2(g) - \Theta_0(g))}}{4\sqrt{2}\omega_0(1 - g^2)^{z\nu+1}}. \tag{E.5}$$

We focus now on the scaling of the residual energy E_r, but same arguments apply in a straightforward manner to other observables. The residual energy, defined as $E_r(g_f) = \langle \psi(g_f)| H_{np}(g_f) |\psi(g_f)\rangle - E_{GS}(g_f)$ can be written as

$$E_r(g_f) = \sum_{n>0} \epsilon_n(g_f) |\alpha_n(g_f)|^2 \approx \epsilon_2(g_f) |\alpha_2(g_f)|^2 \approx \tau_Q^{-2} \frac{g_f^4}{16\omega_0(1 - g_f^2)^{z\nu+2}}.$$

Therefore, if the quench stops before the critical point $g_f < g_c$, the APT is expected to correctly describe the dynamics, and thus, we attain the scaling $E_r \sim \tau_Q^{-2}$. Note that for nonlinear quenches, the same procedure leads to $E_r \sim \tau_Q^{-2r}$, where r is the nonlinear exponent of the protocol (see Chap. 4). Remarkably, once $g_f \to g_c = 1$, the APT breaks down and one needs to resort to Kibble-Zurek arguments which correctly describe the observed scaling, $E_r \sim \tau_Q^{-z\nu/(z\nu+1)}$, as explained in Chap. 4. Remarkably, from previous expressions $|\alpha_2|^2$ does not show a scaling if $g_f = g_c$ using Kibble-Zurek arguments, which is consistent with the predicted scaling $\tau_Q^{-d\nu/(z\nu+1)}$ for d-dimensional systems.

E.2 Time Evolution of Fidelity in Quench Dynamics

Here we show how to calculate the time evolution of the fidelity between the evolved state and the ground state at the same coupling instant $g(t)$ in the $\Omega/\omega_0 \to \infty$ limit of a QRM. We start considering the effective Hamiltonian for $0 \le g \le 1$, which under a protocol $g(t) = t/\tau_q$ reads

$$H_{\mathrm{np}}(t) = \omega_0 \left(1 - \frac{t^2}{2\tau_Q^2}\right) a^\dagger a - \frac{\omega_0 t^2}{4\tau_Q^2}\left(a^2 + a^{\dagger 2}\right) - \frac{\omega_0 t^2}{4\tau_Q^2} - \frac{\Omega}{2}, \qquad (\mathrm{E.6})$$

where we have made the Hamiltonian to be normal ordered. Then, the Schrödinger equation for the unitary evolution operator reads

$$\frac{\partial}{\partial t} U_{\mathrm{np}}(t) = -i H_{\mathrm{np}}(t) U_{\mathrm{np}}(t). \qquad (\mathrm{E.7})$$

Since the ground state energy term gives simply a phase factor, we split it according to $U_{\mathrm{np}}(t) = e^{i\Lambda} U(t)$, where $\Lambda = \left(\frac{\omega_0 t^3}{12\tau_Q^2} + \frac{\Omega}{2}t\right)$, and the equation of motion for $U(t)$ reads

$$\frac{\partial U(t)}{\partial t} = -i\left[\omega_0\left(1 - \frac{t^2}{2\tau_Q^2}\right)a^\dagger a - \frac{\omega_0 t^2}{4\tau_Q^2}(a^2 + a^{\dagger 2})\right] U(t) \qquad (\mathrm{E.8})$$

Note that $U(t)$ is a function of a and $a^\dagger$, that is, $U(t) = U(a, a^\dagger, t)$. Taking the diagonal elements in a coherent state basis, $U^n(\alpha, \alpha^*, t) = \langle\alpha| U(a, a^\dagger, t) |\alpha\rangle$, the equation becomes

$$\frac{\partial U^n(\alpha, \alpha^*, t)}{\partial t} =$$
$$-i\left[h_0(t)\alpha^*\left(\alpha + \frac{\partial}{\partial\alpha^*}\right) + h_1(t)\left(\left(\alpha + \frac{\partial}{\partial\alpha^*}\right)^2 + \alpha^{*2}\right)\right] U^n(\alpha, \alpha^*, t) \qquad (\mathrm{E.9})$$

where we have used

$$|\alpha\rangle \langle\alpha| a = \left(\alpha + \frac{\partial}{\partial\alpha^*}\right) |\alpha\rangle \langle\alpha| \tag{E.10}$$

and defined

$$h_0(t) = \omega_0 \left(1 - \frac{t^2}{2\tau_Q^2}\right), \quad \text{and} \quad h_1(t) = -\frac{\omega_0 t^2}{4\tau_Q^2}. \tag{E.11}$$

Once we obtain $U^n(\alpha, \alpha^*, t)$, we can get $U(a, a^\dagger, t)$ by replacing α and α^* to a and $a^\dagger$, respectively, in a normal ordered fashion. Since Eq. (E.9) contains only quadratic terms in α and α^*, we rely on an Ansatz of the form

$$U^n(\alpha, \alpha^*, t) = e^{a(t) + b(t)\alpha^{*2} + c(t)\alpha^2 + d(t)\alpha^*\alpha} \tag{E.12}$$

from where it follows

$$\frac{\partial}{\partial t} U^n(\alpha, \alpha^*, t) = -i \left[(h_0(t) + 4h_1(t)b(t))(1 + d(t))\alpha^*\alpha \right. \tag{E.13}$$

$$+ (h_1(t) + 2h_0(t)b(t) + 4h_1(t)b^2(t))\alpha^{*2} \tag{E.14}$$

$$\left. + (h_1(t) + 2h_1(t)d(t) + h_1(t)d^2(t))\alpha^2 + (2h_1(t)b(t))\right] U^n(\alpha, \alpha^*, t) \tag{E.15}$$

while the left hand side reads

$$\frac{\partial}{\partial t} U^n(\alpha, \alpha^*, t) = \left[\dot{d}(t)\alpha^*\alpha + \dot{b}(t)\alpha^{*2} + \dot{c}(t)\alpha^2 + \dot{a}(t)\right] U^n(\alpha, \alpha^*, t). \tag{E.16}$$

Therefore, the coefficients obey the following set of differential equations,

$$\dot{a}(t) = -i2h_1(t)b(t) \tag{E.17}$$

$$\dot{b}(t) = -i \left(h_1(t) + 2h_0(t)b(t) + 4h_1(t)b^2(t)\right) \tag{E.18}$$

$$\dot{c}(t) = -i \left(h_1(t) + 2h_1(t)d(t) + h_1(t)d^2(t)\right) \tag{E.19}$$

$$\dot{d}(t) = -i \left(h_0(t) + 4h_1(t)b(t)\right)(1 + d(t)) \tag{E.20}$$

Since the equation for $b(t)$ is decoupled from others, we solve $b(t)$ first and then $a(t)$ and $d(t)$, followed by $c(t)$. We can construct the unitary evolution operator from these solutions,

$$U(a, a^\dagger, t) = \mathcal{N}[U^n(\alpha, \alpha^*, t)] = e^{a(t)} e^{b(t)a^{\dagger 2}} \mathcal{N}[e^{d(t)a^\dagger a}] e^{c(t)a^2} \tag{E.21}$$

where $\mathcal{N}$ is a normal ordering operator. Note that the coefficients $a(t)$, $b(t)$, $c(t)$ and $d(t)$ have to define an unitary operator $U(a, a^\dagger, t)$ for any time t. Including the phase factor, we have

$$U_{np}(t) = e^{i\Lambda} U(t) = e^{a(t)+i\Lambda} e^{b(t)a^{\dagger 2}} \mathcal{N}[e^{d(t)a^{\dagger}a}] e^{c(t)a^2}. \tag{E.22}$$

Since we take as initial state the vacuum, $|\psi(0)\rangle = |0\rangle$, the evolved wave function simplifies to

$$|\psi(t)\rangle = U_{np}(t)|0\rangle = e^{a(t)+i\Lambda} e^{b(t)a^{\dagger 2}} |0\rangle. \tag{E.23}$$

The wave function can be expressed in terms of the instantaneous eigenstates

$$|\psi(t)\rangle = \sum_{n=0}^{\infty} \alpha_n(t) \left| \varphi_{np}^n(g(t)) \right\rangle, \tag{E.24}$$

and thus, the probability of finding $|\psi(t)\rangle$ in the instantaneous ground state is simply $|\alpha_0(t)|^2$, that is

$$P_0(t) = |c_0(t)|^2 = |\langle \varphi_{np}^0(g(t))|\psi(t)\rangle|^2$$

$$= e^{2\mathrm{Re}[a(t)]} \left| \left\langle 0 \left| S^{\dagger}[r(t)] e^{b(t)a^{\dagger 2}} \right| 0 \right\rangle \right|^2 \tag{E.25}$$

Calculating $\left\langle 0 \left| S^{\dagger}[r(t)] e^{b(t)a^{\dagger 2}} \right| 0 \right\rangle$ seems to be rather involved, but we can exploit the following property for a squeezing operator [4, 5],

$$S(z) = \exp[\frac{1}{2}(za^{\dagger}a^{\dagger} - z^*aa)]$$

$$= \exp[\frac{1}{2}(e^{i\theta}\tanh|z|)a^{\dagger}a^{\dagger}] \exp[-\ln(\cosh|z|)(a^{\dagger}a + \frac{1}{2})] \exp[-\frac{1}{2}(e^{i\theta}\tanh|z|)aa] \tag{E.26}$$

where $z = |z|e^{i\theta}$. Therefore,

$$\exp[\frac{1}{2}(e^{i\theta}\tanh|z|)a^{\dagger}a^{\dagger} = S(z)\exp[\frac{1}{2}(e^{i\theta}\tanh|z|)aa]\exp[\ln(\cosh|z|)(a^{\dagger}a + \frac{1}{2})], \tag{E.27}$$

which allows us to identify $b(t) = \frac{1}{2}e^{i\theta(t)}\tanh|z(t)|$, or equivalently, $|z(t)| = \mathrm{arctanh}(2|b(t)|)$ and the angle $\theta(t) = \arctan\left(\frac{\mathrm{Im}[b(t)]}{\mathrm{Re}[b(t)]}\right)$. Therefore,

$$\left\langle 0 \left| S^{\dagger}[r_{np}(g(t))] e^{b(t)a^{\dagger 2}} \right| 0 \right\rangle = \langle 0 | S^{\dagger}[r_{np}(g(t))]S[z]| 0 \rangle \cosh^{\frac{1}{2}}|z| = \langle r_{np}(g(t))|z \rangle \cosh^{\frac{1}{2}}|z|,$$

which can be now computed making use of the overlap between two squeezed states (see Eq. (3.25) in Ref. [6]),

$$S^{\dagger}[\xi]aS[\xi] = a\cosh r + a^{\dagger}e^{i\phi}\sinh r \tag{E.28}$$

with a definition of the squeezing operator $S(\xi) = \exp[\frac{1}{2}(\xi(a^\dagger)^2 - \xi^* a^2)]$ and $\xi = re^{i\phi}$. The overlap reads

$$\langle \xi_2 | \xi_1 \rangle = (\cosh r_1 \cosh r_2 - e^{i(\phi_1 - \phi_2)} \sinh r_1 \sinh r_2)^{-\frac{1}{2}} \tag{E.29}$$

and for our particular case,

$$\langle r(t) | z \rangle = (\cosh |z| \cosh r(t) - e^{i\theta} \sinh |z| \sinh r(t))^{-\frac{1}{2}}. \tag{E.30}$$

Finally, we obtain the probability $P_0(t)$,

$$P_0(t) = e^{2\mathrm{Re}[a(t)]} \left| \left\langle 0 \left| S^\dagger[r(t)] e^{b(t)a^{\dagger 2}} \right| 0 \right\rangle \right|^2 \tag{E.31}$$

$$= \frac{e^{2\mathrm{Re}[a(t)]}}{\cosh r_{\mathrm{np}}(g(t)) \left| \left(1 - 2e^{i\theta} |b(t)| \tanh r_{\mathrm{np}}(g(t))\right) \right|}, \tag{E.32}$$

while the fidelity given in Eq. (4.8) corresponds simply to $F(t) = \sqrt{P_0(t)}$.

E.3 Heisenberg Equation of Motion

Here we provide the derivation of the Heisenberg equation of motion for the $\Omega/\omega_0 \to \infty$ limit first, namely, Eqs. (4.17) and (4.18) in Chap. 4. Then, we consider a finite Ω/ω_0 value. Finally, we give the necessary information to obtain residual energy formulas used in Chap. 4 for the discussion on thermal states Sect. 4.1.3 and sudden quenches Sect. 4.1.4.

E.3.1 $\Omega/\omega_0 \to \infty$ limit

In order to solve the critical dynamics under the effective Hamiltonian of the QRM in the $\Omega/\omega_0 \to \infty$ limit, H_{np}, we resort to the Heisenberg equation, $i\dot{a}_H(t) = \left[a_H(t), H_{\mathrm{np},H}(t)\right]$ where the subscript indicates operators in the Heisenberg picture. The bosonic operators are now expressed as $a_H(t) = u(t)a + v^*(t)a^\dagger$ with $u(0) = 1$ and $v(0) = 0$ (as a Bogoliubov transformation). In addition, to satisfy the commutation relation $\left[a_H(t), a_H^\dagger(t)\right] = 1$, the coefficients must fulfill $|u(t)|^2 - |v(t)|^2 = 1$. The Hamiltonian reads

$$H_{\mathrm{np},H}(g(t)) = \omega_0 a_H^\dagger(t) a_H(t) - \frac{\omega_0 g^2(t)}{4} \left(a_H(t) + a_H^\dagger(t)\right)^2 - \frac{\Omega}{2}, \tag{E.33}$$

and thus, the Heisenberg equation results in

$$
i\dot{u}(t)a(0) + i\dot{v}^*(t)a^\dagger(0) \tag{E.34}
$$

$$
= \omega_0\left[\left(1 - \frac{g^2(t)}{2}\right)u(t) - \frac{g^2(t)}{2}v(t)\right]a(0) + \omega_0\left[\left(1 - \frac{g^2(t)}{2}\right)v^*(t) - \frac{g^2(t)}{2}u^*(t)\right]a^\dagger(0)
$$

which upon taking a commutator with $a(0)$ and $a^\dagger(0)$, we obtain the Eqs. (4.17) and (4.18).

E.3.2 Finite Ω/ω_0 Case

We calculate here the equation of motion when $\Omega/\omega_0 < \infty$, including the first-order correction in ω_0/Ω, namely, using

$$
H_{\mathrm{np}}^{\Omega}(t) = \omega_a a^\dagger a - \frac{\omega_0 g^2(t)}{4}(a + a^\dagger)^2 - \frac{\Omega}{2} + \frac{\omega_0^2 g^4(t)}{16\Omega}(a + a^\dagger)^4. \tag{E.35}
$$

The procedure is straightforward, however, we need to obtain $i\dot{a}_H(t) = \left[a_H(t), H_{\mathrm{np}}^{\Omega}\right]$ which involves $\left[a_H(t), \left(a_H(t) + a_H^\dagger(t)\right)^4\right]$. Introducing again $a_H(t) = u(t)a + v^*(t)a^\dagger$ with $u(0) = 1$ and $v(0) = 0$, and using $\left[a, \left(a + a^\dagger\right)^4\right] = 4a^{\dagger 3} + 4a^3 + 12a^{\dagger 2}a + 12a^\dagger a^2 + 12a^\dagger + 12a$, we arrive to the Heisenberg equation (to ease the notation we denote $a \equiv a(0)$ and $a^\dagger \equiv a^\dagger(0)$)

$$
i\dot{u}(t)a + i\dot{v}^*(t)a^\dagger = \left[a_H(t), H_{\mathrm{np}}(t)\right]
$$

$$
+ \frac{\omega_0^2 g^4(t)}{16\Omega}\Big\{ 4\Big(v(t)^3 a^3 + u^*(t)v(t)^2 a^2 a^\dagger + u^*(t)^2 v(t)a^\dagger a
$$

$$
+ u^*(t)v(t)^2(2a^\dagger a + 1)a + u^*(t)^2 v(t)(2a^\dagger a + 1)a^\dagger + u^*(t)^3 a^{\dagger 3}\Big)
$$

$$
+ 4\Big(u(t)^3 a^3 + v^*(t)u(t)^2 a^2 a^\dagger + v^*(t)^2 u(t)a^{\dagger 2}a + v^*(t)^3 a^{\dagger 3}
$$

$$
+ v^*(t)u(t)^2(2a^\dagger a + 1)a + v^*(t)^2 u(t)(2a^\dagger a + 1)a^\dagger\Big)
$$

$$
+ 12\Big(v(t)^2 u(t)a^3 + |u(t)|^2 u^*(t)a^{\dagger 2}a + |u(t)|^2 v(t)(2a^\dagger a + 1)a
$$

$$
+ +|v(t)|^2 v(t)a^2 a^\dagger + u^*(t)^2 v^*(t)a^{\dagger 3} + u^*(t)|v(t)|^2(2a^\dagger a + 1)a^\dagger\Big)
$$

$$
+ 12\Big(v(t)u(t)^2 a^3 + |v(t)|^2 v^*(t)aa^{\dagger 2} + |v(t)|^2 u(t)a(2a^\dagger a + 1)
$$

$$
+ |u(t)|^2 u(t)a^\dagger a^2 + u^*(t)v(t)^2 a^{\dagger 3} + |u(t)|^2 v^*(t)a^\dagger(2a^\dagger a + 1)\Big)
$$

$$
+ 12\left(v(t) + u(t)\right)a + 12\left(v^*(t) + u^*(t)\right)a^\dagger\Big\}. \tag{E.36}
$$

We now keep only linear terms in a and $a^\dagger$ in the normal-ordered Heisenberg equation, which is an approximate procedure and valid as long as $\Omega/\omega_0 \gg 1$. In this manner, applying the commutator with a or $a^\dagger$ we achieve the equations of motion,

$$\frac{d}{dt}u(t) = -i\omega_0 \left[\left(1 - \frac{g^2(t)}{2}\right) u(t) - \frac{g^2(t)}{2} v(t) \right] - i \frac{3\omega_0 g^4(t)}{4\Omega} \left(u(t) + v(t)\right) |u(t) + v(t)|^2 ,$$

$$\frac{d}{dt}v(t) = i\omega_0 \left[\left(1 - \frac{g^2(t)}{2}\right) v(t) - \frac{g^2(t)}{2} u(t) \right] + i \frac{3\omega_0 g^4(t)}{4\Omega} \left(u(t) + v(t)\right) |u(t) + v(t)|^2 ,$$

which are those presented in Eq. (4.31). Then, the residual energy follows from

$$E_r^\Omega(g) = \langle \psi_0 | H_{np,H}^\Omega(t) | \psi_0 \rangle - E_{GS}(g) \tag{E.37}$$

where $|\psi_0\rangle = |0\rangle$ is the initial state at $g = 0$ (note that the spin state is simply $|\downarrow\rangle$ and not explicitly written).

E.3.3 Residual Energy and Initial States

In general, for an initial state $|\psi_0\rangle = \left|\varphi_{np}^n(g_0)\right\rangle = S[r_{np}(g_0)] |n\rangle |\downarrow\rangle$ at coupling constant $g_0 < g_c$ we can calculate the residual energy $E_r(n, g)$ as

$$E_r(n, g(t)) = \langle \Psi_0 | H_{np,H}(t) | \Psi_0 \rangle - E_n(g(t)). \tag{E.38}$$

In terms of u and v, we obtain,

$$\langle \psi_0 | a_H^\dagger a_H | \psi_0 \rangle =$$
$$(2n + 1) \left(|u + v|^2 \frac{\cosh(2r_0)}{2} - \text{Re}\left\{uv^*\right\} \cosh(2r_0) - \text{Re}\left\{uv\right\} \sinh(2r_0) \right) - \frac{1}{2}, \tag{E.39}$$

$$\langle \psi_0 | \left(a_H^\dagger + a_H\right)^2 | \psi_0 \rangle =$$
$$(2n + 1) \left(|u + v|^2 \cosh(2r_0) - \left[\text{Re}\left\{uv\right\} - \text{Re}\left\{u^2\right\} - \text{Re}\left\{v^2\right\}\right] \sinh(2r_0) \right). \tag{E.40}$$

Therefore, the residual energy is achieved from

$$E_r(n, g(t)) = (2n + 1) \left[\omega_0 \left(|u + v|^2 \frac{\cosh(2r_0)}{2} - \text{Re}\left\{uv^*\right\} \cosh(2r_0) - \text{Re}\left\{uv\right\} \sinh(2r_0) \right) - \right.$$
$$\left. - \frac{\omega_0 g^2(t)}{4} \left(|u + v|^2 \cosh(2r_0) - \left[\text{Re}\left\{uv\right\} - \text{Re}\left\{u^2\right\} - \text{Re}\left\{v^2\right\}\right] \sinh(2r_0) \right) \right] -$$
$$- (2n + 1) \frac{\omega_0 \sqrt{1 - g^2(t)}}{2}$$
$$= (2n + 1) E_r(0, g(t)). \tag{E.41}$$

where $E_r(0, g(t))$ corresponds to the residual energy when the initial state is the ground state. This result is used in Sect. 4.1.3 for thermal states and in Sect. 4.1.4 for sudden quenches. In particular, Eq. (4.29) follows when considering $u = 1$, $v = 0$ and $g = g_c = 1$ with $g_0 < g_c = 1$.

References

1. G. Rigolin, G. Ortiz, V.H. Ponce, Beyond the quantum adiabatic approximation: adiabatic perturbation theory. Phys. Rev. A **78**, 052508 (2008). https://doi.org/10.1103/PhysRevA.78.052508
2. A.K. Chandra, A. Das, B.K. Chakrabarti (eds.), *Quantum Quenching, Annealing and Computation* (Springer, Berlin Heidelberg, 2010)
3. C. De Grandi, V. Gritsev, A. Polkovnikov, Quench dynamics near a quantum critical point: application to the sine-Gordon model. Phys. Rev. B **81**,224301 (2010). https://doi.org/10.1103/PhysRevB.81.224301
4. R.A. Fisher, M.M. Nieto, V.D. Sandberg, Impossibility of naively generalizing squeezed coherent states. Phys. Rev. D **29**, 1107 (1984). https://doi.org/10.1103/PhysRevD.29.1107
5. D.R. Truax, Baker-Campbell-Hausdorff relations and unitarity of SU(2) and SU(1,1) squeeze operators. Phys. Rev. D **31**, 1988 (1985). https://doi.org/10.1103/PhysRevD.31.1988
6. H.P. Yuen, Two-photon coherent states of the radiation field. Phys. Rev. A **13**, 2226 (1976). https://doi.org/10.1103/PhysRevA.13.2226

Appendix F
Orstein-Uhlenbeck Process

In this Appendix we provide some details of the Orstein-Uhlenbeck (OU) stochastic process used in Chap. 5 to model realistic magnetic-field fluctuations in a trapped-ion platform.

An stochastic process ξ represents an OU noise [1, 2] when ξ is continuous and memoryless (a continuous Markov process) whose time evolution follows an explicit and analytic closed-form expression,

$$\xi(t+dt) = \xi(t)e^{-t/\tau} + \left[\frac{c\tau}{2}\left(1 - e^{-2t/\tau}\right)\right]N(t), \tag{F.1}$$

where c and τ completely determine the OU process ξ, and $N(t)$ represents a normal random variable, $\overline{N(t)} = 0$ and $\overline{N(t)N(t')} = \delta(t - t')$, where the overline denotes stochastic average. Recall that a continuous Markov process fulfills i) the increment is memoryless, that is, given $\xi(t) = \zeta$, then $\Theta(dt; \zeta, t) = \xi(t+dt) - \xi(t)$ depends solely on t, dt and ζ, ii) $\Theta(dt; \zeta, t)$ depends smoothly on dt, ζ and t and iii) $\xi(t)$ satisfies the continuity condition, i.e., $\lim_{dt\to 0} \Theta(dt, \zeta, t) \to 0 \ \forall t, \zeta$. See [3] for further mathematical details of this stochastic process.

This Gaussian noise depends on two variables, namely correlation or relaxation time τ and diffusion constant c. In this context, an important property is the so-called spectral density $S(f) \equiv \lim_{T\to\infty} 2|\hat{\xi}(f)|^2/T$ which quantifies the portion of noise intensity at a particular frequency, with $\xi(t) = \int df \hat{\xi}(f)e^{-2\pi i f t}$ the Fourier transform. It is then defined in terms of the auto-covariance $C(t')$ of a stationary ξ, which for an OU process has the following form [3, 4]

$$C(t') \equiv \overline{\xi(t)\xi(t+t')} = \frac{c\tau}{2}e^{-t'/\tau} \quad \text{for } t' \geq 0. \tag{F.2}$$

Then, it can be shown that $S(f)$ can be written as

$$S(f) = 4\int_0^\infty dt' \, C(t')\cos(2\pi f t') = \frac{2c\tau^2}{1 + 4\pi^2\tau^2 f^2}. \tag{F.3}$$

© Springer Nature Switzerland AG 2018
R. Puebla, *Equilibrium and Nonequilibrium Aspects of Phase Transitions in Quantum Physics*, Springer Theses, https://doi.org/10.1007/978-3-030-00653-2

Additionally, it is worth noting the total intensity of the noise is given by $\overline{\xi^2(t)} = C(0) = c\tau/2$ or equivalently, by $\int_0^\infty df\, S(f)$. The noise spectrum decays as $1/f^2$ for large frequencies (Brownian noise), while it features an approximately constant spectrum at small frequencies (white noise). The characteristic frequency $f_{cr} = 1/(2\pi\tau)$ provides a good estimate of this crossover between these regimes.

For the following purposes, it is convenient to introduce the stochastic average of the integral of $\xi(t)$, namely, $\Xi(t) = \int_0^t dt'\, \xi(t')$. In particular, we are interested in $\overline{\Xi^2(t)}$ whose expression can be written as [3]

$$\overline{\Xi^2(t)} = c\tau^2 \left(t - \frac{3\tau}{2} + 2\tau e^{-t/\tau} - \frac{\tau}{2} e^{-2t/\tau} \right). \tag{F.4}$$

F.1 Coherence Time T_2

The coherence time T_2 corresponds to the time it takes for an initial state, say $|\uparrow\rangle_x$, $\sigma_x |\uparrow\rangle_x = +|\uparrow\rangle_x$, to reduce its coherence a factor $1/e$. For longer times, the coherence is lost and the state evolves towards $\overline{|\psi\rangle\langle\psi|} \approx |e\rangle\langle e| + |g\rangle\langle g|$ due to the presence of pure dephasing noise. This situation can be modeled by a stochastic Hamiltonian

$$H = \frac{\xi(t)}{2}\sigma_z \tag{F.5}$$

that effectively captures pure dephasing noise and where ξ stands for an OU process, as aforementioned. Therefore, the expectation value of σ_x after a single run and time t reads

$$\langle \sigma_x(t)\rangle = \cos\left(\Xi(t)\right) \tag{F.6}$$

which upon stochastic average becomes

$$\overline{\langle \sigma_x(t)\rangle} = \overline{\cos\left(\Xi(t)\right)} = e^{-\frac{1}{2}\overline{\Xi^2(t)}}, \tag{F.7}$$

where we have assumed that the stochastic noise ξ is Gaussian, that is, the moments fulfill $\overline{\xi^{2n}} = \overline{\xi^2}^n (2n)!/(2^n n!)$, as is the case for an OU noise. Finally, $\overline{\Xi^2(t)}$ can be obtained analytically as a function of τ and c, given in Eq. (F.4) [4]. Thus, we can obtain a relation between c, τ and T_2 since $\overline{\langle \sigma_x(T_2)\rangle} = e^{-1}$, leading to

$$c = \frac{2}{\tau^2 \left(T_2 - \tau \left(\frac{3}{2} - 2e^{-T_2/\tau} + \frac{1}{2} e^{-2T_2/\tau} \right) \right)}. \tag{F.8}$$

In the special of a short correlated noise, $\tau \ll T_2$, $c \approx 2/(T_2\tau^2)$, which gives an exponential decay of the coherence, $\overline{\langle\sigma_x(t)\rangle} \propto e^{-t/T_2}$ as observed experimentally [5]. To the contrary, when $\tau \gtrsim T_2$, $\overline{\langle\sigma_x(t)\rangle}$ features a Gaussian decay, $e^{-(t/T_2)^2}$. Both scenarios have been illustrated in Fig. 5.5b.

References

1. G.E. Uhlenbeck, L.S. Ornstein, On the theory of the Brownian motion. Phys. Rev. **36**, 823 (1930). https://doi.org/10.1103/PhysRev.36.823
2. M.C. Wang, G.E. Uhlenbeck, On the theory of the Brownian motion II. Rev. Mod. Phys. **17**, 323 (1945). https://doi.org/10.1103/RevModPhys.17.323
3. D.T. Gillespie, The mathematics of Brownian motion and Johnson noise. Am. J. Phys. **64**, 225 (1996a). https://doi.org/10.1119/1.18210
4. D.T. Gillespie, Exact numerical simulation of the Ornstein-Uhlenbeck process and its integral. Phys. Rev. E **54**, 2084 (1996b). https://doi.org/10.1103/PhysRevE.54.2084
5. D.J. Wineland, C. Monroe, W.M. Itano, D. Leibfried, B.E. King, D.M. Meekhof, Experimental issues in coherent quantum-state manipulation of trapped atomic ions. J. Res. Natl. Inst. Stand. Technol. **103**, 259 (1998). https://doi.org/10.6028/jres.103.019

Printed in the United States
By Bookmasters